WAGE LABOUR AND UNFREEDOM IN AGRICULTURE

WIDER

Studies in Development Economics embody the output of the research programmes of the World Institute for Development Economics Research (WIDER), which was established by the United Nations University as its first research and training centre in 1984 and started work in Helsinki in 1985. The principal purpose of the Institute is to help identify and meet the need for policy-oriented socio-economic research on pressing global and development problems, as well as common domestic problems and their interrelationships.

WAGE LABOUR AND UNFREEDOM IN AGRICULTURE

An Indian Case Study

V. K. RAMACHANDRAN

CLARENDON PRESS · OXFORD

This book has been printed digitally and produced in a standard specification in order to ensure its continuing availability

OXFORD
UNIVERSITY PRESS

Great Clarendon Street, Oxford OX2 6DP

Oxford University Press is a department of the University of Oxford.
It furthers the University's objective of excellence in research, scholarship,
and education by publishing worldwide in

Oxford New York

Auckland Cape Town Dar es Salaam Hong Kong Karachi
Kuala Lumpur Madrid Melbourne Mexico City Nairobi
New Delhi Shanghai Taipei Toronto
With offices in
Argentina Austria Brazil Chile Czech Republic France Greece
Guatemala Hungary Italy Japan South Korea Poland Portugal
Singapore Switzerland Thailand Turkey Ukraine Vietnam

Published in the United States
by Oxford University Press Inc., New York

Reprinted 2011

ISBN 978-0-19-828647-9

Printed and bound in Great Britain by CPI Antony Rowe,
Chippenham and Eastbourne

For P. Sundarayya

FOREWORD

From its inception, a major theme of WIDER's research has been the study of the world's poorest billion, most of whom live in South Asia and sub-Saharan Africa. This research has been conducted at several levels, embracing conceptual issues of hunger and living standards and empirical work on food and nutrition, as well as short-term issues of famine and crisis and long-term issues of endemic hunger. The empirical studies have been conducted at different levels of aggregation—global, regional, and national, and also at the level of individual villages. The present study represents the fruit of years of patient inquiry, from the vantage point of a village in India, into the evolving conditions of living and employment of hired agricultural labourers, the class that clearly constitutes the poorest segment of the world's poorest billion.

Unlike most parts of sub-Saharan Africa, India has made quite impressive progress in agriculture in the recent past, despite the closing of land frontiers in the early 1960s. Progress has not been even across the country, but certain regions have made quite spectacular advances. What effect has this advance in the forces of production had on the socio-economic characteristics of agricultural labourers, traditionally the most deprived section of the rural community? This is the question that the author addresses by focusing his attention on a relatively advanced agrarian region in the state of Tamil Nadu, and on a single village within that region.

In inquiring into the evolving conditions of life and labour of agricultural labourers, this study goes beyond the usual concerns with such indices of living standards as income, consumption, etc. In addition, and more importantly, a novel attempt has been made to look at the freedom of labourers to sell their labour power. As the author explains in the book, the focus on freedom was motivated by two concerns. In the first place, the freedom of employment is an important component of positive freedom. In his research on the conceptual foundations of hunger and living standards, at WIDER and elsewhere, Amartya Sen has argued that the standard of living must be judged by the degree of positive freedom enjoyed by the people. Certainly, the freedom of employment, together with other aspects of freedom such as freedom from hunger and freedom from avoidable morbidity and mortality, provides a basic measure of positive freedom, and one that is relevant for judging the living standards of the poorest of the people.

The second aspect of the concern with the question of freedom relates to the classical, particularly Marxist, concern with the freedom of the worker as being (I quote the author) an 'indicator of the direction of historical change and of progressive socio-economic transition'. The 'unfree' nature of Indian labour, weighed down by centuries of poverty, by caste and other social restrictions, and by gender discrimination, has often been thought to be a serious impediment to the development of the forces of production. To what extent has the new wave of progress succeeded in breaking down the old fetters of 'unfreedom', and thus secured for itself the basis of further progress? This important question has been the second motivation behind the concern with freedom.

The relevance of the issues with which this study is concerned and the originality of the manner in which these issues have been investigated—by combining the methods of economics with those of other disciplines in the social sciences—should prove an important departure in the field of research on the living conditions of the poor.

Lal Jayawardena
Director, WIDER

PREFACE

The primary data for this book were gathered from 1977 to 1986 in Gokilapuram and the Cumbum Valley. For most of that period, from 1979 to 1986, I was at the Madras Institute of Development Studies. My first debt is to C. T. Kurien, who was Director of the Institute, for his supervision of the doctoral work of which this book is an extension, and for his patience during the long periods of time that I spent in the village and Valley. From 1986 I have been with the World Institute for Development Economics Research (WIDER) in Helsinki. WIDER gave me the facilities to collect additional secondary information on agricultural labour in India, and to work my material into a book. I thank Lal Jayawardena warmly for his concern and for his tolerance of work in different fields of enquiry.

The final changes to the manuscript of this book were made during my year as a visitor at the Harvard Center for Population Studies. I am grateful to Lincoln Chen and the Center for the welcome and the facilities that I have received.

My work and this book owe a great deal to Susan Ram, who was the main organizer of the 1977 survey in Gokilapuram, and who gave me the material on the Cumbum Valley that she had collected over more than four years of research. N. Ram was involved with the research in its early stages, and many of the issues and views in this book were discussed with him at the time they were being formulated. I have drawn heavily on Madhura Swaminathan's help and her research experience in Madurai district. S. Alamelu, M. Balaji, V. S. Balakrishnan, K. Bharathan, T. R. Govindarajan, K. Jayashree, K. Nagaraj, V. Namasivayam, R. Prabhakar, R. Rajaram, Jeemol Unni, and Sridhar Venkataraman worked with me in the collection and processing of the primary data. Kaushik Basu, Sheila Bhalla, K. Bharathan, G. A. Cohen, S. Guhan, John Harriss, T. N. Krishnan, Saraswathi Menon, R. Ramaraj, Paul Seabright, Sunil Sengupta, A. K. Shiva Kumar and S. Subramanian commented on my work at different stages of its preparation. Over the years my parents, N. Ram and Madhura Swaminathan have been sources of invaluable support and encouragement.

WIDER's friendly staff helped to make my stay there a very pleasant one, and Timo Matoniemi, Juhani Holm, Martina Jägerhörn, Brenda Baker and Liisa Roponen contributed to this book in different ways. Amit Bhaduri provided detailed criticism of the manuscript and encouraged its completion, and Carlo Fonseka, vigilant blue pencil in hand, read the manuscript with concern and care. Amartya Sen helped the most to make the transition from thesis to book, and gave generously of his time to read successive drafts of the manuscript and provide criticism of the content and presentation of its argument. K. Nagaraj and Siddiq Osmani gave me the most detailed advice and guidance that I have received with regard to this study, and although, in a purely legal sense, I imagine that I am responsible for all the views expressed in this book, in respect of most of it a reasonably good case of joint liability can be made against one or the other of them.

The efforts and co-operation of friends in Gokilapuram and the Valley, whose friendship I cherish, made this book possible. Rayar Karuppana Thevar and V. Subramaniam, old and respected agricultural labourers of Gokilapuram, both of whom are now dead, shared—with patience, good humour, and great generosity—the experiences of lifetimes of labour and toil. P. Sundarayya—teacher, comrade, and exemplar for many people in India—encouraged us, for many years, to study the agrarian question in India.

I am deeply grateful to all of them.

V.K.R.

Cambridge, Mass.
1 December 1989

CONTENTS

APPENDICES

LIST OF MAPS

LIST OF TABLES

ABBREVIATIONS

AERC	Agro-Economic Research Centre
EPW	*Economic and Political Weekly*, Bombay
GOI	Government of India
GOTN	Government of Tamil Nadu
ICRISAT	International Crop Research Institute for the Semi-Arid Tropics
IJAE	*Indian Journal of Agricultural Economics*
ILO	International Labour Office
ILO-ARTEP	ILO Asian Regional Training and Employment Programme
IRDP	Integrated Rural Development Programme
MIDS	Madras Institute of Development Studies
NSS	National Sample Survey
NSSO	National Sample Survey Organisation
RLE	Rural Labour Enquiry

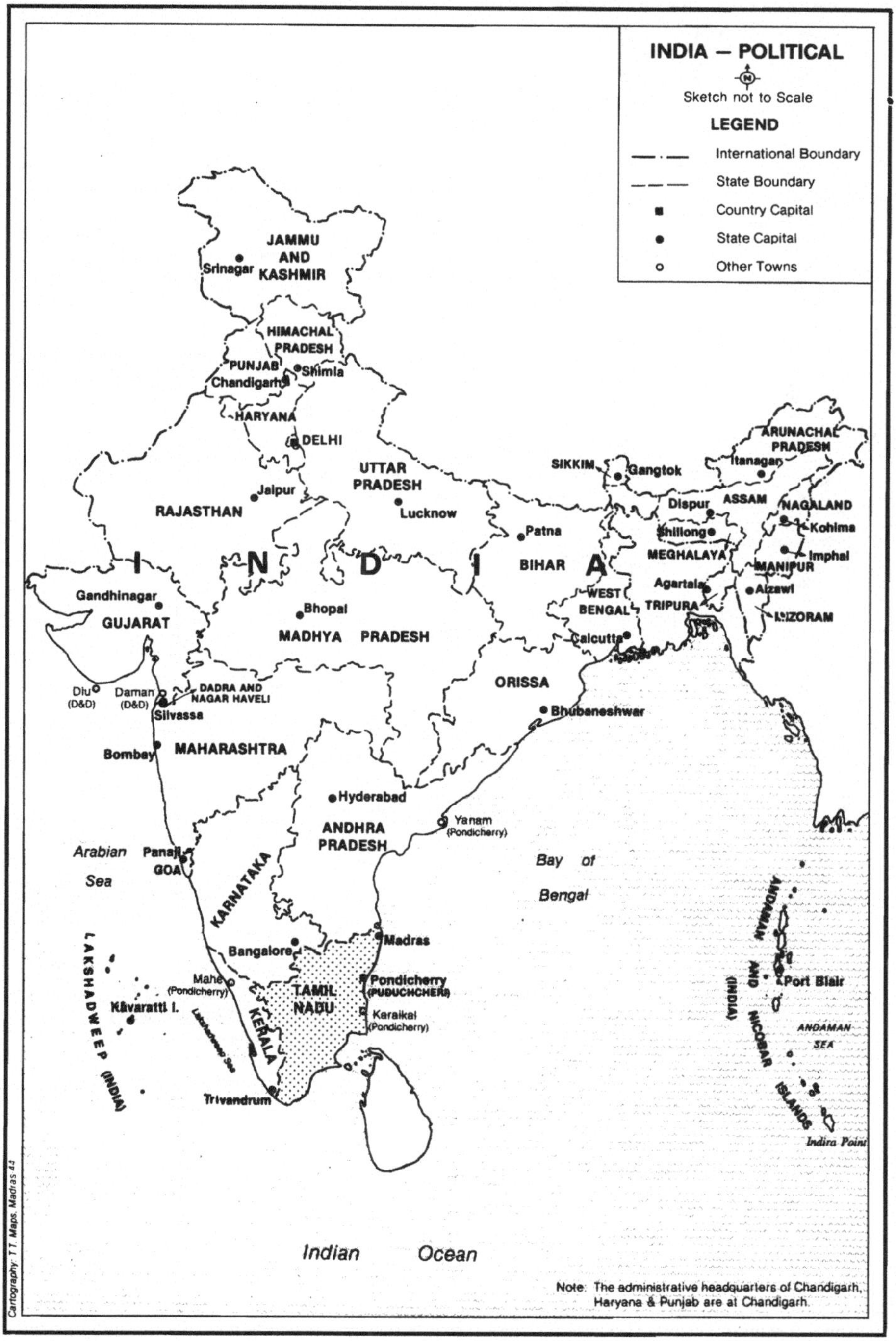
INDIA — POLITICAL
Sketch not to Scale
LEGEND
International Boundary
State Boundary
Country Capital
State Capital
Other Towns
JAMMU AND KASHMIR
Srinagar
HIMACHAL PRADESH
PUNJAB
Chandigarh
Shimla
HARYANA
DELHI
UTTAR PRADESH
Lucknow
RAJASTHAN
Jaipur
I N D I A
SIKKIM
Gangtok
ARUNACHAL PRADESH
Itanagar
ASSAM
Dispur
NAGALAND
Kohima
Shillong
MEGHALAYA
MANIPUR
Imphal
Agartala
TRIPURA
Aizawl
MIZORAM
Patna
BIHAR
WEST BENGAL
Calcutta
Gandhinagar
GUJARAT
Bhopal
MADHYA PRADESH
ORISSA
Bhubaneshwar
Diu (D&D)
Daman (D&D)
DADRA AND NAGAR HAVELI
Silvassa
Bombay
MAHARASHTRA
Hyderabad
ANDHRA PRADESH
Yanam (Pondicherry)
Arabian Sea
Panaji
GOA
KARNATAKA
Bay of Bengal
Bangalore
Madras
Mahe (Pondicherry)
TAMIL NADU
Pondicherry (PUDUCHCHERI)
Karaikal (Pondicherry)
KERALA
Kavaratti I.
LAKSHADWEEP (INDIA)
Trivandrum
ANDAMAN AND NICOBAR ISLANDS (INDIA)
Port Blair
ANDAMAN SEA
Indira Point
Indian Ocean
Note: The administrative headquarters of Chandigarh, Haryana & Punjab are at Chandigarh.
Cartography: T.T. Maps, Madras 44

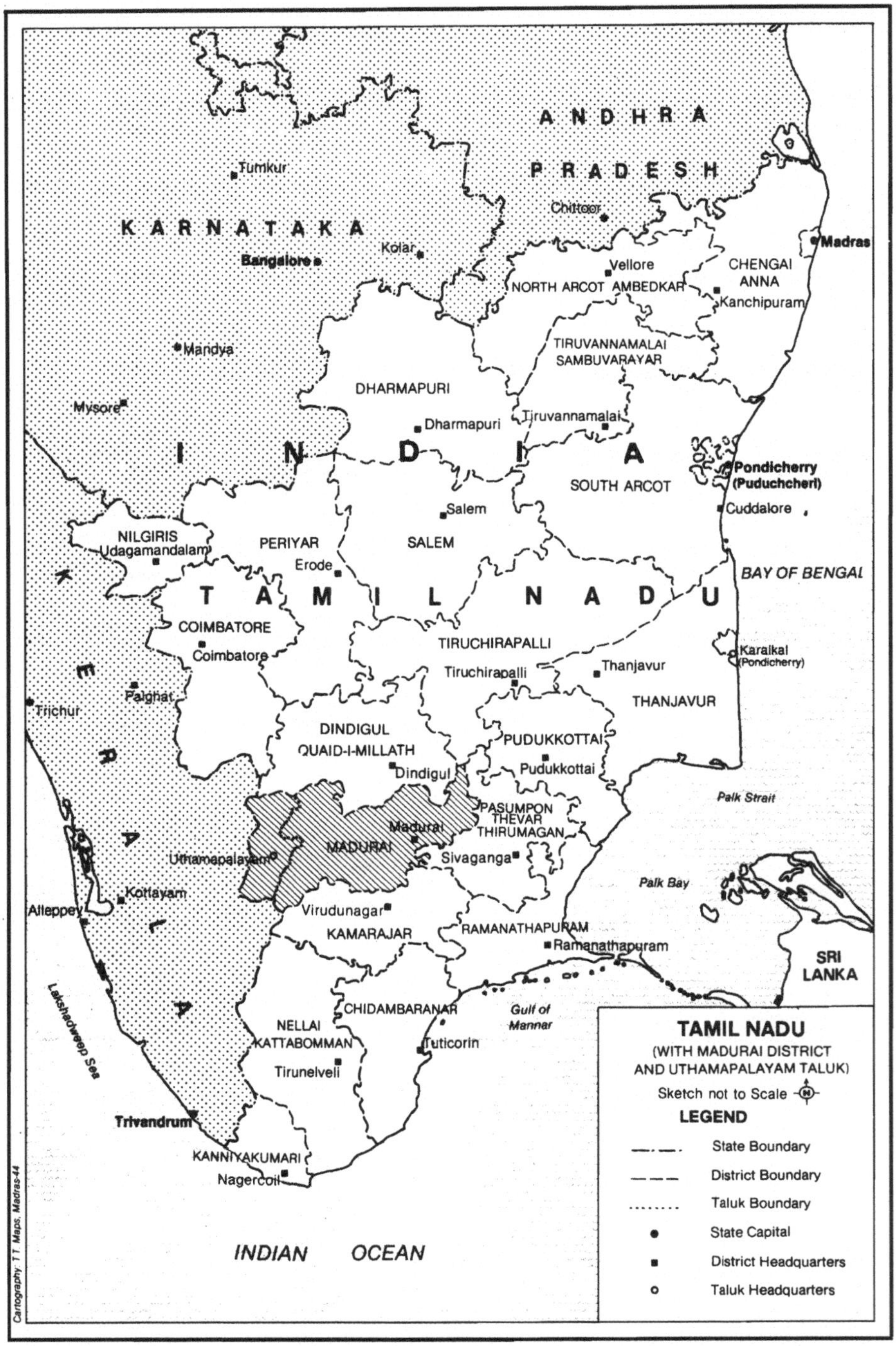
ANDHRA
PRADESH
Tumkur
KARNATAKA
Chittoor
Madras
Kolar
Bangalore
Vellore
CHENGAI
ANNA
NORTH ARCOT AMBEDKAR
Kanchipuram
Mandya
TIRUVANNAMALAI
SAMBUVARAYAR
DHARMAPURI
Mysore
Tiruvannamalai
Dharmapuri
INDIA
Pondicherry
(Puduchcheri)
SOUTH ARCOT
Salem
Cuddalore
NILGIRIS
Udagamandalam
PERIYAR
SALEM
Erode
BAY OF BENGAL
KERALA
TAMIL NADU
COIMBATORE
TIRUCHIRAPALLI
Karaikal
(Pondicherry)
Coimbatore
Thanjavur
Tiruchirapalli
Palghat
THANJAVUR
Trichur
DINDIGUL
QUAID-I-MILLATH
PUDUKKOTTAI
Pudukkottai
Dindigul
Palk Strait
PASUMPON
THEVAR
THIRUMAGAN
Madurai
MADURAI
Uthamapalayam
Sivaganga
Palk Bay
Kottayam
Alleppey
Virudunagar
KAMARAJAR
RAMANATHAPURAM
Ramanathapuram
SRI
LANKA
Lakshadweep Sea
CHIDAMBARANAR
Gulf of
Mannar
NELLAI
KATTABOMMAN
Tuticorin
Tirunelveli
Trivandrum
KANNIYAKUMARI
Nagercoil
INDIAN OCEAN
Cartography: T T Maps, Madras-44
TAMIL NADU
(WITH MADURAI DISTRICT
AND UTHAMAPALAYAM TALUK)
Sketch not to Scale
LEGEND
State Boundary
District Boundary
Taluk Boundary
State Capital
District Headquarters
Taluk Headquarters

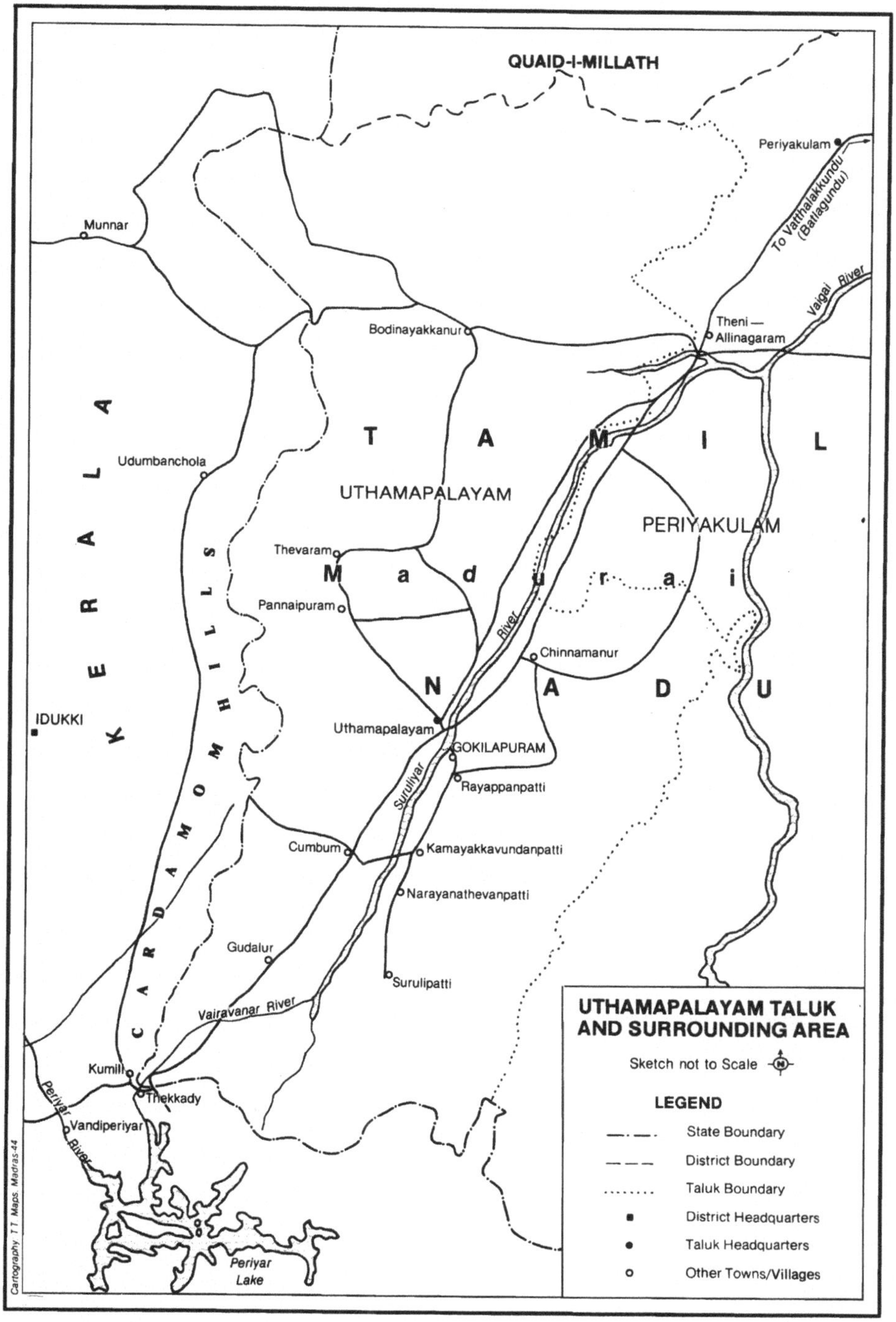
QUAID-I-MILLATH
Periyakulam
To Vatthalakkundu (Batlagundu)
Munnar
Vaigai River
Theni — Allinagaram
Bodinayakkanur
KERALA
TAMIL NADU
Madurai
Udumbanchola
UTHAMAPALAYAM
PERIYAKULAM
Thevaram
Pannaipuram
River
Chinnamanur
IDUKKI
Uthamapalayam
GOKILAPURAM
Rayappanpatti
Suruliyar
CARDAMOM HILLS
Cumbum
Kamayakkavundanpatti
Narayanathevanpatti
Gudalur
Surulipatti
Vairavanar River
Kumili
Thekkady
Periyar River
Vandiperiyar
Periyar Lake
Cartography T T Maps Madras-44
UTHAMAPALAYAM TALUK AND SURROUNDING AREA
Sketch not to Scale
N
LEGEND
State Boundary
District Boundary
Taluk Boundary
District Headquarters
Taluk Headquarters
Other Towns/Villages

1
Introduction

The aim of this book is to explore the development of wage labour in agriculture. The study of wage labour in the book is based on an analysis of agricultural growth and of the impact of that growth on the conditions of life of agricultural labourers. At the core of the analysis is quantitative and qualitative material gathered over nine years in the Cumbum Valley, an area of appreciable agricultural modernization in Tamil Nadu, India. And in the Valley, intensive surveys and case studies were conducted in Gokilapuram village, the story of whose working people is a concern that pervades this book.

While field studies in Gokilapuram and in the Valley as a whole are the major component of the data base of this work, the book also draws on relevant literature that relates to the rest of India and, more particularly, to the agrarian scene in Tamil Nadu.

For the purposes of this book, wage labour must be distinguished from the general category of hired labour. A wage labourer or proletarian is a hired labourer who is propertyless and is free to sell his or her labour power to the employer of his or her choice. This is in contrast, for instance, to a bonded labourer, who, though a hired labourer all right, is unfree to choose his or her employer. In practice, the freedom of the wage labourer to sell his or her labour power is often subject to constraints, and between proletarian labour and the labour of dark bondage intervene many gradations, nuances. The crucial distinction for this book, then, is the difference between the freedom of workers who can sell their labour power to the employers of their choice, and the unfreedom of those who cannot, and the extension of free wage labour is dealt with as an indicator of evolving socio-economic relations.

The emergence and development of wage labour as a basic indicator of capitalist development is, of course, a crucial analytical theme in classical Marxist writing.[1] Marx distinguishes between (to use the terms used by Jon Elster) the *formal freedom* of the worker under capitalism and the *real unfreedom* of workers in pre-capitalist systems: 'the freedom of a worker to change employers makes him free in a way not found in earlier modes of production'.[2]

[1] And this book draws on the discussions in Marx (1977, particularly pts. vi and viii), as well as on Lenin, particularly (1972*a*).

[2] The characterization is Elster's (1985, p. 208), from his discussion of this idea in the work of Marx and others. It must be stressed that the 'freedom' of the worker referred to here is only in relation to the condition of workers in socio-economic environments in which workers are unfree to sell their labour power. There is a separate and extensive literature on the incompleteness of the freedom of the proletarian under capitalism itself, for instance, on the fact that workers under capitalism are forced, as a class, to sell their labour power (Marx 1973, p. 464; Cohen 1985, 1986, 1987; Elster 1985, particularly ch. 4). Of Marx's contribution to this discussion, Elster writes that

The study of the development of wage labour in agriculture is important from another perspective as well. The extension of the freedom of workers in a society to sell their labour power is an enhancement of their positive freedom, which is, in turn, an important measure of how well that society is doing.

So far as the question of the development of wage labour is concerned, the book addresses two major questions. First, does hired labour in Indian agriculture represent wage labour, the existence of a work-force whose members are propertyless and free to sell their labour power? Secondly, is the influence of non-contractual or, broadly speaking, other than economic factors on the freedom to sell labour power an important influence on production relations and the socio-economic status of the agricultural labourer? What is perhaps the major conclusion of the book may as well be revealed at once: in Gokilapuram village and the Cumbum Valley 'free' wage labour predominates in major production tasks in agriculture. However, the basic relation of free wage labour has not emerged from the ashes of the old; on the contrary, the modern wage labour relation has emerged in a swamp of pre-capitalist relations and forms of labour. Despite the generalized occurrence of the wage labour form in actual production, extreme poverty, the extension of the land monopoly into the present, staggering levels of inequality in the village, and a general absence of democratization of village life are fetters on the creation of a completely free rural work-force.

In exploring the development of wage labour in agriculture it was also necessary to consider the role of agricultural growth and its impact on agricultural labour households in some detail. This entailed an analysis of the development of the productive forces in agriculture and of the distribution of land and other assets among different classes in the population; it also includes a study of changes in the size and composition of the labour force and in labour absorption in agriculture as well as an analysis of the material conditions of life of agricultural labourers. There is a wide and disparate literature on these issues in India on which to draw. One major conclusion of this literature is that the path of development has been one in which the rural poor and agricultural labourers have not shared in the fruits of growth in any substantial—or just and equitable—way. The book attempts to study the validity of this conclusion with reference to a specific agrarian region and village.

In the literature on long-term changes in the structure of the Indian economy over the period of planned development and over a longer historical period analyses of shifts in the sectoral configuration of the work-force have come to occupy a significant place (this literature is itself partly derivative from pioneering studies of other countries, notably by Simon Kuznets: see Kuznets

when Marx refers to the freedom of the worker under capitalism, 'he unfailingly adds that (i) the worker depends on capital even if he does not depend on any particular capitalist and (ii) that the independence in the latter sense hides the real dependence in the former sense' (1985, p. 208).

1971; Krishnamurty 1984). A striking conclusion of the literature is that the secular trend in the distribution of the work-force is one of a dominance of workers in the primary sector, in particular, in agriculture.[3] Concurrent research has also shown, however, that while the share of agriculture in the aggregate work-force remained relatively stable,[4] there has been a remarkable and relatively rapid restructuring of the work-force within agriculture itself, an intrasectoral change that has manifested itself in a sharp increase in the proportion of agricultural labourers in the rural work-force.[5] Kalpana Bardhan, reviewing the all-India data from the Censuses of 1961 and 1971, concludes from the Census primary abstract data as well as the Census resurvey of occupational categories of 1974 that over the period under consideration 'the proportion working primarily as agricultural labourers went up considerably'.[6] That this trend was particularly marked in changing Tamil Nadu agrarian society was an important result thrown up by C. T. Kurien's (1981) research; in subsequent papers, whose major results are reported below, it has been shown that this trend of increase in Tamil Nadu, when disaggregated to the taluk level, displays systematic intra-State patterns (Ramachandran 1980*a*; Subramanian and Ramachandran 1982).[7]

In Tamil Nadu, a state where there has been a relatively major—although uneven—development of the productive forces and spread of the new technology in agriculture, persons who are engaged in agricultural tasks (the Census categories of cultivators and agricultural labourers together) as a proportion of the total work-force was over 76 per cent in 1981 (Census of India 1981 undated).[8] At the same time, among the most important concomitants of the process of rural transformation in Tamil Nadu since the 1960s is the large-scale increase in the number of agricultural labourers and of the proportion of agricultural labourers in the work-force. Data from successive Censuses indicate the significance of this trend and its regional variation in Tamil Nadu.

The Census of India is the most comprehensive source of data—and the only one of its kind—on agricultural labourers and the work-force in the State, and here we have used the primary census abstracts from the Censuses of 1961,

[3] Krishnamurty (1984) notes interesting intrasectoral developments behind the aggregates and regional variations in development, although his discussion of change relates mainly to the secondary and tertiary sectors.

[4] At about 70% between 1881 and 1951, by Krishnamurty's (1984) data.

[5] For an earlier period, see Patel (1952); Kurien (1981); and Patnaik (1985*a*).

[6] There is an extensive literature on increases in the size of the agricultural labour force in different parts of India; see e.g. Breman (1974); Bhalla (1976); Nayyar (1977); Beteille (1980); Gough (1981); Parthasarathy and Pothana (1983); C. R. Reddy (1985); Patnaik (1987*a*); and others.

[7] Indeed, almost the only paper that argues the contrary trend, of decline in the proportion of agricultural labourers, and is based on an analysis of quantitative data is Sundaram (1977); this paper is discussed in Ramachandran (1986), where I try to show that Sundaram's conclusions are based on a comparison of non-comparable sets of data and cannot be taken as reliable for Tamil Nadu.

[8] A more detailed analysis of Statewide trends is in Ramachandran (1980*a*) and Subramanian and Ramachandran (1982).

1971, and 1981. While the definition of 'agricultural labourer' at the three Censuses has remained basically unaltered,[9] the precise comparability of data on agricultural labourers across Censuses has been marred by changes in the definition of the general category of 'workers', in the criteria for the inclusion of women in the work-force, as well as in the actual collection of data. These and problems of Census enumeration and classification have been discussed elsewhere (Ramachandran 1980*a*; Subramanian and Ramachandran 1982); in general, although their combined effect is to make it difficult to use Census results for precise quantitative measurements (particularly in specific localities), Census data do allow for the possibility of spatial and intertemporal comparisons of orders of magnitude and, as it happens, these are striking enough at the taluk, district, and state level in Tamil Nadu over the intercensal periods from 1961 to 1981.

Our method here is to measure, by taluk, the incidence of agricultural labourers in the working population (agricultural labourers as a percentage of total workers in 1961 and 1971 and agricultural labourers who were main workers as a percentage of all main workers at the Census of 1981)[10] and change (as a percentage) in incidence. The measurements are of the whole work-force (male and female) and the agricultural labour force in the rural segment of each taluk in the State, and we have calculated the incidence and change for each taluk over the three Censuses.[11]

The results give us a broadbrush illustration of what has happened in Tamil Nadu. There was a large-scale and general increase in the incidence of agricultural labourers in the work-force of the State over the period 1961–81. This incidence increased from 22 per cent in 1961 to 40 per cent in 1981 (Table 1.1). This trend was also general, in that no taluk displayed a contrary trend of decline in the proportion of agricultural labour. Disaggregating this increase into its two intercensal components, the data suggest that the change in incidence was much steeper in the decade 1961–71. In the second decade, although there was an increase in the agricultural labour force, there was a fall-off in incremental change. This reflects the higher base in 1971 compared with 1961 levels as well as the change to a more restrictive definition by the Census authorities of the category of 'main worker' in 1981. The extent to which this

[9] An agricultural labourer is a person who works on another person's land for wages in cash, kind or share and who does not have a right of lease or contract on the land which he or she works and does not share in the risks of cultivation. For a very interesting survey of the definitions of agricultural labour in labour legislations from different parts of the world, see International Labour Conference (1981, particularly ch. 2).

[10] The Census of 1981 defined a 'main' worker as a worker who worked for at least 180 days in a year, and a 'marginal' worker as one who worked for less than 180 days in a year. The breakdown by occupation of marginal workers at the Census of 1981 is not available; in any case, by the Census results, they constitute only 8.1% of the labour force in Tamil Nadu.

[11] Where there have been changes in taluk or district boundaries, necessary adjustments have been made in the data. Madras district (which is wholly urban), the taluks of the Kanniyakumari and Nilgiris districts (for which data were not comparable), the urban taluk of Saidapet, and three hill taluks, Shencottah, Yercaud, and Kodaikkanal, have been excluded when calculating rank correlations and coefficients of variation, as these areas have special characteristics.

Table 1.1 *Agricultural labourers in the working population of rural Tamil Nadu, incidence and change, 1961, 1971, 1981* (%)

Year(s)	Incidence	Change
1961	21.8	—
1971	38.1	—
1981	39.9	—
1961–81	—	82.9
1961–71	—	74.7
1971–81	—	4.7

Source: Census of India.

reflects an actual process on the ground requires further and more detailed study.

In terms of the spatial distribution of change that occurred over the period 1961–81, two tendencies can be discerned. First, the hierarchy of taluks ranked by the incidence of agricultural labourers did not alter radically: the coefficient of rank correlation for taluks ranked by the proportion of agricultural labourers in the work-force between 1961 and 1971 was 0.8 and between 1971 and 1981 was 0.9.

The second tendency, which asserted itself strongly over the period, was characterized by faster growth over the two decades of the agricultural labour force in areas where incidence was relatively low in the base period, resulting in a closing of the gap between taluks in respect of the component of their working population represented by agricultural labourers. A simple summary measure of this is the coefficient of variation of incidence over taluks, which has declined steeply, from 0.47 in 1961 to 0.26 in 1981 (see Table 1.2).[12]

For the decade 1961–71, the Census data on agricultural labourers have been set alongside results from the World Agricultural Census, 1970–1, for Tamil Nadu (see Ramachandran 1980*a*).[13] Stated very briefly, the results showed that the incidence of agricultural labourers in 1961 and 1971 was very high in areas where canal irrigation is high, such as the Cauvery-irrigated areas of Thanjavur district, and in certain areas of irrigation from large tanks, such as in Chengalpattu district. Change in the proportion of agricultural labourers in the working population was not high in areas of traditional canal irrigation or in predominantly tank-irrigated areas (where increase was, in fact, relatively sluggish).[14] Such change was very high in areas where irrigation by wells

[12] Similarly, data on Bangladesh and West Bengal covering a longer historical period and analysed by van Schendel and Faraizi indicate that regional differences in the proportion of agricultural labourers in the population, which were pronounced in the 19th cent., became blurred by the 1970s (van Schendel and Faraizi 1984, p. 46).

[13] This could not be done for the decade 1971–81 since the results of the Agricultural Census for 1980–1 were not available.

[14] See Subramanian (1983) for a review of Census data pertaining to agricultural labourers in tank-irrigated regions of the State.

Table 1.2 *Coefficient of variation across taluks of incidence of agricultural labourers, Tamil Nadu, 1961, 1971, 1981* (%)

Year	Coefficient of variation
1961	46.8
1971	27.5
1981	25.6

Source: Census of India.

constituted a higher than average proportion of irrigation. Rates of growth were steepest in taluks where a higher than average area under canal irrigation was supplemented by a higher than average area under well irrigation.

An important component of the changes in the agrarian scene in the 1960s was the development of minor irrigation, particularly irrigation from wells equipped with electric motor pumps. The transformation of single-cropped unirrigated land growing, for instance, coarse cereals, to irrigated garden land that is in use the whole year round or for most of the year and on which a variety of crops—such as cotton, groundnut, fruit, and high-yielding cereals—can be grown involves a sharp increase in the amount of labour power required for cultivation. The Census data indicate that the increased demand for labour in such areas was accompanied by a rapid increase in supply as well.[15] A large section of the working population, from among cultivators and others, was pulled into the hired labour force, particularly the agricultural labour force (that is, the labour force captured within the Census category 'agricultural labourers').[16]

The large-scale increase in the absolute and relative size of the labour force in agriculture represents not merely quantitative change, but has an important qualitative aspect as well[17] and, as has been said, it is the qualitative aspect, the

[15] While intra-State rural–rural migration is an important phenomenon, aggregate permanent migration into the rural areas of the State from outside the State can be assumed to be low; see Nagaraj (1981).

[16] Correspondingly, there was a decline in the incidence of the Census category of 'cultivators' in the State, from 42% in 1961 (taluk average: 32%) to 29% in 1981 (taluk average: 18%). The RLEs of 1964–5 and 1974–5 suggest a comparable trend of increase. Agricultural labour households as a proportion of all rural households in Tamil Nadu increased from 28% in 1964–5 to 38% in 1974–5. While all rural households increased from 6,193,000 in 1964–5 to 6,304,000 in 1974–5, or an increase of 1.8%, agricultural labour households increased from 1,735,000 in 1964–5 to 2,399,000 in 1974–5, an increase of 38.3% (RLE 1974–5 (1978)).

[17] In a discussion of qualitative changes in present-day village society, Béteille (1980) refers to the increase in the size of the labour force, male and female, as 'probably the most significant feature of the Indian village of the present' (p. 118).

socio-economic characteristics of agricultural labourers in the contemporary period, that is the major concern of this study. Before entering the literature in economics that attempts to deal directly with the question of wage labour in contemporary agriculture, I shall note, very briefly, certain historical features of unfreedom in the socio-economic status of agricultural labourers, focusing mainly on some of the evidence for Tamil Nadu.

A feature of India's agrarian history, and one of continuing relevance to production relations in agriculture, is that the creation of a propertyless labour force in Indian agriculture, a 'specially repressed' class of agricultural labourers, considerably pre-dates capitalist relations in agriculture and the onset of colonial rule. Irfan Habib dates the emergence of agricultural labour as a separate class to the first millennium AD, the period which saw, he says, 'the completion of the great division between the peasantry and landless labour.' He says that this section of the people was 'largely created out of the food-gatherers and forest folk who had been already converted into ostracized *jatis* during the five centuries before Christ' (Habib 1982, p. 21). (In his agrarian history of Mughal India, Habib points to the sharp seasonal fluctuations in the demand for labour in cultivation as having necessitated a large reserve supply of labour (Habib 1963, pp. 121–2, cited in 1982, p. 21).) Noboru Karashima's work on two Tamil Nadu villages in the Cauvery Valley in the ninth and tenth centuries shows that even in the village where individual land holding had not emerged and land was cultivated in common by members of the *ur* or village community, the village was stratified in that there was a class of agricultural labourers who were not members of the *ur* (Karashima 1984, pp. 12, 43–55).

The emergence of a labour force comprised of casual workers, with wage labour at a low stage of development, began well before Independence. In the area of study, the Cumbum Valley, this had certainly taken place by the late nineteenth century. The size and qualitative features of the labour force in agriculture, a labour force assembled in the pre–1947 period, changed rapidly with the economic expansion following Independence.

Saraswathi Menon's research (1983, particularly chs. 3, 6)[18] provides a valuable assessment of changes in agrarian relations during the colonial period in Tamil Nadu and, specifically, in the socio-economic characteristics of the class of agricultural labourers in Thanjavur district over the nineteenth century. She documents the transition from the beginning of the century, when the class of agricultural labourers was in a condition of corporate bondage, to the period at the end of the century, when agricultural labourers were bonded to individual *mirasdars*. By the end of the century, a group of

[18] For a description of the place and role of landless labourers in agrarian society in South India from the 6th to 17th cent., see Sastri (1955, pp. 327–8); for analyses of the labour force in agriculture in Tamil Nadu in the 17th and 18th cent. see P. Menon (1986, pp. 183–8); see Kumar (1965) and Hjelje (1967) for research on the 19th cent.; see also the survey in Petra Heidrich's recent monograph (Heidrich 1988, particularly ch. 4).

'occasional labourers' (the quantitative dimensions of which she has not been able to estimate), precursor of the casual labour force of this century, began to emerge in the Cauvery delta. The relevance of her research to this study also lies in providing a description and analysis of the historical antecedents of many of the obstacles to freeing the labour force and of forms of institutionalized unfreedom that continue to exist in contemporary Tamil Nadu agriculture.

An outstanding feature of the class of labourers at the beginning of the nineteenth century was its corporate identity and condition of corporate bondage; agricultural labourers were bound as a group to the cultivation of wet land in a village:

> [I]t is clear from the information from late eighteenth and early nineteenth century Thanjavur that the essential form of labour was still set within corporate structures. The labourers formed a social group not only in the act of labour, but also in their relations to the village as a whole and to the group of mirasdars in particular. (*ibid.*, pp. 155–6)

The caste system played a pivotal role in ordering social relations and establishing the social status of the labouring class; from the evidence it appears that on the wet-land regions of the Cauvery delta, this class was drawn exclusively from the so-called untouchable castes. Combined with the absence of social mobility was occupational immobility and the lack of freedom of the labourer to choose between alternative forms of employment. And to this was added geographical immobility: legal and social sanctions provided for 'absconding' labourers to return to 'cultivate the land to which they belong' (*ibid.*, p. 141).[19]

Agricultural labourers on the Cauvery wet lands were clearly unfree to sell their labour power; they were also propertyless, denied any rights over agricultural land, stock or implements in the village.

The basic remuneration of agricultural labourers also reflected their group identity: it was a share of the gross produce of the village (the share varied from village to village) which was handed over to the group. Agricultural labourers were given house sites and backyard plots, and they received cloth and money from the community of village proprietors at stated periods such as festivals.

Certain other features of the conditions of agricultural labourers in the early nineteenth century are of interest here. The first is that the differences that continue to exist in the forms of labour on types of land with differing irrigation and cropping patterns have long historical roots.[20] There were important qualitative differences in forms of labour and in relationships

[19] 'Most of the sources identify a group (often a particular sub-caste) of labourers with a specific region where their ancestors had laboured and within which they were bound to labour' (S. Menon 1983, pp. 140–1).

[20] Variations in forms of labour on different irrigation types of land in Gokilapuram and the Cumbum Valley in the contemporary period are dealt with in subsequent chapters, particularly ch. 9.

between the community of proprietors and agricultural labourers on unirrigated land and on canal-irrigated paddy land in the Cauvery delta. An elaborate and rigid structure of control over the labourer characterized relationships in the surplus-producing paddy region, while relationships that were less elaborate and restrictive characterized cultivation on unirrigated land (although the precise nature of relationships on dry land is not explored). Secondly, the inferior traditional offices of watchmen (*thalayari*), sweepers, and cleaners (*thotti*) were held by harijans, a practice that continued statutorily until 1981 and, in transmuted form, continue to the present day in the village of study. Thirdly, although at the village level the control of the labour force was still that of a class under corporate bondage, the concentration of land control in the hands of a few *mirasdars* provided a basis for transition to forms of individual control of the labour force.

While landless labour existed before British rule, fundamental changes in the character and the size of the labour force occurred after its imposition. Changes in the agrarian regime in the nineteenth century—the erosion and dissolution of older forms of ownership, particularly corporate forms; the increase in the numbers of agricultural labourers, particularly as a result of the revenue demands of the colonial power on the peasantry; and new avenues of employment for agricultural labourers, particularly as plantation workers overseas—made for important changes in the forms of employment of the labouring people of rural Thanjavur.

Certain conspicuous features of the agrarian scene at the end of the nineteenth century are of importance here. First, in a development that paralleled the erosion of corporate structures in the sphere of ownership, agricultural labour was increasingly employed—in a relationship of bondage—by the individual proprietor or *mirasdar*. Remuneration was still a share, but with the important difference that it was a share of the product of the land owned by the individual employer, not of the aggregate produce of the village. Other elements in the maintenance of labour power were also part of the individual labourer–landowner relationship. Debt bondage, of which there were instances at the beginning of the century, became a form of tying labourers to particular employers that was practised widely. Secondly, increased emigration and the increase in numbers of the labouring population of the village took place concurrently with the creation of alternative sources of employment as well as changes in the composition of the labour force. An important source of demand for rural labour—forced labour—was at public works, particularly irrigation works. Thirdly, since the concentration of economic power in individual hands was not accompanied by the destruction of other socio-political institutions that characterized the *corporate* structure of the village community, aspects of the client status of the labouring community *vis-à-vis* the community of village proprietors also continued to exist. Traditional village servants were still drawn from the same castes (and often the same families); the mode of payment of farm servants, including payment at festivals, was still influenced strongly by traditional practices; and the members of the so-called untouchable castes still predominated in the labour force:

> The labourers were still drawn mainly from the untouchable population of the village and were therefore lineal descendants of the labourer in pre-colonial times. This . . . appeared to give the labouring population a particularly timeless character [which] seemed further strengthened by the repetitive and constant cycle of paddy cultivation . . . (*ibid.*, p. 479)

Of great significance for future development was the emergence from among agricultural labourers of a contingent of 'occasional labourers', whose members, though still confined by the traditional village system, were not bound, as were farm servants, to employment by an individual employer. This group was the forerunner of the casual labour force of this century.

In recent years, authoritative work on agrarian relations in post-Independence India by two social scientists working in different parts of India has paid careful analytical attention to the fact that in the relationship between employer and employee in agriculture, history dies hard. Jan Breman's (1974) research on the decline of the *hali* system of attached labour in Gujarat represents an important, and paradigmatic, case study of the emergence of 'free' labour in Indian agriculture.[21] The decline in the prevalence of this system, once deep-rooted in parts of Gujarat, was not swift or precipitous, but took place over a prolonged period:

> There was certainly no sudden turn in the relationship . . . nor is it readily conceivable that such an important institution in the social and economic life of a large part of the agrarian population should come to an end within such a short period of time. (*ibid.*, p. 68)

Breman describes different factors contributing to this decline. He argues against the contention that it was a result of population growth and against the position that the decline of *hali* was associated with an alleged decline in the economic status of the landlord-employer class. He sees the disintegration of the system as being rooted in changes in the scale and type of agriculture pursued; among other things, 'because of a drastic change in the crop system, the landlords no longer needed to use permanent labour' (*ibid.*, p. 74). He also links the decline of the *hali* system to migration, to the decline in the system of providing marriage loans, to factors of a socio-political nature (including legislation, the role of the judiciary, and the role of the freedom movement), as well as to new aspirations among agricultural labourers. Complete dependence on the employer, once identified with personal security, has become an onerous liability in the present day.

It is an important feature of the status of an agricultural labourer that unfreedom is manifested not only in the labourer's relation with a specific

[21] For an earlier set of studies on agrarian relations in western India by economists that discusses the existence and erosion of the *hali* system, see Vyas (ed.) (1964).

employer or his or her status in a specific contract, but in his or her place and role in village society as a whole. Unfreedom, in this context, has been conceptualized within the framework of *jajmani* relations and patron–client ties, and the use of these categories has been the subject of controversy in recent peasant studies.[22] Current research also documents the erosion of patron–client ties as a result of the emergence of new forms and levels of production and the breakdown of the isolation of the individual village community. Breman describes the declining importance of patron–client relations over this century, and identifies such ties as being a specifically feudal relationship. Another noteworthy point that emerges from Breman's work in Gujarat, Gough's work in Thanjavur district, and Nagaraj's village studies in Dakshina Kannada district in Karnataka is that the withering away of patron–client relations is distinctly asymmetrical as between the patronage and clientage aspects of the relation: the landlord who ceases to extend patronage still expects, and is often able to secure, services that derive from the client status of the labourer (on this, see also Saldanha 1989).

Kathleen Gough conducted village studies in Thanjavur in the early 1950s and the late 1970s (Gough 1981, 1987). The first study was conducted in the early years of independent India and the second after an important phase of agricultural change, and each after signal periods of activity among the rural poor by Communist organizers in the villages of Thanjavur, particularly east Thanjavur. Gough's studies provide landmarks in our understanding of historical transition in the socio-economic status of labourers in Tamil Nadu village society. Her discussion of conditions in Thanjavur as they existed in the early 1950s deals with the district as well as two villages where Gough did field-work, Kumbapettai in west Thanjavur and Kirippur in the east. Gough employs an analysis of production relations as well as social relations to classify agricultural labour, providing information in dense descriptive detail on distinctions of caste and social status, and distinctions in tasks performed and in levels and forms of remuneration. While the thrust of her argument is that the trend is one of the transformation of the work-force towards proletarian status, she is careful to note and describe the impress of history, of traditional village society, and of pre-capitalist forms of oppression on the conditions of life and labour of agricultural workers.

Gough identifies three broad types of agricultural labour in the villages of Thanjavur in the early 1950s. The first were the *pannaiyals*, farm servants of the former 'untouchable' castes, lineal descendants of the landless labourer of the historical past. Although this was still the dominant type, the number of farm servants was decreasing every year. Gough's description of the dominance of landlords over socio-economic life in the village is particularly important; farm servants were bonded by debt to landlords and, since the system of bonded

[22] See Brass (1986*a*) for a criticism of explaining bondage in terms of patron–client relationships.

labour still had legal sanction, farm servants could free themselves and gain some degree of mobility only by repaying the debt, having it paid by a new master, or being 'forgiven' it. Bonded labourers who ran away could be brought back by other landlords or the police. The period of service of a labourer with a particular landlord or landlord family continues to be a hardy indicator of the labourer's unfreedom: in Thanjavur in the early 1950s, while some farm servants had served the same employer's family over generations, with the increasing instability in the system a larger number had changed employers over their working life. Farm servants, who were employed by individual landlords, continued to have duties towards the village community, particularly in respect of irrigation and temple tasks (Gough 1981, pp. 50 and *passim*).[23] They also had rights to certain common property resources: 'mud and palmyra leaves to build their huts, fuel, small fish, crabs and rats from the fields and irrigation channels, carrion beef and the hides of dead cattle, and gifts of cooked food at village temple festivals' (*ibid.*, p. 51).

The second group of agricultural labourers was also composed of farm servants, but distinguished from the first by the fact that they were (middle- or low-ranking) caste Hindus. Differences in caste also made for differences in the types of duties that they performed, in their privileges, and in levels of remuneration.

The third type were casual workers. Gough describes some of the factors in the development of this section of the labour force, factors increasing the supply of labour while working to erode the system of employing permanent farm servants. She sees a 'permanent superfluity of labour' as having emerged after the 1930s. In addition, there were concurrent processes of population increase (enhanced by return migration, particularly from overseas) as well as the fragmentation (and presumably the loss) of small land holdings. Regular hiring was also associated with a decline in the share of grain payments in the total earnings of hired labourers. An important factor in the increase in the casual labour force was legislation passed in the wake of struggles led by the Communist Party in the region.

Casual labourers had no customary rights, although in practice in some cases employers did provide house sites, building material, and fuel. Gough summarizes the condition of this section of the working people in the early 1950s in this way:

> Although they were free to leave the master, and usually worked more limited and regular hours than did *pannaiyals*, coolies had no one who accepted responsibility for

[23] Compulsory duties at temple festivals, such as pulling the temple chariot, by sections of the agricultural labour community continue to this day in many parts of Tamil Nadu; in Gokilapuram in 1977, as we shall see, irrigation workers still had special tasks at the annual temple festival. The association between irrigation and compulsory labour has been particularly close: the substantive clauses of the Compulsory Labour Act, enacted in the Madras Presidency in 1858, deal almost exclusively with commandeering labour for work on channels, tanks, and embankments.

their survival throughout the year. In short, they were casual labourers of the most depressed variety. (*ibid.*, p. 55)

Gough points to two important features of change in the period between her two studies of Kirippur and Kumbapettai. The first was a 'dramatic rise' in the category of landless labourers in both villages; the second was that the numbers and proportions of attached labourers declined greatly while casual labour greatly increased. With poverty and unemployment as endemic features of the modern agricultural labourer's situation, the process of 'depatronization' continued; at the same time, corporal punishment and similar forms of abuse by the landlords declined under the impact of social pressure and the influence of the Communist movement. 'In short', writes Gough, 'the agricultural labour population had become less pre-capitalist and more proletarianised' (1987, pp. 289–90).

The study of hired labour–employer relationships in agriculture by economists was off to something of an early start in independent India. Agricultural wages and systems of payment was one of the themes for discussion at the 1948 conference of the Indian Society of Agricultural Economics (*IJAE* undated) and in the 1950s, two pioneering works in this field were published: Surendra Patel's *Agricultural Labourers in Modern India and Pakistan* (Patel 1952) and Daniel and Alice Thorner's paper 'Employer–Labourer Relationships in Indian Agriculture' in 1957 (Thorner and Thorner 1962). Although their work relied heavily either on secondary data or on a somewhat impressionistic data base and are today rather out of date, they were of lasting methodological value: employer–labourer relationships in agriculture were seen in terms of the relative freedom of agricultural labourers to sell their labour power to employers (and, particularly in the case of Patel's work, in terms of the freedom of the agricultural labourer from ownership of the means of production as well). In a passage that is applicable to the study of agrarian relations in the present day (and provides a frame of reference of a great deal of the discussion in this book), the Thorners wrote:

For the purposes of analysis, the most important distinction to be made is between those arrangements in which the labourer contracts freely and those in which his bargaining power is abridged. A free labourer is one who is able to accept or reject the conditions and wages offered by the employer. If he wishes he may refrain altogether from working. Once having taken a job, he can decide to give notice and quit. Economic stringency may indeed compel a free labourer to agree temporarily to terms he does not consider favourable. But his basic right to refuse work or to seek alternative employment remains uncompromised.

An unfree, or bond, labourer, by contrast, is one whose bargaining power is virtually non-existent, or has been surrendered. Such a labourer does not possess the right, or has yielded the right, to refuse to work under the terms set by his master. Through custom, compulsion or specific obligation, the bond labourer is tied up to his master's needs. He can neither quit nor take up work for another master without first receiving permission. (*ibid.*, p. 21)

The work of the Thorners and Patel, which refer to the 1950s and earlier, pre-date the major technical and institutional changes that have taken place in Indian agriculture. The discussion by economists of forms of production and the nature of the capitalist tendency in Indian agriculture in recent years gave rise to different streams of research literature; one of these was the discussion of the socio-economic characteristics of the labour force in agriculture. The compendium of relationships encompassed by a system of production seems to be summarized, as it were, at that point where worker meets employer. This has impelled economists to focus their attention on employer–employee relationships in agriculture with renewed research interest. Some writers have attempted to conceptualize the question of the socio-economic underpinnings of the relationship between employers and hired labourers in the context of the production system as a whole. There is also an extended literature relating to the labour force in contemporary agriculture and of relevance to the question of its proletarianization. This literature derives from work on class differentiation in rural society and forms of hired labour, on the supply of and demand for hired labour in rural labour markets, on labour migration and the role of village boundaries (particularly their role as a factor in wage determination). It also deals with types of labour contracts and the labour process in agriculture as well as with the material conditions of life of the rural poor.

A remarkable feature of the labour market in Indian agriculture (as well as in other countries in South and South-East Asia) is the complex diversity of labour contracts and the forms of employment of agricultural labour, as well as the diversity of patterns of change under the impact of new methods of production in these forms. Institutional forms of labour hire, of the mode of payment of wages, and of the gender division of labour differ widely across the country and across regions. As we shall see, even a single village can accommodate many different ways of organizing the labour process at different tasks and use a wide range of local names, often village-specific, to describe them.[24] Alongside this phenomenon is a clear association between agricultural growth and the increased prevalence of casual labour, a move away from long-term contracts to daily-paid and piece-rated work in different parts of the country.[25] (An important exception here comes from the work of Sheila Bhalla (1976),

[24] There is, again, a very extensive literature on this subject. See e.g. S. Bhalla (1976); Clay (1976); Hirashima (1977); Chattopadhyay (1979); Bardhan and Rudra (1981, 1983); Gough (1981); Nagaraj (1981); Binswanger and Rosenzweig (eds.) (1984); and others.

[25] Kandaswamy (1964); Djurfeldt and Lindberg (1976); Dasgupta (1977); Parthasarathy (1977); Mencher (1978); Chattopadhyay (1979); Nagaraj (1981); Nakamura (1982); Parthasarathy and Pothana (1983); Hatti and Raagard (1985); C. R. Reddy (1985); Drèze *et al.* (1986); Drèze and Mukherjee (1987); Gough (1987); B. Harriss (1987); van Schendel and Faraizi (1984); and others. Hobsbawm and Rudé point to the shortening of the period of hire as an important feature of the conditions of employment of farm-hands in England in the mid-19th cent. They write, 'In a word, the farmhand became essentially a casual labourer, hired and dismissed at will, and lacking even the guarantee, as he set out in a misty dawn, that he would return home that night with any earnings at all' (Hobsbawm and Rudé 1973, p. 24).

whose data from Haryana from the early 1970s show a predominant proportion of workers on long-term contracts in one region of Haryana.)

In dealing with some of the major contributions to the literature that relates directly to the freedom of the labour force, it may be useful to start by setting out some opinions at opposite ends of the spectrum. At one end, there are economists who state that non-contractual, or 'extra-economic', forms of coercion do not constitute significant barriers to the free sale of labour power. This has been argued by Pranab Bardhan and Ashok Rudra for West Bengal (Bardhan and Rudra 1981, 1983, particularly ch. 1), and by others for other parts of India. (Among these sources is a detailed and useful study by Rammanohar Reddy of Varhad in Maharashtra, a predominantly unirrigated region which, nevertheless, has a high degree of commercial cotton farming (C. R. Reddy 1985).) Bardhan and Rudra recognize that there can be various obligations due to the employer by a worker, that a worker may have to be at the beck and call of a loan-giving or land-allotting employer for a specified period of time, and that a worker may or may not be free to sell his or her labour power to other than a specific employer at particular periods of time. The contention of Bardhan and Rudra's that has received a great deal of attention, particularly in the literature on interlocked markets, is that a major constraint on the free sale of labour power arises where workers pledge their labour to an employer during the peak season in return for privileged employment provided by that employer in the lean season. For all this, nevertheless, Bardhan and Rudra consider to be incorrect the characterization of constraints on the free sale of labour power as non-contractual (that is, outside the sphere even of implicit, not necessarily formal, contracts) or extra-economic forms of coercion. In their collaborative work they emphasize 'the absence of extra-economic coercion by the employer and the mutual advantage of the employer–employee contracts' (Bardhan and Rudra 1981, p. 105).

At the other extreme, we have the view that the development of the labour force in agriculture into a recognizable agrarian proletariat has not taken place at all, that labourer–employer relationships in agriculture have little or no affinity with capitalist relations of free hire. An extreme version of this view is that the degree of development of a labour force on capitalist principles is precisely zero, and this view is, ironically, expressed in the work of Ashok Rudra, in his report on a study of the conditions of labour hire in Birbhum district, West Bengal (Rudra 1982, particularly ch. 11). The general 'model' at the end of Rudra's study has it that agricultural labourers, being subjected to a host of 'extra-economic coercive forces', are not free to sell their labour power, that the basis of exchange on the labour market is personal knowledge between employers and labourers, that labour power is not commoditized, and that 'free labour is totally absent in the labour power of an agricultural labourer' and, in general, that

> the way the entire community of labourers in a village remains at the mercy of the handful of employers in the village and gets dominated in their entire life process by

even fewer leading employers and the way the life process and the labour process gets conditioned by the restrictions and taboos of social custom imposed by the Village Society are examples of extra-economic coercion *par excellence* which rob the labour relations of any trace of capitalist character. (*ibid.*, p. 116)

The Soviet scholar Rastyannikov, in his assessment of the position of hired labour in agriculture, poses the question in this way:

Who are they, the unpropertied population of the Eastern countryside that forms a colossal reserve of the nation's labour force? What are the social types among them? What is the character of the changes that have taken place in their system of social links? (Rastyannikov 1976, p. 67)[26]

Can the labour force be characterized as proletarian? Rastyannikov first discusses the nature of the production units on which hired labourers expend their labour power and the question of whether hired labour in agriculture produces surplus value at all. In his view, the principal factor in the mass employment of hired labour in agriculture is the struggle for sources of subsistence in petty production, that is, that the major demand for hired labour comes from an impoverished peasantry, which is involved basically in subsistence farming. From this point of view, hired labourers do not, in general, work in a production system where surplus value is produced, and they do not, in general, constitute a force of wage labourers or an agrarian proletariat:

The hired labour drawn into the agricultural structures functioning on the basis of simple reproduction evidently cannot be regarded as *wage labour* in the political economy sense of the word (i.e., as labour producing surplus value for the sake of its further growth). The hired labourers bringing their labour force into these structures are a type of *agrarian protoproletariat*. In India's multiform agrarian structure people of this type form the largest contingent among the sellers of labour power. The hired labourer of this type does not create surplus value (while in huge sectors of the agrarian economy he does not create even a simple exchange value) and he is unable to provide himself by his labour with the necessary minimum subsistence. He is therefore a *pauper in the fullest and absolute meaning of the word*. The scale of pauperism in the agrarian society of one region or another may be most fully gauged by precisely the numerical strength of the agrarian protoproletarians and the scale of their personal consumption. (*ibid.*, p. 69)

A second part of Rastyannikov's assessment of the proletarian (or protoproletarian) character of the labour force relates to the freedom of the worker to sell his or her labour power. Here he sees the continuing sway of forms of group dependence (the caste system and untouchability) and individual dependence (forms of debt bondage) as formidable and as yet unsurmounted obstacles to the freedom to sell labour power and as contributing to the underdevelopment of a proletariat in Indian agriculture. Wage labour as a category is still at a low

[26] His arguments are presented in more elaborate form in Rastyannikov (1975, 1981).

stage of development in Indian agriculture, according to Rastyannikov, and its development is marked by unevenness across regions.[27] He sees hired labourers, therefore, not as a single class, but as a '*conglomerate of lower social groups*, whose economic image is determined by the character of the economic structures in which they are employed' (*ibid.*, p. 73).

Utsa Patnaik, in an overall evaluation of the extent and nature of the development of capitalist tendency in Indian agriculture, analyses the trend in the following way:

> the capitalist path in India's agriculture is one dominated by a socially narrowly-based 'landlord capitalism' with semi-feudal features, of caste subordination of workers, which is capable of raising the level of productive forces only under certain exceptional conditions, and which acts as a long-run fetter on agricultural growth, and hence on the overall growth of the economy. (Patnaik 1985*a*, p. 45)

'Landlord capitalism', in this view, is marked by semi-feudal features, it 'preserves and utilises servile, customary, caste-based relations in the village' (Patnaik, 1987*b*, p. 8), and increasing investment in agricultural production proceeds on a narrow base, socially as well as spatially. In a recent paper Patnaik has argued that there has been a trend of rising investment in agriculture only in areas where a 'substantial middle and rich peasantry has existed' (*ibid.*) over historical time. The empirical basis for Patnaik's characterization of the labour force in agriculture derives mainly from Sheila Bhalla's data on Haryana in the early 1970s[28] and, in Patnaik's assessment, while the employment of hired labour is an integral part of the agrarian regime, it is also a development that is characterized, in part, by a strengthening of relations of dependence:

> As the expansion of output with adoption of 'green revolution' technology raises overall labour demand, and imposes more stringent requirements with respect to the time-distribution of labour inputs, it becomes important for the landlords to have labour-power 'on tap' in the same way that irrigation water is 'on tap'. Methods of labour-control begin to change in the direction of longer-term contracts which carry, built into them, debt-bondage of a new type. *Given the increasingly desperate condition of rural labour, debt-bondage may be expected to increase, not decrease as the capitalist tendency develops. A reinforcing of servility and bondage is a logical corollary of the socially retrogressive character of landlord capitalism.* (Patnaik 1985*b*, p. 28; my italics)

Do continuing obstacles to the free sale of labour power represent survivals of pre-capitalist forms of exploitation of the labour force? In her introduction to a volume on bonded labour in India, Utsa Patnaik makes a forceful case for a positive answer to the question:

> India is a country of the survival and adaptation to new functions, of social forms which elsewhere have become extinct. The various forms of subordination of one class of

[27] It is worth noting that the data base of Rastyannikov's writings of the 1970s relates mainly to the 1950s and early 1960s, and derives from secondary sources.

[28] The material is analysed and discussed in S. Bhalla (1976).

human beings by another, have not been swept away here by any revolutionary upheaval; on the contrary their characteristics have accumulated, and have mutated sometimes under the impact of capitalism to produce an extraordinary amalgam of the modern and the archaic. (Patnaik 1985*a*, p. 1)

An opposing viewpoint comes from the school of thought that holds that—to paraphrase—it's all capitalism now (as in Gunder Frank 1969). The general contention of the members of this school is that the bourgeois revolution in the Third World is essentially complete and within this framework, by means of something of a definitional trick, they convert even the most backward forms of rack-rent and bonded labour into essentially capitalist forms of exploitation. Sudipto Mundle's work on bonded labour in Palamau district in Bihar bears theoretical affinity to this literature. His study is an attempt by an economist to research a truly backward area and bonded labour within it; it contains very useful material on forms of labour bondage in contemporary India and on levels of living of bonded labourers. The brief case studies by Mundle's associate Bikramaditya Singh at the end of the book provide some of the most vivid descriptions in the literature of what must be among the worst forms of contemporary agrarian bondage.

According to Mundle, the *kamiuti* system of bondage—which rests on the indebtedness, at usurious rates of interest, of the labourer on the employer—originated in Palamau at the end of the nineteenth century and signified a clean break with previous forms of exploitation, most notably the system of *begari* (forced labour). The agrarian regime in nineteenth-century Palamau was characterized by extremely low productivity and a shortage of labour in agriculture. Wages were kept down through the system of bondage, and 'the entire system of labour relations came to rest, therefore, on the arrangement of debt bondage' (Mundle 1979, p. 5). Thenceforth surplus was extracted by increasing absolute surplus: 'the increasing physical exploitation of labour, not its rising productivity, formed the principal basis of extracting the surplus product' (*ibid.*, p. 91).[29]

Do the worst forms of non-contractual, extra-economic coercion indicate a survival of pre-capitalist forms of exploitation with debt bondage? According to Mundle, they do not. He draws a barrier between the *kamiuti* system and systems of forced labour that preceded it, distinguishing what he calls the 'new' relation of *kamiuti* from the 'traditional' system of forced labour or *begari* (*ibid.*, pp. 95–6). (And here he also uses Marx's theoretical category of the formal subordination of labour by capital, arguing that the structure of formal subordination under capital persists in Palamau.)[30]

[29] There are, of course, limits on increasing output through an unlimited expansion of absolute surplus value in agriculture as compared with industry (see e.g. Georgescu-Roegen 1976), because of the differences in the nature of the production process in each of them—there are just so many days, for example, in the space of which a paddy seedling can be transplanted. For that reason, there are limits as well on drawing a parallel on these grounds between bonded labour in agriculture and sweated labour in a workshop or manufactory, as Mundle does.

[30] Here is an example of the application of Marx's category; the example derives almost wholly

Mundle's methodology in dealing with changes in agrarian relations that took place in the nineteenth century is very different from the one that Saraswathi Menon applies to a historical study of labour relations in Thanjavur district in the nineteenth century, as well as from those of Kathleen Gough and Jan Breman. The important difference is that in their analyses these three authors not only record the forms of exploitation that gained in dominance, they also document the way in which evolving new relations were influenced deeply by traditional forms.

Differing assessments of *empirical* reality by Bardhan and Rudra, Rudra, Rastyannikov, and Patnaik are essential components of their differing assessments of the nature and extent of wage labour in India. In Bardhan and Rudra's assessment, extra-economic forms of coercion are absent from employer–employee contracts. In Rudra's assessment, their influence is total. Rastyannikov sees forms of group dependence and individual dependence as being major obstacles to freeing labour power. Patnaik's analysis is among the few that considers the relations of production in agriculture in the context of changes in the economy as a whole. A specific feature of her assessment is that she expects that debt bondage will increase with the development of the capitalist tendency in Indian agriculture, and that landlord capitalism reinforces bondage and servility.

Rastyannikov argues that hired labourers in Indian agriculture do not constitute a proletariat because they do not create surplus value (and often do not even produce exchange values). This argument has an empirical aspect: is it a fact that the bulk of hired labourers do not create surplus value? It also has a definitional aspect: it suggests that only workers who create surplus value are to be termed proletarians. This is open to discussion,[31] but, for present purposes, I shall simply reiterate the definition of the wage worker used in this book: the wage labourer or proletarian is a worker who is propertyless and is free to sell his or her labour power to the employer of his or her choice.

from Ponomarev *et al.* (1980). Marx distinguishes a stage of capitalist development in which, say, a cottager takes in yarn and gives back cloth to a capitalist entrepreneur, still retaining features of a small household economy but having no direct access to the capitalist market, from, say, the stage of the manufactory, where production is more socialized and draws many workers under the same roof. The former broadly represents a *formal* subordination of labour by capital, while the latter is a move towards *real* subordination, and 'the root cause of this difference lies in the extent of alienation of . . . the means of production and the means of subsistence from the direct producer and in their opposition to him as alien property, as capital. The greater the alienation, the more complete is the formal subordination of labour to capital, which is, in turn, the precondition and basis for real subordination' (Ponomarev *et al.* 1980, i. 68). The analysis provides a very useful way of distinguishing different phases of the development of the labour process; it is not clear, however, that it is applicable to the situations to which it has often been applied in Indian agriculture, or that it can be commandeered to 'prove' that where labour has been formally subordinated by capital, pre-capitalist forms of exploitation cease to be an important component of production relations.

[31] Marx, for instance, takes into account the category of wage labourers who do not produce exchange-value. Cohen (1978) cites Marx's example of the proletarian seamstress employed to produce clothing for the capitalist's wife, who 'is retained so that use value, not exchange value, will accrue' (Cohen 1978, p. 182).

By contrast, in Sudipto Mundle's perception, where labour has been formally subsumed by capital (as he asserts is the case in Palamau), the relation between employer and employee is essentially a capitalist relation and a worker is essentially a wage worker *even when he or she is unfree* and, indeed, even when the worker is bonded. The definition of wage labour used in this book is clearly different from Mundle's.

The foregoing is not a comprehensive review of the issues involved; it is an attempt to give a flavour of the debate and of the complexity of the factors that mould the socio-economic characteristics of agricultural labourers. The debate provides rich ground for research, and for polemic; the assessments presented in the preceding paragraphs, as well as the contributions of others to specific aspects of the study of agricultural labour in India and Tamil Nadu are discussed in later sections of this book.

The process of proletarianization can only be identified as the product of a wide set of socio-economic circumstances. To list those that I have attempted to deal with in this study:

- the level of development of the productive forces in agriculture; irrigation, cropping patterns, and work participation; levels of productivity in agriculture and developments in farming practices (Chapter 2);
- the position of different classes in the economy and their relative position as owners of land and other assets and as employers and workers; occupational structures and the mediation of the occupational pattern by the caste system; literacy and work participation among different strata of the village population (Chapter 3);
- the expansion of the labour force and the sources of expansion (broadly speaking, issues involving the 'supply of labour') (Chapter 4);
- labour absorption: the demand for labour generated by new levels of development of the productive forces and by the general diversification of economic activity (Chapter 5);
- the work calendars of agricultural labourers; underemployment and seasonality in employment patterns (Chapter 6);
- the material conditions of life of agricultural labourers (in particular, wages, earnings, and indebtedness) and, in general, the (adverse) conditions in which labour power is reproduced in rural society (Chapter 7);
- the existence of remnants of older socio-economic formations and the existence of relations of unfreedom which involve the socio-economic dependence of the hired labourer on his or her employer (Chapter 8); and
- the nature of the labour process itself, in different crops and at different operations; the specificities of the task of mobilizing workers for different types of work; and, in general, the ways in which the social character of production asserts itself in a modernizing agrarian context (Chapter 9).

This study is an attempt to put together something of a composite picture of the socio-economic characteristics of agricultural labourers in the context of a changing agrarian system on the basis of evidence collected through an area-limited, time-limited field study. The study is based on one village, Gokilapuram, in the Cumbum Valley of the Madurai district of Tamil Nadu. By concentrating on a single village in the region in the contemporary period the aim has been to achieve a degree of detail and to cover a range of issues relating to agricultural labourers that may not otherwise have been possible.

It is certainly not the argument of this book that the individual village stands apart as an isolated socio-economic unit, or that the isolation of the village is not breaking down. Nevertheless, the intricate web of interrelationships that characterize village society continue to distinguish the village ('village' variously defined) in South Asia[32] as an entity distinct from, say, a village that is merely a unit of neighbourhood.[33] And despite the importance of forces crucial to the process of proletarianization that are eroding village boundaries and their role in daily life, if the complexity of employer–employee relations in agriculture is to be rescued from anonymity, study at the village level appears imperative.

The choice of this specific region (if we put aside the various personal and circumstantial factors involved in the researcher's selection of an area for study) has been purposive: the Cumbum Valley stands out in Tamil Nadu as a vanguard agrarian region. It is an area of relative modernization of agriculture, with a relatively stable irrigation system, a diverse cropping pattern, relatively advanced levels of agricultural technique (paddy-yield levels are among the highest in South India and compare favourably with areas of East Asia), with a developed transport and agricultural extension infrastructure. The agriculture of the region draws on a numerically preponderant, largely settled force of agricultural labourers, supplemented by seasonal inflows of migratory labour.

The Cumbum Valley is a distinct geographical and agro-economic region within Madurai district. The Valley is shaped, roughly speaking, like an inverted triangle with a rounded apex, wedged between the Cardamom Hills, whose watershed marks the western and south-western wall of the Valley (and the border between Tamil Nadu and Kerala) and the High Wavy and Erasakkanayakanur Hills in the east and south-east. Its specific features include loamy and sandy soils of comparatively high fertility and assured irrigation for a significant part of the region.

The geographical area of the Valley roughly coincides with the Uthamapalayam taluk in south-western Madurai district (although it extends to some villages in the Theni block of Periyakulam taluk), about 50 villages in all. Uthamapalayam taluk lies between latitudes 9° 34′ and 10° 10′N. and longitudes 77° 10′ and 77° 2′E. and has an area of 1,426 square kilometres.

[32] See Connell and Lipton (1977) for a discussion of the specific features of the village as a unit of study in South Asia, and for 'tentative approaches' to a definition of a village.

[33] On this point, see Gough (1963).

The population of the taluk by the Census of 1981 was 485,899 persons, with an average population density of 341 persons per square kilometre. Over the two preceding intercensal decades, 1961–71 and 1971–81, the annual compound rate of growth of population was lower in the taluk and village than in the district or State (see Table 1.3). The rate of growth of population in Gokilapuram village has been, by Census figures, particularly low, especially in the 1970s. The decadal rate of growth of population in Gokilapuram over the intercensal period 1971–81 was only 6 per cent. The natural rate of increase of population in the rural areas of Tamil Nadu between 1971 and 1981 was of the order of 17 per cent; an important reason for the difference between that and the rate of growth in Gokilapuram was substantial emigration (this also appears to have been the case in the 1960s, although on a smaller scale). The most important emigrant destinations were cardamom and other plantations, on the Kerala side of the border and in Tamil Nadu, to the south of the Valley.

Table 1.3 *Annual compound rate of growth of population, by region, 1961–1971, 1971–1981* (%)

Region	1961–71	1971–81
Tamil Nadu	2.03	1.60
Madurai district	2.06	1.41
Uthamapalayam taluk	1.88	1.07
Gokilapuram village	1.35	0.57

Source: Census of India.

Persons belonging to the scheduled castes are generally a higher proportion of the population in the village (1981: 23 per cent) than in the taluk (1981: 14 per cent), the district (1971: 15 per cent), or the State (1971: 19 per cent) (see Table 1.4).

The taluk has four panchayat unions-cum-development blocks—Cumbum, Uthamapalayam, Chinnamanur, and Bodinayakkanur. There are five

Table 1.4 *Scheduled castes and tribes as a percentage of total population, 1961, 1971, 1981*

Region	1961	1971	1981
Tamil Nadu	18.8	18.5	n/a
Madurai district	15.4	15.1	n/a
Uthamapalayam taluk	12.4	13.3	13.8
Gokilapuram village	20.4	19.0	23.2

Note: n/a = not available.

Source: Census of India.

medium-sized towns (by Census classification), with populations of 20,000 to 60,000 persons each.

Gokilapuram village is in the Uthamapalayam block, about 6 kilometres east of the taluk headquarters at Uthamapalayam. It is off the major highway, although served by long-distance and local bus routes. To the west and south of the village, bounded on the west by the Periyar River and on the east by the Palayamparavu Channel, are more than 600 acres of prime paddy land, irrigated by an intricate network of field channels that branch off from canals and canal-fed tanks. Ten to 15 kilometres to the east of the village, the eastern range rises sharply from the plains. While the land to the east of the village and up to the hills (this tract was formerly within the *zamindari* estate of Erasakkanayakanur) is predominantly unirrigated, it is also marked by a growing, and in economic terms assertive, sector of land irrigated by wells.

When the canals and tanks are full and the paddy is up, Gokilapuram is a village set in a sea of luminous green; it is a plains village set close to the hills, one of open spaces and of tamarind and riverside coconut groves where the shade is dense, a village celebrated in the folk poetry of the region for its rich natural endowments.

Major Sources of Primary Data

I was involved in collecting primary material for this book in Gokilapuram and the Cumbum Valley from 1977 to 1986. Most of the material comes from three sources:

- a census-type survey covering 650 households resident in Gokilapuram, conducted between April and June 1977, for which the reference year for agricultural operations was 1976–7;
- case study and other observational material collected in the village and region between 1977 and 1986; and
- a sample survey of 43 landless agricultural labour households in the village conducted in June 1986, for which the reference year for agricultural operations was 1985–6.

There follows a brief description of the concepts and definitions used in the surveys and in this book;[34] specific aspects of the survey and case-study material, of data, and of data collection, are discussed at different points in the book, and particularly in Appendix 1.

The village. The basic unit of revenue administration, the revenue village, is the village unit used in most official surveys. The World Agricultural Census, 1970–1, for Tamil Nadu is recomputed from revenue records; the Censuses of

[34] Concepts and definitions that relate to specific subsections of the household schedule are dealt with at the relevant places. In general, the definitions used here apply to the whole study, not only to the survey of 1977.

India for 1971 and 1981 deal, in general, with the revenue village; and the National Sample Survey (NSS), for purposes of sample stratification, uses the revenue village. In Gokilapuram, the panchayat village (the basic unit of development administration and of local administration) and the revenue village coincide; this gives us the advantage of having village-level data for the Census of 1961 and successive Censuses that are comparable.[35]

Ours was a survey of households resident in Gokilapuram village.[36] A consequence of covering households resident in the village is that the ownership and operational holdings covered by the survey lies (in administrative terms) in the revenue village of Gokilapuram and in adjacent revenue villages, particularly Anamalaiyanpatti (across the canal and to the east of Gokilapuram) and Uthamapalayam, to the west. The survey does *not* cover households resident outside Gokilapuram who have land holdings within the administrative jurisdiction of the village. For this reason, our survey appears to have taken a cross-sectional snapshot, one that (in the event) captures rather better the irrigation regime and cropping pattern of the Valley as a whole, and the labouring pattern of agricultural labourers, while at the same time not being complete in its coverage of the agricultural land of an overwhelmingly canal-irrigated village, as Gokilapuram is.

Household. The basic unit of investigation was the household, which, following the Census of India and the NSS, was defined as a group of persons normally living together (and normally belonging to the same family) and eating food from a common kitchen. It was assumed that members of a household do not merely live together, but conduct their operations and socio-economic activities together.

This concept of household is very useful as a unit of investigation; its definition is one that respondents in the village related to very quickly. The household was also convenient as a unit of classification when we dealt with size categories of land holdings and asset holdings, details of agricultural production and aggregate income, and consumption and indebtedness. There were problems, however, with using the household as a unit of classification when we

[35] The Census of India 1961 provided data for panchayat villages in the District Census Handbooks. In order to provide comparability between the data for villages in 1961 and 1971, the Census of India 1971 gives lists by taluk of the location code numbers of each village and town in 1971, with location code numbers for the same village or town for 1961 given alongside. However, when the returns by village and town for the Cumbum Valley were actually compared using the conversion table used in the 1971 District Census Handbooks, there was a divergence (of varying degrees of magnitude) in the area of villages at the two Censuses for a large number of villages.

[36] The Census of India 1971 records that the number of households resident in the village was 733, and the survey, as mentioned, covers 650 households in the village (the actual number surveyed was 652; two schedules were discarded). Among the omissions were three major landlord households (one refused to give us an interview, the other two always managed to slip out on us on the many occasions that we took unfilled or part-filled schedules to interview them), one prominent trader household, and some agricultural labour-cum-plantation worker households that left the village for the estates (some during the survey period). For statistical purposes, the 650 households covered by the survey are considered to be the population of the village and not a sample.

attempted a socio-economic classification of households. Single households can, and in many cases do, consist of persons with different occupations; correspondingly, the sources of employment and income of its members may be quite independent and diverse.

Ownership holdings of households. Here we follow the NSS definition—land held by a household in ownership (with the right of permanent heritable possession) or owner-like possession. Specifically, the household ownership holding consists of land owned and operated with family and/or hired labour (referred to in some sources as land under 'direct cultivation' or 'self cultivation') plus land leased out and land mortgaged out.

Operational holdings of households. This is the land actually operated by, and under the management of, a household. Specifically, it consists of land owned and cultivated by family and/or hired labour, land leased in, and land mortgaged in.

The Tamil Nadu government has renamed districts and subdivided them in recent years. References in the text to district level data for Madurai prior to 1985 are to the old Madurai district, which included the present Madurai and Dindigul Quaid-i-Millath districts. 'Chengalpattu' in the text refers to the present Chengai Anna district, 'North Arcot' to the present North Arcot Ambedkar and Tiruvannamalai Sambuvarayar districts, 'Ramanathapuram' to the present Pasumpon Thevar Thirumagan, Kamarajar and Ramanathapuram districts, and 'Tirunelveli' to the present Nellai Kattabomman and Chidambaranar districts.

2

Some Features of the Agrarian Economy of the Cumbum Valley

In terms of achieving high levels of agricultural production, the Cumbum Valley must be counted as one of Tamil Nadu's most successful agricultural regions. This chapter concentrates on two key aspects of the development of the productive forces in its agriculture: on changes in irrigation and cropping patterns and on the unevenness of these changes across different types of irrigated land.[1] The chapter also looks at secondary data on changes in the work-force in agriculture in the Cumbum Valley in the recent period. While the discussion here focuses on changes that have taken place in agricultural technique in the 1960s and after, it must also be remembered that modern agriculture in the Valley, as in many other pockets of intensive agricultural development in the country, has built on certain historical advantages in terms of soil, climate, irrigation, and a settled labour force.

Since secondary data on irrigation and cropping patterns are available for the administrative unit of the taluk, the taluk is used in this chapter as proxy for the Cumbum Valley.

The distinction between three types of cultivation, with the source of irrigation as the major determinant, is of great significance to the agro-system of the Cumbum Valley. The implications of such a classification relate to distinct variations in the scale and type of agriculture. These variations include, among other things, differences in the number and type of crops grown on the land and methods of cultivation used to produce them, in the degree to which the productivity of labour has developed and the intensity and forms of employment of agricultural labour, and in the extent to which the produce of the land is sold on the market.

Well-entrenched in popular perception, the classification is as follows:

Nanjai, or surface-irrigated land, fields that are irrigated by canals and tanks,[2] of which the bulk of the acreage in the Valley is under the Periyar system (described below). The predominant crop on surface-irrigated land is high-yielding variety (HYV) paddy, a large part of which goes to the market. In some of the low-lying villages, betel, sugar-cane, and other minor crops are grown, although on a small scale. Some surface-irrigated land is planted with green manure and fodder crops during the dry months from March to June.

[1] The unequal impact of development on different classes is discussed in subsequent chapters, particularly ch. 3. Changes in farming practices involving changes in labour absorption are dealt with in ch. 5.

[2] This refers to land actually under surface irrigation and not merely those lands registered in the revenue records as 'wet'.

Punjai, or unirrigated, rain-fed land—land that 'looks up to the sky'. The quality of unirrigated land in the Valley varies, from the more productive groundnut and sesame-growing *punjai* in the south of the Valley near Gudalur, on which cultivation has acquired a substantial degree of commercialization, to what is the most familiar type of rain-fed land—dry and low-yielding land. In such tracts, unirrigated land represents the backward, traditional sector of agriculture. In years when the rain fails, agriculture is reduced to being pursued as traditional craft, in a state of near-ruin—the soil is scratched with a plough, seeds broadcast, and little care is taken of the crop until the time comes for a meagre harvest. Households whose holdings consist entirely of unirrigated land must allocate their labour time elsewhere—generally to labouring out for others—for subsistence. The crops grown on unirrigated land are mostly coarse grains of traditional varieties—*cholam*, *ragi*, *varagu*, and *samai*, and, on some fields, unirrigated groundnut. *Cumbu* is also grown, particularly in the north-west of the Valley. Pulses are planted as intercrops.

Thottam, groundwater-irrigated or garden land, where wells are the source of irrigation. *Thottam* is planted with a diverse variety of commercial crops (such as banana, cotton, high-yielding and traditional *cholam* and *ragi*, vegetables, chilli and fruit) cultivated on an increasingly intensive scale.

While there are variations within these three types that have to do with the quality and reliability of irrigation (particularly important in the case of groundwater-irrigated land), soil types and fertility, and the location and topography of fields, the main *nanjai–punjai–thottam* classification helps to stamp cultivation (and the landlords and peasants involved in different sectors of cultivation) with distinct labels.[3]

Rainfall and Irrigation

Local rainfall and its annual distribution affects unirrigated land primarily and irrigated land secondarily; as in many parts of Tamil Nadu, the major determinant of irrigation in the Cumbum Valley is rainfall on the windward side of the hills, in the catchment area of the Periyar, which is the major source of surface irrigation and an important source of recharge for groundwater. For that reason, irrigation in the Valley is not so much dependent on rainfall in the Valley itself as on the south-west monsoon in the hills of southern Kerala.

Rainfall in the Uthamapalayam Agricultural Division (which more or less coincides with the taluk and the Valley) was about 745 millimetres per annum over the period 1973–4 to 1983–4, compared with the long-term average of 943 millimetres per annum for Tamil Nadu as a whole (Guhan 1984; Uthamapalayam Agricultural Division 1985). In Tamil Nadu, this precipitation

[3] The survey also covered *thope*, or groves, land on which tree crops are raised. This included coconut and tamarind groves and also some planted with fruit, mainly mango and jackfruit.

derives mainly from the north-east monsoon (October–November), even more so in the Cumbum Valley than in the State as a whole (see Table 2.1).

It is a distinguishing feature of the Cumbum Valley that it draws together in a single, distinct, and reasonably compact geographical zone a diversity of irrigation types (surface irrigation and lift irrigation, drawing on groundwater) of cultivated land and farming conditions.[4]

Table 2.1 *Rainfall in different seasons as a percentage of annual rainfall, Tamil Nadu and Uthamapalayam Agricultural Division*

Region	NE monsoon	SW monsoon	Other summer and winter rain
Tamil Nadu	47.6	32.5	19.9
Uthamapalayam Agricultural Division	50.2	28.7	21.1

Sources: Guhan (1984); Uthamapalayam Agricultural Division (1985).

The Periyar System

Most of the area under surface irrigation in the Cumbum Valley is part of the Periyar system of irrigation, which irrigates a total of 166,880 acres in Madurai and other districts in Tamil Nadu.

In 1885, a masonry dam built across the west-flowing Periyar River was completed and the water of the river diverted to the east from the lake at Thekkady (7 kilometres from the border with Tamil Nadu) through a tunnel cut through the watershed. The river descends from the lake into the Cumbum Valley, which is the first and best-served part of the command area of the Periyar system. In the Valley, the river follows the main watercourse of the system that existed before the Periyar project, the Suruliyar-Vairavanar system.

The Periyar system is a canal and tank system, one that developed on foundations that were laid in the pre-Periyar period. Channels branch out from the main river. There are openings to crop land directly from the channels, and the command area (*ayacut*) of such vents or sluices is called the 'direct' command area of the system. Channels also lead into storage tanks, which are the reservoirs of the Periyar system, and the area irrigated by the tanks is the system's 'indirect' command area. There are 16 barrages (*anaikattu*) across the Periyar in the Cumbum Valley subdivision; 21 channels branch off from these barrages. These have a total command area of 14,784 acres of double-cropped land. There are 19 systems tanks fed by these channels; the indirect command area constitutes 27 per cent of the total command area in the Valley (PWD 1981). The period of release of the water—the intensity varies over the year—is generally from 1 June to 1 March.

[4] Some of the issues here are treated in greater detail in Ramachandran (1983, pp. 6–16; s. 1, 2).

It has been recorded that even in pre-Periyar times, the 'Suruliyar was probably the most reliable irrigation source in the district', irrigating about 12,000 acres in the Cumbum Valley (Boag 1915, p. 26).[5] Despite that, irrigation as a whole was less reliable than it is today and a substantial portion of the command area was single-cropped. (It was estimated in 1915 that not more than half the command area in the pre-Periyar period could have been irrigated for two crops (Boag 1915).)[6] There was no source of irrigation to provide for the dry months from June to September and, over much of the land where the irrigated paddy crop is now planted in June, fields were either left uncultivated or planted with a dry crop.

Of the surface-irrigated land in Uthamapalayam taluk, about 65 per cent is under the Periyar system. The rest lies mainly in Bodinayakkanur block and consists of river and tank irrigation. This is a much poorer irrigation system than the Periyar system, with most of the command area irrigated for only one crop season in the year.

Groundwater Irrigation

The land irrigated by the Periyar forms the rich green core of the Cumbum Valley; to the east and west is its groundwater-irrigated land, land irrigated by wells that are mostly fitted with electric motor pumps. (Interspersed with the garden land, and stretching beyond it to the base of the hills that form the boundaries of the Valley, are the unirrigated crop lands of the Valley.)

The major stimulus to the spread of well irrigation came in the 1950s and 1960s, with the extension of rural electrification. The process involved the conversion of land previously irrigated by wells with water drawn manually or by means of bullocks to electric pump irrigation, and the conversion of land that was unirrigated to well irrigation, accompanied, in most cases, by the consolidation[7] and improvement (including levelling) of dry-land plots.

The process of land improvement and conversion to modern garden land involves, of course, heavy private investment. The hugely increased and uncoordinated exploitation of groundwater resources has also caused a fall in the water table, particularly in the western tracts of the Valley. In one case reported from Pannaipuram, to the west of Uthamapalayam, the water level fell from about 20 feet in 1953, when electric pumps were first installed, to

[5] Under the land settlement of 1885, many of the wet lands in Periyakulam taluk, of which the Cumbum Valley comprised a sizeable portion, were compulsorily registered as double-crop lands, at 1.5 times the single-crop rate; this was not done in other taluks.

[6] Although a substantial acreage of dry land and waste land was converted into surface-irrigated land after the Periyar project, a major consequence of the project in the Cumbum Valley seems to have been the provision of a wholly secure water supply for the *existing* wet land, unlike, for instance, in the Nilakkottai and Madurai taluks, where the major consequence was the wholesale extension of canal-irrigated land.

[7] Consolidation here is effected by periodic purchases of, and foreclosures on, contiguous plots of groundwater-irrigated land and dry land and is not consolidation by legislation. Tamil Nadu does not have any legislation on the consolidation of land holdings.

more than 80 feet in 1977. This has also led to high levels of investment in well-deepening, which in many cases must be undertaken every two years. Another consequence of the present pattern of extension of lift irrigation has been disuniformity in the quality of well irrigation available in different parts of the Valley and to different cultivators. There are maintenance costs as well: electricity bills must be met and pumps kept in good repair. The growth of the highly commercialized sector of garden cultivation has led to significant changes in agricultural production and technique and in the nature of production relations in the region.

Data on irrigation and cropping patterns in Uthamapalayam taluk over the decade ending 1981 provide a useful picture of how tendencies observed for an earlier period in Tamil Nadu (see Kurien 1981, particularly ch. 3) worked themselves out in a specific region during the 1970s.

Trends in Irrigation, 1970–1971 to 1980–1981
(see Tables 2.2 and 2.3)

One problem with the statistics on irrigated cultivation is that they deal with the extent of area irrigated—its nominal acreage—and not with the availability of water and the quality of irrigation; no distinction is made, for instance, between a plot of land that depends for its irrigation on a rain-fed tank in a dry zone and one for whose irrigation there is a dependable source of water.[8]

Table 2.2 *Area under different sources of irrigation, Uthamapalayam taluk, 1970–1971, 1980–1981*

Irrigation type	Years 1970–1 (acres)	Years 1980–1 (acres)	Decadal variation (%)
Canals and tanks	17,234	19,623	13.9
Wells	23,669	39,024	64.9
Others	1,023	0	−100.0
Net irrigated area	41,026	58,647	39.9
Net unirrigated area	101,917	87,760	−13.9
TOTAL SOWN AREA	143,843	146,407	1.8

Sources: GOTN (1977); Uthamapalayam taluk, revenue records, 1980–1.

[8] The limitations of data on irrigation that do not provide information on the volume of water available from individual sources of irrigation are particularly acute when dealing with land irrigated by wells. There are also problems of undercounting the number of energized wells and recording the area served by wells inaccurately. (On the incorrect recording of irrigated land in village records, see Nakamura 1976, p. 33.)

Table 2.3 *Irrigated and unirrigated land as a percentage of net sown area, Uthamapalayam taluk,* 1970–1971, 1980–1981

Irrigation type	1970–1	1980–1
Canals and tanks	12.0	13.4
Wells	16.5	26.7
Others	0.7	0.0
No irrigation	70.8	59.9
TOTAL	100.0	100.0

Sources: GOTN (1977); Uthamapalayam taluk, revenue records, 1980–1.

Although the degree of irrigation (net sown area under irrigation as a proportion of total net sown area) in Uthamapalayam taluk is only about 40 per cent, which is roughly the same as that for the State as a whole, the quality of irrigation in the taluk is relatively high. This is because of the dominance of the Periyar system in surface-irrigated agriculture (and, as has been said, the Cumbum Valley is the first and best-served command area of the Periyar) and also because wells whose artesian sources are fed by rainfall and are recharged by the Periyar canals account for a much larger proportion of irrigation than in the State as a whole. As it happens, roughly two-thirds of the irrigated area of the taluk are irrigated by wells as against only one-third in the State (although the quality of well irrigation may no longer be on a rising curve). The significance of irrigated land is also indicated by the fact that while the irrigated area was 40 per cent of net sown area in 1980–1, it was about 51 per cent of the gross cropped area.

While the data show an increase in area under irrigation of about 40 per cent between 1970–1 and 1980–1, it must also be remembered that, amidst this, the bulk of net sown area still remains without any form of irrigation—a large unirrigated and relatively backward sector that constitutes a major drag on agricultural development.

In Tamil Nadu, systems of canal irrigation that draw water from riparian sources are generally associated with large and medium surface-irrigation projects, most of which were completed by the 1960s. In the case of the Cumbum Valley, while work has been done since Independence to modernize the irrigation system—repairing the barrages on the river, extending and lining channels, regulating water outlets, de-silting and repairing tanks—the source of surface water, the Periyar River and Periyar Lake in Kerala, has not changed, and, consequently, the total volume of water has remained roughly the same over the last century. It is not surprising, then, that the area under surface irrigation in Uthamapalayam taluk (65 per cent of which is Periyar-irrigated), which occupied about 13 per cent of net sown area in 1980–1, increased only marginally from 1970–1 to 1980–1. This 13 per cent is, however, a particularly weighty segment of total agricultural land, since it receives water for nine months (enough for two paddy crops) a year and irrigation is reliable. Land in the Periyar tract is the highest priced in the Valley; the importance of surface-irrigated land is also indicated by the fact that paddy, mostly grown on these tracts, occupies the largest share of any crop in the taluk's gross cropped area.

The increase in area irrigated by wells is the highest among different irrigation types of land, indicating a continuation of the trend of extension of irrigation through wells that began in an earlier period. The extension of well irrigation cannot, however, be an untrammelled process, particularly given the problem of a falling water table in the Valley and the widely differing quality of wells as irrigation sources.

A partial indicator of the quality of irrigation is the index of multiple cropping, or gross cropped area divided by net sown area, on irrigated land. (The conclusions here derive from Table 2.2 read with Table 2.4.) This was fairly constant for canal- and tank-irrigated land over the 1970s: it was 1.68 in 1970–1 and 1.66 in 1980–1. All land under regular Periyar irrigation is irrigated for two crops (encroachment land in the Periyar command area may be cultivated only for a single season) and the average index here probably reflects the poorer quality of surface irrigation in the Bodinayakkanur block. The trends in the multiple-cropping index on groundwater-irrigated land are interesting: it declined from 1.67 in 1970–1 to 1.39 in 1980–1. This decline may reflect the increasing importance of long-duration crops (such as banana, coconut, and grape);[9] it also suggests a possibility of a decline in the quality of well irrigation concurrent with its relative expansion.

The poor quality of cultivation on unirrigated land is also confirmed by the data. The gross cropped unirrigated area of 82,771 acres was *less* than the net sown unirrigated area of 87,760 acres, a shortfall of 6 per cent. This may be a confirmation from official data of an observed phenomenon—that some sown land on these dry and unproductive tracts is not cropped at all. If we assume that the overall multiple-cropping index on unirrigated land is greater than unity, since some unirrigated land is double-cropped, the area that is sown but not cropped is likely to be higher.

Trends in Cropping Patterns, 1970–1971 to 1980–1981
(see Tables 2.4 to 2.7)

Paddy

Paddy production in Tamil Nadu has been transformed where semi-dwarf varieties, chemical fertilizers, and pesticides have been used in combination

[9] Although even excluding 'orchards and condiments', which include the major long-duration crops, the ratio dropped from 1.88 to 1.53 for land irrigated by wells.

with adequate and controlled water supply.[10] The Periyar tract in the Cumbum Valley, where paddy farming produces the highest yields in Tamil Nadu, is an area of intensive paddy cultivation; in the six-stage ranking order of intensity (in ascending order of intensity) of rice cultivation proposed by Barker *et al.* (1985, p. 27)[11] the Periyar tracts occupy the third rank, and parts of it occupy the fourth rank. Paddy is grown in two seasons, the first (*kodai*) from June to October, and the second (*kalam*) from October to February. Although mechanization is relatively limited (see Chapter 5), and herbicides are not used, paddy farming on well-cultivated farms in the Valley conforms to Francesca Bray's lucid summary of the 'highly productive' system of rice farming:

> A highly productive system would include careful water control, based on an adequate irrigation network; the fields would be accurately levelled, carefully tilled and manured; selected seed would be sown in a separate seed-bed, to be transplanted in the main field in straight lines suitably spaced, the right amount of fertilisers would be applied and weeding would take place at frequent intervals, and herbicides and pesticides would be used to safeguard the crop. The success of the Green Revolution is premissed on such practices . . . (Bray 1986, p. 45)

Area planted to paddy increased over the 1970s and, by 1980–1, paddy was the dominant and specialized crop on surface-irrigated land, to the almost complete exclusion of other crops. Other surface-irrigated crops, mainly sugar-cane and betel, are localized and occupy a very small share of gross cropped area. In 1980–1, paddy occupied the largest cropped area among cereals (38 per cent of area under cereals and 73 per cent of all irrigated cereals) and the largest area of any single crop in the taluk (19 per cent of gross cropped area and 37 per cent of irrigated gross cropped area). A very important feature of paddy cultivation in the Valley is that by the 1980s the entire area under

[10] This has been documented for Tamil Nadu as a whole in Kurien (1981) and at the micro-level. John Harriss (1986*a*), for instance, while pointing to constraints that tend to limit the dynamism of the paddy-producing sector of agriculture in North Arcot district, also cites evidence that 'lends credence to official estimates of increased average yields of paddy from 1620 kg/ha in 1966–67 to 2635 kg/ha in 1976–77 and 3215 kg/ha in 1980–81 . . . Paddy production has been calculated to have increased by about 50 per cent between 1951–66 and 1967–84, the most important source of this growth having been the widespread adoption of HYVs' (Harriss 1986*a*, p. 8). Kathleen Gough presents valuable micro-level data on gross paddy yields in Kumbapettai village in the fertile Cauvery delta in Thanjavur district in the early 1950s. Paddy yield 'in a very good year' varied from 3.5 to 3.8 tonnes/ha. in the *kuruvai* (harvested in Oct.) season and was 4.6 tonnes/ha. in the *samba* (harvested in Feb.–Mar.) season. In 1951–2, the average gross yield of paddy in Kumbapettai was 2.7 tonnes/ha. in the *kuruvai* season and 1.9 tonnes/ha. in the *samba* season (Gough 1981, p. 234). for a description of step-by-step increases in yield in Gokilapuram, culminating in the changes of the late 1960s and early 1970s, see App. 2.

[11] The six types are: rice/fallow; rice/dry land; rice/rice; rice/rice/dry land; rice/rice/rice; and rice/upland/rice/upland. In most areas of the Valley, a dry-land fodder or green-manure crop is grown after two crops of rice. In some villages, irrigation water is released throughout the year, although its volume is greatly reduced from February to May. A key factor in making this intensity of cultivation possible is, of course, the reliability of the irrigation source (for an assessment of the role of the quality of irrigation from different sources in rice cultivation in India, see Kurian 1986, p. 58).

Table 2.4 *Gross cropped area, Uthamapalayam taluk, by crop, 1970–1971 and 1980–1981* (acres)

Crop category	Gross cropped area 1970–1			Gross cropped area 1980–1		
	Irrigated	Unirrigated	Total	Irrigated	Unirrigated	Total
Rice	29,079	59	29,138	32,493	0	32,493
Cholam	11,239	39,906	51,145	7,268	24,710	31,978
Cumbu	521	1,039	1,560	2,164	976	3,140
Ragi, other cereals	10,278	15,402	25,680	2,475	14,924	17,399
Pulses	101	3,906	4,007	964	8,185	9,149
Total food-grain	51,218	60,312	111,530	45,364	48,795	94,159
Condiments	2,729	1,292	4,021	1,707	1,978	3,685
Cotton	7,868	1,227	9,095	15,548	1,209	16,757
Orchards	2,864	2,390	5,254	8,644	1,872	10,516
Sugar-cane	54	46	100	375	0	375
Vegetables	1,208	98	1,306	4,195	113	4,308
Coconut	876	464	1,340	3,364	187	3,551
Groundnut	1,342	29,547	30,889	6,747	18,496	25,243
Sesame	6	228	234	38	5,287	5,325
Other oil seeds	7	33	40	18	63	81
Others	441	4,249	4,690	834	4,771	5,605
Total non-food-grain	17,395	39,574	56,969	41,470	33,976	75,446
TOTAL	68,613	99,886	168,499	86,834	82,771	169,605

Sources: GOTN (1977); Uthamapalayam taluk, revenue records, 1980–1.

paddy was planted with high-yielding varieties.[12] The use of these varieties is not confined to a small stratum of the population; at our survey, even small peasants cultivating a fifth of an acre of surface-irrigated land used high-yielding varieties. The cultivating poor cannot, of course, afford to use chemical fertilizers, farmyard manure, and pesticides in the same amounts as the rich do; they have less access to credit and are more in debt than the rich and they are unable to replenish their seed stocks as frequently. The transition to sowing high-yielding varieties, however, is complete in the region.

As for the village, Gokilapuram is endowed with a combination of natural conditions—soil, climate, topography, and the availability of water—greatly suited to the cultivation of paddy. At the harvest of the first crop in 1981–2 (by my field notes), paddy production on the better wet-land tracts was about 8.1 tonnes per hectare and over most of the surface-irrigated land about 6 tonnes per hectare, or between 4 and 5.4 tonnes of rice per hectare. Yields compare well with other parts of Asia (including the Japanese average of 5.1 tonnes of rice per hectare in 1971 and the central Taiwan average, in 1972, of 5.2 tonnes per hectare (Barker *et al.* 1985, table 9.2; see Table 2.8). Yields are very high in

[12] Another area which showed a very high rate of HYV utilization was North Arcot district; in the district as a whole 90% of the area under paddy was planted with HYVs (Harriss 1986*a*, p. 8).

comparison with other parts of the district and State; in the official crop-cutting survey estimates, the standard yield of rice in kilograms per hectare over all tracts in 1981–2 was 2,799 kilograms for Madurai district and 2,265 kilograms for Tamil Nadu (GOTN 1986, table V-A). An important feature of paddy cultivation, dealt with in greater detail in Chapter 5, is that while physical output per hectare is high, productivity measured in terms of kilograms of rice produced per person-day of hired labour employed is very low by international standards.

Crop Diversification: Irrigated Non-Cereal Crops

A major feature of the cropping system in the Cumbum Valley has been diversification, a shift from coarse cereals and pulses and low-grade oil-seed production to irrigated non-cereal crops. This process of crop diversification has occurred through changes in the crop mix on the existing cropped area, and not through an extension of cropped area, which has not really changed. Diversification is a process that has run concurrently with the development of

Table 2.5 *Gross cropped area, Uthamapalayam taluk, by crop, 1970–1971 and 1980–1981* (%)

Crop category	Gross cropped area 1970–1			Gross cropped area 1980–1		
	Irrigated	Unirrigated	Total	Irrigated	Unirrigated	Total
Rice	42.4	0.1	17.3	37.4	0.0	19.1
Cholam	16.4	40.0	30.4	8.4	29.8	18.9
Cumbu	0.8	1.0	0.9	2.5	1.2	1.8
Ragi, other cereals	15.0	15.4	15.2	2.8	18.0	10.3
Pulses	0.1	3.9	2.4	1.1	10.0	5.4
Total food-grain	74.7	60.4	66.2	52.2	59.0	55.5
Condiments	4.0	1.3	2.4	2.0	2.4	2.1
Cotton	11.4	1.2	5.4	17.9	1.5	10.0
Orchards	4.2	2.4	3.1	10.0	2.2	6.2
Sugar-cane	0.1	0.1	0.1	0.4	0.0	0.2
Vegetables	1.7	0.1	0.8	4.8	0.1	2.5
Coconut	1.3	0.5	0.8	3.9	0.2	2.1
Groundnut	2.0	29.6	18.8	7.8	22.3	14.9
Sesame	n	0.2	0.1	n	6.4	3.1
Other oil seeds	n	n	n	n	0.1	0.1
Others	0.6	4.2	2.8	1.0	5.8	3.3
Total non-food-grain	25.3	39.6	33.8	47.8	41.0	44.5
TOTAL	100.0	100.0	100.0	100.0	100.0	100.0

n = negligible
Sources: GOTN (1977); Uthamapalayam taluk, revenue records, 1980–1.

Table 2.6 *Gross cropped area, Uthamapalayam taluk, changes of crop between 1970–1971 and 1980–1981* (%)

Crop category	Decadal change in gross cropped area		
	Irrigated	Unirrigated	Total
Rice	11.7	0.0	11.5
Cholam	–35.3	–38.1	–37.5
Cumbu	315.3	–6.1	101.3
Ragi, other cereals	–75.9	–3.1	–32.3
Pulses	854.4	109.5	128.3
Total food-grain	–11.4	–19.1	–15.6
Condiments	–37.4	53.1	–8.4
Cotton	97.6	–1.5	84.2
Orchards	201.8	–21.8	100.1
Sugar-cane	594.4	–	275.0
Vegetables	247.3	15.3	229.9
Coconut	284.0	–59.7	165.0
Groundnut	402.8	–37.4	–18.3
Sesame	533.3	2218.8	2175.6
Other oil seeds	157.1	90.0	102.5
Others	89.1	12.3	19.5
Total non-food-grain	138.4	–14.1	32.4
All crops	26.6	–17.1	0.7

Sources: GOTN (1977); Uthamapalayam taluk, revenue records, 1980–1.

Table 2.7 *Gross cropped area under irrigated crops as a percentage of total gross cropped area, Uthamapalayam taluk, 1970–1971, 1980–1981*

Gross cropped area (GCA)	Year	
	1970–1	1980–1
GCA under irrigated crops as percentage of total	40.7	51.2
GCA under irrigated food-grain crops as percentage of GCA under all food-grains	45.9	48.2
GCA under irrigated non-food-grain crops as percentage of GCA under all non-food-grain crops	30.5	55.0

Sources: GOTN (1977); Uthamapalayam taluk, revenue records, 1980–1.

well irrigation and, as mentioned, is marked by an increase in the number of irrigated non-cereal crops planted and of the area under such crops. Over the 1970s, gross cropped area under irrigated non-cereal crops increased by 138 per cent from 17,395 to 41,470 acres and the share of these crops in irrigated gross cropped area increased from 25 to 48 per cent. Almost every subcategory within this group of crops—cotton, fruit, vegetables, groundnut, and coconut —reported absolute increases in gross cropped area over the decade.

These are some examples of farming activity in the cultivation of irrigated non-cereal crops, examples that attempt to illustrate the pace and nature of change in this sphere. The area under irrigated cotton (generally considered a price-responsive crop) showed a sharp increase over the decade, although it fluctuated over the period. One reason for this fluctuation was a pest attack for a two-year period beginning about 1978. A noteworthy feature of the cropping pattern of the Valley in the 1970s is the remarkable rapidity with which the cultivation of two new varieties of banana—Dwarf Cavendish (introduced in the taluk in 1966–7) and Robusta (introduced in 1972–3)—has advanced since the 1960s. The area under banana increased more than 10 times in the six-year period from 1969–70 to 1975–6,[13] the rich Udayar landlord families of Rayappanpatti village south of Gokilapuram having been particularly noteworthy 'innovative farmers' in banana cultivation. The area planted with irrigated groundnut increased sharply (by 40 per cent) over the decade. (There was a concurrent fall, by 37 per cent, in the area under unirrigated groundnut.) Vegetable crops are short-term crops, generally involving short-term

Table 2.8 *Paddy yields and productivity in Gokilapuram and selected locations in East and South-East Asia, 1971–1981*

Location	Year(s)	Yield (tonnes per hectare)	Productivity (kilograms per man day of hired labour)
Gokilapuram[c]	1976–81	4.0–5.4	10.4–15.6[d]
Central Luzon, Philippines[b]	1966–74	2.9	35.6
Central Taiwan[a]	1972	5.2	61.9
Hwangsunggan, South Korea[b]	1974	7.2	51.8
Japan[a]	1971	5.1	36.2
Java[a]	1977–80	3.7	25.6
Laguna, Philippines[b]	1966–74	3.5	33.1
Suphanburi, Thailand[b]	1971–72	3.4	29.3

Note: The data are, in each case, for modern varieties of rice.

[a] Derived from Barker *et al.* (1985, table 9.2, p. 126).

[b] Derived from Barker *et al.* (1985, table 9.4, p. 127).

[c] From survey data.

[d] Assuming a yield of 5.4 tonnes per hectare of rice and 346–519 person-days of hired labour per hectare.

[13] Data from the offices of the Deputy Agricultural Officer (DAO), Cumbum Block, the DAO (Banana), Chinnamanur, and the DAO (Banana Package Scheme), Uthamapalayam.

decisions in crop management. The increase in gross cropped area under vegetables is marked by quick changes in crop mix, noticeable even over relatively short periods; in 1980–1, the major vegetable crops were onion, tomato (a new variety, PKM–1, developed in nearby Periyakulam, had been introduced), beetroot, cabbage, and aubergine. The area under grape, a crop that requires high investment, particularly initially, and intensive care over the crop year, occupied a small though increasing area on groundwater-irrigated land in the Valley, particularly in its eastern and southern villages.

There has been a steady increase in the area under coconut in the taluk, a trend that accelerated in the late 1970s and in the 1980s. These trees are planted not only in the traditional grove areas, but increasingly on garden land converted to coconut fields, and are generally inter-cropped with vegetables for the first two years. While the initial investment on this presently highly profitable tree crop is high, it is seen as less demanding than other crops, in the long run, in respect of water management in groundwater-irrigated areas, and as a crop that will severely cut down on the demand for labour and, correspondingly, on labour costs on groundwater-irrigated land.

Certain features of commercial cultivation are most pronounced on groundwater-irrigated land on which non-cereal crops are grown. Many such crops require high initial investment—banana, grape, and coconut are examples. Most such crops involve high running costs, including costs of seed, fertilizer, pesticides, and of high inputs of labour—cotton, banana, and grape come in this category. Vegetable crops and flowers also draw heavily on hired labour, particularly for digging irrigation basins, weeding, and harvesting. The area planted to irrigated non-cereal crops such as cotton generally shows a marked response to product prices and are responsive as well to trends in demand. One crop that appeared and receded during the decade was *nityakalyani* (*Vinca rosea*), an ingredient in drugs being developed for the chemotherapy of certain cancers. Cultivation of irrigated non-cereal crops can also be risky: plant disease at the last stages of, say, a banana or cotton crop can ruin a small cultivator.

An important factor in the diversification of cropping patterns has been the development of transport. Many crops are perishable—such as flowers (which are sent to Madurai, Tiruchi, and to smaller temple towns), banana (marketed mainly in Bangalore and Tiruchi), as well as other fruit and vegetables—and their cultivation has developed in areas to which truck transport now has access.

The type of cultivation that has developed on groundwater-irrigated land—characterized by heavy investment and labour absorption, diversifying cropping patterns and crop schedules, year-round farming and production that is often price responsive and risk-exposed—has contributed towards making the minimum size of plot actually farmed larger than on surface-irrigated land.[14]

[14] The structure of groundwater-irrigated holdings, their distribution, and relative size are discussed in Ramachandran (1983).

Decline in Irrigated Coarse Cereals

The increase in irrigated non-cereal crops has been at the expense, and strikingly so, of irrigated coarse cereals (*cholam, ragi* and other cereals), the area under which declined 46 per cent between 1970–1 and 1980–1. Over the same period, the share of irrigated coarse cereals in gross cropped area declined from 32 per cent to 14 per cent.[15] From my observations in the Cumbum Valley—there are no specific data on the subject—it appears that 'irrigated' cultivation of coarse grain often represents a very poorly irrigated crop, grown on lift-irrigated land that has a well from which the water supply is irregular or scanty, or cultivated by a household that cannot afford to plant crops that need high investment. More abundant water supply and capital generally make the way for diversification to non-cereal crops.

Unirrigated Crops

Among unirrigated crops, coarse grain and oil-seeds remain dominant, and, except in the south of the Valley, where groundnut and, to a lesser extent, sesame, are grown on a commercial basis, their cultivation remains backward and low yielding. There has been a decline in the area under unirrigated coarse cereals; it was of the order of 15,737 acres over the 1970s and accounted for 92 per cent of the decline in total unirrigated cropped area over the period. This can be partly accounted for by the fact of erstwhile dry land being newly irrigated and by shifts to other dry crops, such as sesame and even to some fodder crops.

Another important feature of cropping in the Cumbum Valley is the large-scale commercialization of tamarind production. Gokilapuram is the centre of the Valley's tamarind trade, and the employment thus generated is dealt with in greater detail in subsequent chapters.

Irrigation and Cropping Patterns in Gokilapuram

Table 2.9 shows data on irrigation patterns in Uthamapalayam taluk and Gokilapuram revenue village (both of these are from official revenue records

[15] For a study of the decline in acreage planted to coarse cereals in Tamil Nadu, see Poduval (1987). For an account of stagnation and decline in unirrigated groundnut production in Tamil Nadu and selected districts (inc. Madurai), see Krishnakumar (1989). Corroborative material for the Cumbum Valley is also available from a study conducted in 1981 by the Dept. of Agricultural Economics, Agricultural College and Research Institute, Madurai, into changes in the traditionally established crop-mix in the Uthamapalayam, Cumbum, and Chinnamanur blocks. In a first report on the conclusions from the study it is indicated that over the nine-year period from 1970–1 to 1979–80 the area under irrigated *cholam* declined by 91% and irrigated *ragi* by 89% in Uthamapalayam block; the area under irrigated *cholam* declined by 79% and irrigated *ragi* by 60% in Chinnamanur block; and the area under irrigated *ragi* declined by 60% in Cumbum block. The report suggests that in addition to reducing the supply of coarse grain this decline will also have an adverse impact on the supply of fodder, already short in the Valley.

Table 2.9 *Irrigation patterns for the taluk and revenue village, 1980–1981, and the survey village, 1976–1977* (acres)

Irrigation type	Uthamapalayam taluk (1980–1)		Gokilapuram revenue village (1980–1)		Gokilapuram survey village (1976–7)			
					Ownership holdings		Operational holdings	
	NSA	% of column total	NSA	% of column total	NSA	% of column total	NSA	% of column total
Nanjai	19,623	13.4	685	89.4	174.99	22.2	202.21	23.9
Thottam	39,024	26.7	33	4.3	148.88	18.9	189.14	22.3
Total irrigated	58,647	40.1	718	93.7	323.87	41.1	391.35	46.2
Punjai	87,760	59.9	48	6.3	464.41	58.9	455.83	53.8
TOTAL	146,407	100.0	766	100.0	788.28	100.0	847.18	100.0

Note: NSA: net sown area.

Sources: Uthamapalayam taluk, revenue records, 1980–1; survey data.

for 1980–1) and for the survey village. The table shows that about 90 per cent of the net sown area of the revenue village is, from official records, Periyar-irrigated wet land. While the total extent of land holdings covered by the survey roughly coincides with the total net sown area and current fallows in the revenue village (current fallows were recorded as 82 acres in 1980–1),[16] in the survey there is a large unirrigated sector and a substantial groundwater-irrigated sector, both features that are more representative of the Valley as a whole.

Table 2.10 matches Table 2.9 in that it shows the cropping patterns of Uthamapalayam taluk and Gokilapuram revenue village and our survey village. Here too a comparison of gross cropped areas indicates that the survey village is more typical of the taluk as a whole than the revenue village. Considering the cropping pattern of Gokilapuram itself, while the share of crops for the survey village mirrors irrigation patterns, the last column shows that paddy production on surface-irrigated land is by far the largest component of the gross value of production of field crops. Another noteworthy feature of the table is the extremely low value per unit of cropped area of dry cereal crops, which occupied 39 per cent of gross cropped area and accounted for a meagre 3 per cent of the value of gross production.

So while the survey presents us with a more varied agronomic canvas than the near-unbroken surface-irrigated landscape of the revenue village, it must also be differentiated from the more backward villages where unirrigated cultivation is dominant and from those where lift irrigation has advanced a great deal and is the overwhelmingly dominant sector of field cultivation.

Trends in the Nature of the Work-Force

Work Participation

The high-yielding, diversifying cropping system of the Cumbum Valley is sustained by labouring households that have very high work participation rates.

At the Censuses of 1971 and 1981, the work participation rates (workers as a percentage of the population) in Uthamapalayam taluk and Gokilapuram were higher than the rates for the district and the State (see Table 2.11). This was particularly true of the village; in 1971, the work-force was 52 per cent of the total population (State average: 36 per cent) and in 1981, the rate of work participation, including marginal workers, was 58 per cent (State average: 43 per cent).

[16] Revenue returns, Gokilapuram village, 1980–1.

Table 2.10 *Gross cropped area for the taluk and revenue village, 1980–1981, and the survey village, 1976–1977* (%)

Crop	Gross cropped area under a crop as a percentage of total gross cropped area			Gross value of production of crop as a percentage of value of all crops:
	Uthamapalayam taluk (1980–1)[a]	Gokilapuram revenue village (1980–1)[a]	Gokilapuram survey village (1976–7)[b]	Gokilapuram survey village (1976–7)[b]
Paddy	19.1	94.4	39.3	61.8
Other cereals (unirrigated)	22.2	0.0	38.7	2.7
Other cereals (irrigated)	7.0	0.1	7.2	3.3
Banana	n/a	n/a	3.1	11.9
Cotton	10.0	0.0	4.5	9.8
Vegetables	2.5	0.0	5.1	6.8
Groundnut (unirrigated)	4.0	0.0	0.6	0.5
Groundnut (irrigated)	10.9	0.0	1.3	0.8
Others	24.3[c]	5.5[c]	0.2	2.4

[a] Derived from Uthamapalayam taluk, revenue records, 1980–1.
[b] From survey data.
[c] Includes banana.
Note: n/a: not available

Table 2.11 *Percentage of workers in the population, by region, 1961, 1971, 1981* (%)

Region	1961	1971	1981 (main workers)	1981 (main + marginal workers)
Tamil Nadu	45.6	35.8	39.2	42.6
Madurai district	45.5	36.9	42.3	48.8
Uthamapalayam taluk	43.9	40.0	46.9	46.0
Gokilapuram village	42.0	52.0	48.3	58.1

Source: Census of India.

Agricultural Labourers in the Taluk and Village: Census Returns

Uthamapalayam taluk is an area where the proportion of agricultural labourers in the working population has been higher than the State average at all three Censuses and where the increase in this proportion was higher than in the State as a whole over the two intercensal decades.

The incidence of agricultural labourers in the work-force of the taluk increased from 22 per cent in 1961 to 57 per cent in 1981 (the State average for rural and urban areas together in 1981 was 32 per cent) (see Subramanian and Ramachandran 1982). At the Census of 1981, the incidence of agricultural labourers in the rural segment of the taluk was 61 per cent, which was the second highest among all taluks in the State, after Thiruvarur taluk in Thanjavur district. With reference to the village, in 1961 there was a significant misclassification of workers who were agricultural labourers, many of whom (particularly women) were apparently wrongly classified as workers in 'other services'. At the Censuses of 1971 and 1981, this misclassification appears to have been corrected; the incidence of agricultural labourers was recorded as 61 per cent and 60 per cent.

The high incidence of agricultural labourers is particularly marked in the case of female workers. In 1971, the incidence of female agricultural labourers in the female work-force for Tamil Nadu was 54 per cent; for the taluk it was 76 per cent and for the village 76 per cent. At the Census of 1981, the corresponding figures for the taluk and village (considering only main workers) were 74 per cent and 75 per cent, against a State average of 52 per cent.

Agricultural Labourers in Cumbum Valley Towns

A feature of the rapid growth of agricultural labour in the Valley is that it has occurred not only in its villages, but in its towns as well. There has been a steady rise in the agricultural component of the work-force of the towns of the Valley, with the Censuses reporting a sharp increase in agricultural labourers in the urban work-force. There is no factory industry in the Valley, and the functions of its towns (outside direct crop production) relate mainly to services and retail trade. Diversified urban development—including the growth of urban functions that serve the needs of the Valley—has been largely concentrated in Theni, which is at the northern entrance to the Valley, although geographically just outside it.

In terms of the Census classification of towns, there were seven towns in the Valley in 1961, six in 1971, and five in 1981. The data on the functional classification of towns in 1961 and 1971[17] reflect the steep rise of agricultural labourers in the urban work-force (Table 2.12). A recent study, based on an analysis of Census data, has shown that the conclusions reported here apply also to the period 1961 to 1981 (Rukmani 1989). Between 1961 and 1981, along with a decline in the proportion of non-agricultural workers in the urban work-force of the Valley, there was an absolute decline, of the order of 14 per cent, in the number of non-agricultural workers in the towns of the Valley during the same period.

Table 2.12 *Urban areas of Uthamapalayam taluk, cultivators and agricultural labourers as a percentage of the work-force, 1961, 1971, 1981*

Year	Cultivators as a proportion of all workers	Agricultural labourers as a proportion of all workers	Total
1961	21.8	14.7	36.5
1971	15.6	49.2	64.7
1981	12.3	53.2	65.4

Source: Census of India.

In 1961, one town was bifunctional (Chinnamanur); three towns were monofunctional service towns[18] (Cumbum, Bodinayakkanur, and Uthamapalayam), and three were monofunctional primary-activity towns (Thevaram, Kombai, and Gudalur). By 1971, on account of the increase in the agricultural labourer component in their working populations, all the towns in the Valley were monofunctional primary-activity towns, and two places classified as towns in 1961 (Kombai and Thevaram) ceased to be classified as urban altogether.

[17] The Census of 1971 introduced a functional classification of towns, for which the nine Census occupational categories were regrouped into five 'functional' categories: Census occupational groups I–IV were regrouped into the functional category 'Primary activities'; V*a*, V*b*, and VI into 'Industry'; VII into 'Trade and commerce'; VIII into 'Transport'; IX into 'Services'. When the percentage of workers in any of the five categories is 40% or more of total workers, the town is 'monofunctional'; if less than 40%, the next predominant category is taken into account and, if the percentage is then greater than 60%, the town is 'bifunctional'; if the total of the first two functions is less than 60%, the town is considered 'multifunctional'.

[18] This may, in part, reflect the misclassification of workers in occupational categories.

In his work on the regional dimensions of urban growth in South India, K. Nagaraj sums up the data on urban areas in the Cumbum Valley with the perceptive comment that

> the scale and nature of the agricultural transformation in the [Valley] is such that it introduces a 'scale factor' in the process of urbanisation also: at a particular stage in its transformation, the needs of a vanguard agricultural economy would perhaps be better served by a large, fast-growing town, rather than a number of towns dispersed throughout the region. And in a situation like this, one town grows at the expense of the rest in the region. In the specific case of the Valley, Theni [was] located at a transport node where it could serve both the Valley . . . and the plantation economy . . . and hence, grew at the expense of towns in the Valley. (Nagaraj 1985, p. 174)

To sum up, the Census data indicate that the changes in the agrarian situation in the Cumbum Valley in the period 1961–81 were not associated with any large-scale diversification of non-agricultural occupations or with any large-scale increase in the urban population employed at non-agricultural tasks, but with the emergence of agricultural labourers as the most significant occupational category even in the towns. A feature of present development appears to be a blurring of the distinction between the work-force in small and medium towns and in the villages of the Valley.[19]

Conclusions

The Cumbum Valley is something of an agrarian laboratory: it draws together, within a compact geographical and agro-economic zone, a variety of irrigation and soil conditions, field and tree crops (and plantation crops in its periphery), and farming practices. This chapter has attempted to discuss changes in three key aspects of the development of the productive forces in the Valley in the recent period: irrigation, cropping patterns, and work participation.

The Periyar system, which built on the foundations of the Suruliyar–Vairavanar system that existed before the Periyar dam was constructed in the 1880s, dominates surface (canal and tank) irrigation and assures at least two crops a year. The rice-growing tracts of the Valley score high on international scales of intensity of rice cultivation. The coexistence of this system and an expanding sector of groundwater irrigation (a major stimulus of which was the programme of rural electrification) has been the basis for an irrigation system that is manifestly more reliable than in the rest of the State. The expansion and relative stabilization of irrigation has been accompanied by increased commercialization of agricultural production.

In respect of cropping patterns, on surface-irrigated land (*nanjai*), there is an expanded, specialized, and stable sector of high-yielding paddy cultivation

[19] In this context Rukmani (1989) observes that 'the agrarian transformation in the Valley has affected Valley towns in a qualitative manner, giving them a rural complexion'.

and, although physical output per person-day of labour input is low by international standards, yields per hectare are high. The commercialized sector of land irrigated by wells (*thottam*) is the major area of crop diversification. The recent period has seen increased cultivation of a variety of irrigated non-cereal crops—cotton, banana, grape and other fruit, coconut, groundnut, vegetables, flowers, condiments, and other crops—and a decline in the cultivation of irrigated coarse grain. A simple, vivid indicator of crop diversification is the number and range of different crops planted in the fields of a village; in Kamayakavundanpatti, a village five kilometres south of Gokilapuram, where agriculture is centred on groundwater irrigation, there were over 25 crops being grown when we first went there in 1977. Other than a few pockets in the south of the Valley, dry-land (*punjai*) agriculture is agriculture pursued as traditional craft, and, where practised by poor cultivators, agriculture in a state of near-ruin. There has been an overall emphasis on coarse grain and oil-seeds among unirrigated crops, although on dry land, within that, there has been a decline in unirrigated coarse grain.

Although these are the broad trends, the conclusions from decadal data cannot, of course, be taken as final pronouncements for the future; on the contrary, the data show the relative quickness with which changes in the Valley can and have occurred—the expansion of area under coconut, for instance, will be a major factor in cropping patterns in this decade.[20]

An important aspect of the modernization of the agrarian economy of the Valley relates to changes in the nature of the work-force. Work participation rates are high, for the population as a whole and for agricultural labourers in particular. Among the taluks of Tamil Nadu, Uthamapalayam taluk has also been a high-incidence, high-growth taluk in respect of the proportion of agricultural labourers in the working population.

These features of change in the agrarian economy of the Cumbum Valley have important implications for the structure of the work-force as a whole and for aspects of the market for hired labour—for trends in the supply of labour and labour absorption in agriculture, for the actual deployment of labour at different tasks, and for seasonality in agricultural employment. These are issues that will be dealt with in the chapters that follow.

[20] On changes in cropping patterns, see the description in Yanagasawa (1983, pp. 79 and *passim*) of the shift to coconut, banana, and sugar-cane cultivation in a delta village in Tiruchi district.

3

Classes and Property Inequality in Gokilapuram

As elsewhere in India, so in Gokilapuram, landlords and agricultural labourers stand in glaring contrast. These are the two polar agricultural classes in village society; the village is also characterized by a complex pattern of socio-economic differentiation, by the existence of a heterogeneous peasantry and a sizeable proportion of households whose major activities do not involve direct participation in crop production. Before entering on a detailed analysis of agricultural labour, it is useful to establish perspective, to locate the class of agricultural labourers as one stratum in the socio-economic structure of the whole village, and to discuss its place in the system of ownership and control of land and wealth in Gokilapuram.

This chapter deals with a socio-economic classification of households in Gokilapuram, with work participation rates, caste structure, and literacy, and with property inequality between different sections of the population.

Economic Classification of Households

The socio-economic categories into which households in Gokilapuram have been classified are discussed in some detail here. Detail is necessary because despite extensive research based on village-level data, the Indian literature does not provide any ready-made or widely accepted framework or scheme for classifying socio-economic strata within village and rural populations; there is no 'given' morphology of occupations and socio-economic structure in a village.

In most sources of official data, of course, households are classified by the size of household land holdings, operational holdings or ownership holdings. The rationale for such a classification is that land is the primary prerequisite for production in agriculture and the distribution of land between households is an important indicator of their position in the system of agrarian production. It follows that changes in the distribution of land and growing inequality between rural households in the ownership and operation of this fundamental asset is an important concomitant of changes that are occurring in forms of ownership and forms of production.

The utility of the size of land holdings as an index for classifying households is, however, subject to serious limitations: such classification has the obvious implication that equality in the extent of land possessed by households establishes uniformity between them in their essential features. Such an assumption is certainly not valid, as the examples that follow suggest.

A single size category of land holding may conceal wide variations in the physical qualities of land—variations, for instance, in the irrigation and drainage facilities available to the land, the type of soil and its fertility, land utilization and cropping pattern, and so on. A single size category of land holding may group together households that differ in respect of other essential features, such as the terms of possession of a holding.

There are also important differences between plots of similar acreage that derive from differences in the socio-economic status of households owning or cultivating them. A size category can encompass, on the one hand, households that have made technical improvements on the land, have adopted advanced techniques of cultivation, incurred heavy expenditure during the crop season, and received regular surpluses from the land. On the other hand, it can encompass households whose methods of cultivation are still backward, including households that still farm unproductive dry land. A single size category may conceal differences in the sources, agricultural and non-agricultural, of incomes of households, and wide variations are likely to exist in the assets owned by households. Further, given that technical change penetrates agriculture in varied forms, the extent of land holding need not be a direct indicator of even the scale of production in agriculture. An index based on land holdings *per se* does not subsume the all-important questions of the physical participation of members of households in cultivation, the relationship between labour power employed and labour power expended, and the character and conditions of such labour. It is clear, then, that the size of land holding in terms of its physical extent need not necessarily be (and, as agriculture modernizes, may become less and less) an accurate measure of the *economic* size of the farming unit.[1]

The economic classification of households that is presented here uses survey data on the ownership of the means of production by a household, the extent to which it earns its livelihood through the manual labour of its own members or makes a living through the manual labour of others, and the occupation pursued by it as criteria for classifying the household.

The purpose here is not to provide a general scheme, applicable to Indian agriculture as a whole, but to try to apply the existing literature and village-level data and observation to construct a frame of reference for the study of agricultural labour in the village. In this attempt, there were, of course, problems of classification specific to the areas of particular study, Gokilapuram village and the Cumbum Valley. Landlords in the village leased out very little land; they were mainly capitalist landlords pursuing modern methods of cultivation and employing wage labour as well as preserving the economic and socio-political privileges of landlords in village society. There were strata

[1] The theoretical issues are discussed in Lenin (1972*b*, *c*); for a discussion of these issues in the contemporary Indian context, see Patnaik (1972) and, for Tamil Nadu, Ramachandran (1980). For a pattern of stratification whose empirical basis is a size-distribution of operational holdings of land, see Rudolph and Rudolph (1987), ch. 13.

among them—major and medium landlords—and a subcategory of ex-landlords who had fallen by the wayside and had lost out relative to the major landlords. In classifying the peasantry, we were dealing with a very heterogeneous group, and one in which each stratum participates very actively in the market for hired labour, as buyers *and* sellers of labour power. Employing relatively large numbers of hired labourers characterizes sections of the peasantry that are, in an important sense, 'poor peasants'; at the same time, sending out household members to labour on the fields of others extends well into the upper-middle groups of the working peasantry. A substantial part of the time of agricultural labourers in the Valley is spent at tasks outside direct crop production, at non-agricultural hired jobs and other activities. Another significant feature of the economy of the village is that more than a quarter of households derive their main incomes from sources other than direct crop production; in other words, they are households that are not landlord, peasant or agricultural labour.

The 650 households that were surveyed in Gokilapuram in 1977 have been classified into 10 major socio-economic categories. These groups are:

1. Landlords: major and medium landlords
2. Non-labouring small landowners
3. The peasantry, subdivided into strata
4. Agricultural labourers
5. Non-agricultural labourers
6. Households whose main income is from business activities other than direct crop production
7. Households whose major earnings are from artisan activity or work at traditional caste callings
8. Households whose major earnings are from salaried persons and professionals
9. Households whose main or sole income is from small rents, remittances, old age pensions, etc.
10. Unclassified households.

These socio-economic categories are defined and their distribution in the village population is discussed below.[2] It is also to be noted that the word 'class' used in the text refers to the socio-economic categories defined here.

1. *Landlords: major and medium landlords.* Landlords are those who own and control the largest land holdings in the village, do not participate in the major manual operations on the land, and depend entirely on the labour of others for their living. In Gokilapuram, their lands are worked, above all, by hired labour; some of them lease out small plots of land and their tenants are often persons who do odd jobs for them and help in the management of 'self-cultivated' land. Landlords also draw upon various forms of semi-free labour and extract labour services from agricultural workers.

[2] A more detailed classification is in Ramachandran (1983).

There were 21 landlord households in the survey. Major landlord households are a distinct section—they are separated from the others by the level of land holdings of each household and by the very high level of assets and income of each household. In addition, each major landlord household derives income from trees and groves and plantation land and some from varied sources of business activity other than agriculture. Members of medium landlord households also do not participate in the major manual operations on the land that they own and control; the level of income and assets of each household is, however, distinctly lower and the sources of income much less diverse than among major landlord households. All medium landlord households gain income from holdings of trees and groves and five of them have small plantation holdings. Among these landlords there is also one Chettiar household that is active in the tamarind trade, although with a turnover that is lower than that of the tamarind merchants amongst the major landlords.

Major and medium landlord households are entirely from the Maravar and Telugu-speaking Chettiar castes. Of the 21 landlord households in 1977, 13 were Maravar and 8 were Chettiar.

Landlord households own surface-irrigated land (*nanjai*), groundwater-irrigated land (*thottam*), and unirrigated land (*punjai*). The core of the land holdings among landlord households (with one exception, noted later) is composed of inherited holdings. Each of the four Maravar major landlord households has a place established by tradition in the land holding structure of the village; the holding of a household today has its origins in land holdings that can be traced to family holdings held by many successive generations in the village. The Maravars are, by popular account, the caste that originally established the village, and the largest landowners among them have held a specific position of dominance in the socio-political life of the village. Among the Maravar medium landlords, inherited holdings have diminished in size over generations through subdivision.

Chettiar landlord households, particularly the major landlords among them, acquire land assiduously, and have enhanced traditional holdings through purchase and foreclosure on mortgages. Chettiar major landlords are the richest section in the village, and their income from the land is substantially supplemented by income from business activity.

One household that has been included in the group for purposes of convenience of classification is different from others: it is a well-to-do capitalist farmer household which does not have any traditional holdings in the village; it has only unirrigated and groundwater-irrigated land holdings, and these holdings were obtained through purchase in the period after 1960. It is not a peasant household either, since no member of the household participates in manual labour on the land, which is entirely cultivated by hired labour.

2. *Non-labouring small landowners*. These are households that have traditionally held high-quality surface-irrigated land in the village and typically descend from landlord families. The inherited land holdings are now subdivided

between successive generations and for a variety of reasons, while their households are not ruined, they have not really stemmed the rot, and now live mostly off the income from diminished land holdings. Members of these households do not do any manual work; there is also no significant diversification of sources of income among them. Some of them have comparatively high assets. Apart from the value of land, these consist mainly of the old family house and its appurtenances.

3. *The peasantry.* Peasant households are demarcated from landlord households by the fact that their members work, in varying degree, at the major manual operations on the land that they operate. As a group, the peasantry is marked by heterogeneity in the socio-economic status of its constituent households: it is a group stratified into different categories.

Set in a region with a relatively high development of the market for hired labour, the economy of the overwhelming majority of households in Gokilapuram is marked by working as hired labourers on a substantial scale. It is also marked by high levels of demand for hired labour for cultivation by different sections of the peasantry, of which I give some examples from cultivation on surface-irrigated land and groundwater-irrigated land.

On surface-irrigated land, a peasant, however poor, does not perform irrigation or guarding tasks: these are carried out by a group of agricultural labourers chosen annually by households that cultivate surface-irrigated land. With the development of a specific form of labour organization, namely, the contract system, for transplanting and harvesting work, in the transplanting season most peasants, and in the harvest season virtually all of them, have their land worked by contract groups. (My field notes for 1981–2 included even small peasants with surface-irrigated holdings of 0.19 acres, or 0.08 hectares, that employed hired labour on contract for transplantation and harvesting.) Given the need for intensive care of a high-yielding variety paddy crop (in terms of requirements of labour and modern fertilizer and pesticide inputs) and the gradual decline of the institution of exchange labour, many peasants also employ hired workers for ploughing, weeding, and spraying pesticides, and even for threshing the paddy crop.

In the case of a peasant household that operates groundwater-irrigated land, whether the household grows irrigated food-grain, or whether (by cutting down consumption and scraping funds together or borrowing or taking advances on the crop) it is able to diversify or alter its crop mix and plant banana, vegetable crops, or cotton, it requires a substantial input of hired labour in addition to its family labour. This hired labour is needed particularly for weeding and harvesting, but also for field preparation and forming irrigation basins.

There are two sets of circumstances, both of which are present in Gokilapuram, in which peasant, including poor peasant, families may have to employ outside labour. First, a peasant household may have to *supplement* family labour with outside labour. There are many operations which have to be completed within a span of two or three days and which require, even on a plot of land that is less than an acre, a quantity of labour power that is many times that which can be provided by members of the household. Examples of such operations are transplanting and harvesting paddy, picking cotton, and harvesting vegetables. Secondly, the family may *replace* family labour with outside labour. An example of this comes from the irrigation operation on surface-irrigated paddy cultivation. This requires that water be drawn from the canal and distributed to the fields of all cultivators in the command area of a sluice through a network of field channels, and it is done by a group of workers who control water in the command area of each sluice. Peasant farmers employ these workers, who have a detailed knowledge of the irrigation system and the water requirements of the paddy crop, to irrigate their fields.[3]

Given these requirements of hired labour, and the opportunities for employment as hired labourers in an area of modernizing and diversifying agriculture, most peasant households send out men and women to work as hired labourers for wages on the land of others. They do so in order to compensate partially for the rising costs of production; to pay back debts; in many cases to pay rent to landlords; and in general to augment household income. In seeking opportunities for work as hired labourers, households look to agricultural labour and also to tasks outside the sphere of direct crop production, such as tamarind-processing, hand-pounding paddy, and other occupations in and around the village.

It is clear that rich peasants participate in the market for hired labour as employers of labour power and poor peasants as suppliers of labour power. At the same time research has shown that in many regions of South and South-East Asia, in regions characterized by diverse agricultural conditions, poor and middle peasants are active as employers *as well as* suppliers of labour power.[4] In Gokilapuram, all sections of the peasantry are active as employers of labour power, although in differing degree, and the practice of sending family members to work as hired labourers extends beyond the poorest sections of the peasantry and well into its upper-middle strata. The *household* takes decisions concerning the labour of its members: when to send members to take advantage of opportunities for hired labour and when to keep members back to work on the household operational holding.[5]

[3] See the case study of irrigation workers in Gokilapuram in ch. 9.

[4] On this, see Ryan and Ghodake (1984, p. 180) on semi-arid regions of central India; S. V. Rao (1984, p. 766) on Karnataka; Rudra (1982) on West Bengal; Kotovsky (1964, p. 37); Rastyannikov (1981, p. 153); van Schendel and Faraizi (1984) on West Bengal and Bangladesh; Rahman and Das (1982, p. 13) on Noakhali, Bangladesh; Hara (1981, p. 307) on central Thailand; Fredericks (1977, p. 187) on Selangor, Malaysia; Chung (1981, pp. 412–13) on South Korea; and Hart (1986, pp. 117 ff.) on Java.

[5] On labour supply as a family decision, see P. K. Bardhan (1979*b*) and Rosenzweig (1984). See Rudra (1982, p. 28) for an area where both family labour and employment as a hired labourer are present as alternative forms of employment for the worker in a peasant family. In a very interesting study in central Luzon in the Philippines, Takahashi (1977) shows that in peasant households only the male head of household was involved in family labour; sons joined their

Taking into consideration the pressures on a peasant household in the context of a modernizing, increasingly capitalist agriculture, and the phenomenon of members of peasant households working a great deal as hired labourers, a classification of peasant households was formulated on the basis of a 'labour ratio'. This ratio measures the relationship between the total number of days of manual labour performed by a household and the total number of days of hired labour that they employ. The numerator represents the sum of the number of days of family labour of households (labour days of men and women in the household spent in crop production on the operational holding of the household), the number of days spent by members of the household labouring out at tasks connected with crop production (or agricultural tasks), and the number of days of labour spent by members of the household at tasks other than crop production (or non-agricultural tasks).[6] The denominator denotes the number of person-days of hired manual labour employed by the household on its operational holding.

On this basis, peasant households were classified as follows:

(*a*) households with labour ratio $> 0 < 0.75$;
(*b*) households with labour ratio $\geq 0.75 < 1.25$;
(*c*) households with labour ratio ≥ 1.25.

The ratio attempts to classify peasant households on the basis of a labour criterion. In addition to relating the number of days of manual labour expended by members of a household to the number of person-days of manual labour employed by the household[7] the ratio attempts to take into account the significant degree of employment as hired labourers, not only of agricultural labourers and poor peasants (who constitute the bulk of the labour force in the village), but also of wider sections of the working peasantry.

It should be noted that, in the numerator of the labour ratio, days of family labour have been added to days of working as hired labourers. In Mao Zedong's classical definition of classes in the pre-Liberation Chinese countryside, a sharp distinction is made between working on one's own holding and working for others. Writing in October 1933 on establishing criteria for determining class status in rural China, Mao Zedong stated that 'In general a middle peasant does not need to sell his labour power, while the poor peasant has to sell his labour power. This is the principal criterion for distinguishing between a middle and a poor peasant' (Mao Zedong 1967, p. 139). In preceding

fathers in cultivation work on family farms in a very small proportion of households (about one-seventh), and the women of the families, who worked as hired labourers, were never seen at work on the household operational holdings (p. 99). Takahashi also provides a useful classification of those operations that are undertaken through family labour and those (generally employment-intensive) operations that are invariably undertaken by hired labour.

6 The numerator is the total number of days of labour of the family, irrespective of family size.

7 The following are classic works on the differentiation of the peasantry: Mao Zedong (1967, 'Analysis of Classes in the Rural Areas'); Lenin (1972*a*, particularly chs. 2, 3; (1974). Among the contributions to the Indian academic literature on differentiation in which specific criteria for assessing differentiation are proposed are Patnaik (1976) and Sundarayya (1977).

paragraphs I have attempted to show that in the particular conditions figuring in this study, labouring out extends well beyond members of poor peasant households. In a situation of a relatively high development of the market for hired labour, middle and poor sections of the peasantry have to compensate for relatively high levels of hiring labour by working as hired labourers, often intensively, and it would be difficult in these conditions to make the fact of the sale of labour power by households the criterion by which these sections are demarcated. However, it should also be noted that as the labour ratio for households declines, in general so also do the labouring-out components of the numerator; among households at the top, whose labour ratio is very low, hiring out labour power, predictably, tends towards, or is, zero.[8]

An important part of Mao Zedong's definition of classes in the countryside relates to the material status of households, and the levels of living of landlords, rich peasants, middle peasants, and poor peasants play an important role in determining their class status. While it is true of the data from this survey that there is an inverse relationship between income levels and the labour ratio on aggregate, there are significant sections in the middle where this does not really apply, with peasant households pushing overall earnings up by working as hired labourers. In this classification, I have not tried to construct a composite index based on manual labour and incomes. An implicit assumption of an inverse relationship between the two could obscure the reality of wide sections of the peasantry working as hired labourers.

The three peasant categories are referred to in the text as the upper or rich, middle, and lower or poor sections of the peasantry. With the qualification that the numerator of the labour ratio includes the days of family labour as well as the days spent as hired labourers, in respect of a labour criterion alone, households in class 3(*a*), whose own labour is less than 75 per cent of the labour that they hire for work on their land, correspond to the classical definition of the rich peasantry, whose own labour is less than the labour they employ and for whom reliance on the labour of others is of a constant character. The next category, 3(*b*), whose labour ratio is within the range 1 ± 0.25, corresponds to the middle peasantry, whose own labour is roughly equal to the days of labour that they employ. The manual labour of households in 3(*c*) is more than 125 per cent of the labour that they employ, and these correspond to a semi-proletarianized poor peasantry.

4. *Agricultural labourers* and 5. *Non-agricultural labourers*. For purposes of the discussion in subsequent chapters of this book, agricultural labourers have been divided into four subgroups, labourers who

(*a*) own and operate land
(*b*) own but do not operate land
(*c*) operate but do not own land
(*d*) neither own nor operate land.

[8] For landlord households, of course, the entire numerator would be zero; I have not, however, used the ratio for landlord households.

Two problems regarding the identification of the class of agricultural labourers have been discussed in the literature, and both of them are relevant to this study. The first applies only to agricultural labourers with land, and is the one of distinguishing agricultural labourers with land from the last sections of the poor peasantry. Agricultural labour households with land typically have very small holdings and work extensively on other people's land. Biplab Dasgupta argues that 'the boundary between agricultural labour and poor peasant households is bound to be arbitrary' (Dasgupta 1977, pp. 14–15), and this is a problem that featured in my classification as well. In general, in this classification, manual labourers fulfil both an income criterion (the major portion of their income came from hired labour) as well as a work-time criterion (the major proportion of their working time was spent as hired labourers). In addition, in a peasant household, unlike an agricultural labour household, there is a certain centrality about cultivation on the family operational holding in the socio-economic life of the household. In many of the borderline cases, I had to depend on my familiarity with individual households for final classification.

The second problem is that of separating agricultural labourers from the larger pool of manual labourers in the village. There has been diversification in the range of employment opportunities open to agricultural labourers, and many of them spend an important part of their working time at tasks outside direct crop production.[9] In my classification, categories 4 and 5 both consist of households whose earnings are mainly or exclusively from manual labour. Category 4 consists of those whose manual labour includes tasks at crop production (agricultural labour); category 5 consists of those whose manual labour is exclusively at tasks other than crop production (non-agricultural labour). It must be noted at once that this tends to underestimate those whose main occupation is non-agricultural labour and is a more liberal definition than at successive Censuses. This was a problem that was difficult to avoid: given the fact that my data related to the labouring pattern of households for only a single agricultural year, it would have been arbitrary to have classified a household that worked, say, 50 days at agricultural tasks and 51 days at non-agricultural tasks as a 'non-agricultural labour household'. In order to avoid establishing an arbitrary cut-off point for distinguishing agricultural and non-agricultural manual labour, I decided to classify as agricultural manual labourers those who had, during the reference year, exercised the option of working as hired labourers at agricultural tasks despite the fact that on

[9] John Harriss writes of 'the difficulty of assigning individuals to discrete occupational categories in what are, commonly, circumstances of occupational mobility' (J. Harriss 1986*c*, p. 8). Jan Breman refers to the importance of irregular non-agricultural work in the work schedules of rural labourers in Gujarat, 'who remain committed to agriculture but who find insufficient employment within it' (Breman 1985, pp. 290–2). See also Dasgupta (1977, p. 8) and B. Harriss (1987, pp. 8, 36) for a review of the literature on this question. In an interesting study of a village in Thailand, Atsushi Kitahara categorizes rural workers, who are no longer pure farm workers, as 'miscellaneous workers in rural society' (1977, p. 141).

aggregate this would involve an underestimate of households that worked mainly at non-agricultural tasks.

Category 4 has been further subdivided into four subgroups: three groups of households that own and/or operate holdings, and landless agricultural labour households that neither own nor operate holdings. Land holdings among agricultural labourers are almost always small and unproductive plots of unirrigated land; agricultural labour households with operational holdings of land have not been included in the peasantry, within group 3(*c*), because their labour ratio would be absurdly high—the values of the first component of the numerator, family labour, and of the denominator, hired labour, are, in absolute terms, low.

6. *Households whose main income is from business activities other than direct crop production.* Many households in the village earn incomes from entrepreneurial activity outside the direct process of crop production. The business incomes of the major landlord families have been mentioned; in other households as well, persons attempt to take advantage of possibilities for small entrepreneurship. Most of them operate from a low capital base, working, for example, as contractors for tree produce, as retail sellers of fruit or dry provisions, as small tea-shop owners, at parboiling paddy on a small commercial basis, as itinerant vendors of cloth, fruit, and other commodities, at selling fruit or sweets near the gates to the village schools, and so on. On a larger scale, one household had a small rice mill, another owned a brick-kiln, another an eating-place at Cumbum, a fourth ran a chit fund. Of the households in the village, 12 per cent derived most of their income from such activities; they comprise class 6. This group included households with ownership and operational holdings and landless households.

7. *Households whose major earnings are from artisan activity or work at traditional caste callings.* These include households whose major income is from woodwork and blacksmith work (they belong to the Achari caste), making earthenware (belonging to the Potmaker caste), and washing clothes (the Vannar caste). There are also some scheduled caste households in this group: one household does leather work and prepares corpses for funerals in the village; a few others work as village servants, at tasks which have been traditionally assigned to families of their caste. Many households from traditional artisan castes, particularly from among the Potmaker and Achari castes, belong to the peasantry and to the class that covers the largest number of castes, that of agricultural labourers. In this group I have included only those households whose major earnings are from artisan activity or work at traditional caste callings. There are households with land and households without land in this category.

8. *Households whose major earnings are from salaried persons and professionals.* These include households whose major earnings are from the salaries of schoolteachers, clerks, petty government functionaries, and shop assistants, some of whom own land or cultivate it. There were, incidentally, in the village

one lawyer, who was not living in the village at the time of the survey, and one doctor with an MBBS, who practised in Madras—neither are included in the survey data.

9. *Households whose main or sole income is from small rents, remittances, old age pensions, etc.* These households generally comprise elderly families or those with women as heads of households. They derive their incomes mainly from small rents, or remittances from relatives employed outside the village, or from pensions under the Old Age Pension scheme. In this group as well, there are households with land[10] as well as landless households.

10. *Unclassified households.* Seven households, or about 1 per cent of all households, have been left unclassified. These are households where there are overlapping functions which are difficult to classify in any one of the preceding categories.

Distribution of Households between Socio-Economic Categories (see Tables 3.1 and 3.2)

Landlord households constituted 3 per cent of all households in the village. The peasantry in Gokilapuram has been whittled down by processes of differentiation; only 16 per cent of households were classified as peasant households. Of the peasantry, 40 per cent belonged to its upper section, the ratio of whose own labour to labour power they hired was less than 0.75; there was a slim middle section, consisting of 8 per cent of peasant households; and 38 households (52 per cent of peasant households and 6 per cent of all households) were those that corresponded to the poor peasantry, whose days of own manual labour amounted to at least 125 per cent of the days of labour that they hired.

The most numerous class of all is that of agricultural labourers, who constituted 49 per cent of all households and whose numbers have been rising. Of the agricultural labourers, 257 households (81 per cent of households in the class and 40 per cent of all households in the village) were completely landless. The proportion of the landless among agricultural labourers shows considerable variation in different parts of India. In the backward areas of Palamau district studied by Sudipto Mundle, some villages showed a much higher proportion of agricultural labourers *with* land than, say, Gokilapuram did, while in West Bengal, in Ashok Rudra's study, they constituted less than 25 per cent of agricultural labour households (Mundle 1979, p. 77; Rudra 1982, p. 26).

Manual labourers as a whole, agricultural and non-agricultural, constituted 53 per cent of all households.

The all-India data show that new jobs and services have been created as a result of commercialization and the growth of a non-agricultural sector in villages (Dasgupta 1977, p. 16); this is true of Gokilapuram, in contrast with

[10] In Nakamura's study of Abinnimangalam village in Tiruchi district, the majority of such non-cultivating landowners were widows and their children (Nakamura 1982).

Table 3.1 *Distribution of households, by economic classification, Gokilapuram, May 1977*

Class	No. of households	% of column total
1. Landlords: major and medium	21	3.2
2. Non-labouring small landowners	11	1.7
3. Peasants		
Upper (L>0<0.75)[a]	42	6.5
Middle (L≥0.75<1.25)	8	1.2
Lower (L≥1.25)	54	8.3
4. Agricultural labourers		
(*a*) who own and operate land	29	4.5
(*b*) who own but do not operate land	14	2.2
(*c*) who operate but do not own land	18	2.6
(*d*) who neither own nor operate land	257	39.5
5. Non-agricultural labourers		
who own and/or operate land	8	1.2
who neither own nor operate land	21	3.2
6. Main income from business activities other than direct crop production		
who own and/or operate land	23	3.5
who neither own nor operate land	57	8.8
7. Main earnings from artisan activity or traditional caste callings		
who own and/or operate land	13	2.0
who neither own nor operate land	23	3.6
8. Main earnings from salaried persons, professionals		
who own and/or operate land	9	1.4
who neither own nor operate land	24	3.7
9. Main or sole income from small rents, remittances, old age pensions, etc.		
who own and/or operate land	5	0.8
who neither own nor operate land	6	0.9
10. Unclassified	7	1.1
TOTAL	650	100.0

[a] L = ratio of own labour to hired labour.

Source: Survey data.

data from micro-studies in some other areas (see, for instance, Rudra 1982, p. 1; Reddy 1985, pp. 30–2). Despite the low development of industry and urban occupations in the Cumbum Valley, the proportion of households that do not work mainly in full-time field crop production is noteworthy. At the survey, households whose major income was from business activity other than direct crop production comprised, significantly, 12 per cent of all households in the village (80 households). Taken together with the fact that households in other groups also derived income (although subsidiary) from business activity other than agriculture, the data indicate that in the context of the increasingly

Table 3.2 *Population, by economic classification of households, Gokilapuram, May 1977*

Class	Males	Females	Total
Landlords: major and medium	82	90	172
Non-labouring small landowners	36	28	64
Peasants			
Upper	122	88	210
Middle	20	15	35
Lower	169	125	294
Agricultural labourers			
Total	732	737	1,469
Landless	571	587	1,158
Non-agricultural labourers	63	65	128
Main income from business activities other than direct crop production	226	195	421
Main earnings from artisan activity or traditional caste callings	90	91	181
Main earnings from salaried persons, professionals	81	77	158
Main or sole income from small rents, remittances, old age pensions, etc.	15	27	42
Unclassified	15	14	29
TOTAL	1,651	1,552	3,203

Source: Survey data.

commercialized agriculture of the region, a significant number of households have attempted to cash in on the growth of the home market and have taken to forms of entrepreneurship outside direct crop production—in the majority of cases, petty business ventures—as major or subsidiary sources of household income and earnings.

Work Participation Rates: Survey Data

In the previous chapter, Census data were used to show the relatively high work participation rates in the Valley and village. Survey data confirm the high work participation rates in the village (see Table 3.3).

At the survey, we included as a worker any person who had been, over the reference year,

a manual labourer;
a salaried employee; or
self-employed.

In general, this definition is more restrictive than that of the Censuses of 1961 and 1981 (main and marginal workers together at the Census of 1981), at which those engaged wholly in supervisory work were also considered workers. In comparison with the definition of workers at the Census of 1971, the survey definition is, on the one hand, more restrictive, since it excludes persons who were engaged only in supervisory work, and, on the other hand, more liberal in respect of the reference period for work, since a person who had been at work at any time during the reference year was considered a worker at the survey (although it is likely that the definition of worker stipulated in the 1971 Census was not strictly adhered to in the actual enumeration).

Table 3.3 *Workers as a percentage of population, by economic classification of households, Gokilapuram, May 1977*

Class	Work participation		
	Males	Females	Total
Peasants			
Upper	47.5	39.8	44.3
Middle	45.0	60.0	51.4
Lower	62.1	63.2	62.6
Agricultural labourers	61.0	64.1	62.6
who own and operate land	67.9	56.8	62.6
who own but do not operate land	57.1	72.0	65.2
who operate but do not own land	64.4	52.9	59.1
who neither own nor operate land	59.8	65.7	62.8
Non-agricultural labourers	46.0	43.1	55.5
All households	52.2	47.8	50.0
All households except agricultural labourers	45.2	33.0	39.5

Source: Survey data.

Two broad aspects of work participation rates are discussed here: the difference between work participation among agricultural labourers and the rest of the population, and the difference between male and female work participation rates. In Gokilapuram, the difference between the work participation rates of workers in agricultural labour households and in other households was highly significant for all workers and for male and female workers separately. The average work participation rate among agricultural labour households was 61 per cent against 45 per cent for other households; it was 63 per cent for male members of agricultural labour households against 39 per cent for males in all other households; and 64 per cent for female members of agricultural labour households against 33 per cent for all other households. Participation rates were also high among the peasantry, other than among the highest section.

Female participation rates vary very widely across countries (see Boserup 1970, 1976) and across regions of South Asia, and show greater variation than male participation rates in rural India (see Dasgupta 1977, p. 16). Apart from Gokilapuram and other areas of Tamil Nadu, where female work participation rates are relatively high, they have been shown to be high in agro-economic regions as diverse as the dry Varhad region of Maharashtra, Dakshina Kannada district in Karnataka, Haryana, and semi-arid regions of central India.[11] In West Bengal and Bangladesh, on the contrary, female work participation is relatively low; a village study in Uttar Pradesh notes that the employment of women and children for wages is 'very rare' (Drèze *et al.* 1986, pp. 10–11; see also Rudra 1982; Rudra and Bardhan 1983). A comparison between Tamil Nadu and West Bengal is quite striking in this respect. Data from the Census of 1981 show that the proportion of women workers (main workers and marginal workers) to all women was 34 per cent in Tamil Nadu and only 9 per cent in West Bengal. Data from the 38th Round of the National Sample Survey (NSS) show that women whose 'usual status' was that of workers as a proportion of all women was 42 per cent in Tamil Nadu and 11 per cent in West Bengal (Census of India 1981 (1983); *Sarvekshana* 1981, 1986, cited in Nagaraj 1989).

For the country as a whole, the proportion of women among rural manual workers is higher than the proportion in the total rural work-force.[12] An interesting feature of studies conducted by ICRISAT researchers working in semi-arid regions of central India was that in the villages that they studied, male labour predominated in family labour, while female labour predominated in wage labour, a conclusion that was true for Gokilapuram in 1977 (Ghodake and Ryan 1981, p. 38; see also Takahashi 1977, p. 99).[13]

Assessments of inter-state variations in female work participation rates are based on secondary macro-level data; these data have been noted to be problem-ridden in their recording of the type and volume of female labour participation (see Nayyar 1987). An early study by Leela Gulati (1975) found no satisfactory explanation for variations in female labour participation across states in terms of five variables: these were per capita income, cropping patterns, literacy, male participation rate, and the sex ratio. K. Nagaraj's (1989) study of female workers in rural Tamil Nadu provides a new source of detailed information on Tamil Nadu as well as of comparative inter-state data. Work participation was higher among women of the scheduled castes and tribes than among others and the data showed that the coefficient of variation in the female work participation rate across states was lower among people of the scheduled castes and tribes than among others. This coefficient of variation was lowest among scheduled tribes. Further, the National Sample Survey data showed that in states that had a low female work participation rate, the rate fell steadily as the size class of per capita household consumer expenditure

[11] C. R. Reddy (1985): Varhad; Patnaik (1987*a*, p. 174): Haryana; Nagaraj (1981, pp. 95 ff.): Dakshina Kannada; Ghodake and Ryan (1981, p. 38), Ryan and Ghodake (1984, pp. 170–93): semi-arid regions of central India; AERC (undated *d*, village survey no. 49): other areas of Tamil Nadu.

[12] See K. Bardhan (1983, p. 199) and C. R. Reddy (1985, pp. 14–15 and table 5) for Varhad in Maharashtra; Bardhan sees development as being characterized by a growth in female work participation at a low intensity of employment at low rates of wages.

[13] In Gokilapuram, changes since 1977 have worked to decrease female labour absorption.

increased. Differences in the female work participation rates across size classes of household per capita consumer expenditure was lower in states that had a relatively high overall female work participation rate. The all-India secondary data showed a weak positive relation between the proportion of agricultural labourers in the work-force and the female work participation rate (although this positive relation did not hold across districts in Tamil Nadu, a state with a relatively high overall female work participation rate). The Tamil Nadu data also showed that female work participation rates varied with the marital status of women: the rate was higher among widows and divorcees than among others. A study by Rohini Nayyar (1987) concludes that female participation rates are highly correlated with poverty and landlessness in rural India (see also Chatterji 1984; C. R. Reddy 1985);[14] many micro-level studies also conclude that participation in rural agricultural and non-agricultural labour is higher among poor women (Vyas 1964; Jain 1980, tables 11, 13, pp. 100–1; Parthasarathy and Rao 1981, p. 130; Nagaraj 1981, 1989). Another feature of female work participation is that it is greatly influenced by socio-cultural factors (see, for instance, Vyas 1964; Nagaraj 1981, 1989).

In Gokilapuram, the male work participation rate for all households was 52 per cent against a female work participation rate of 48 per cent (the difference was significant at a 1 per cent level); among households other than agricultural labour households the male work participation rate was 45 per cent and the female work participation rate was 33 per cent, a difference that was highly significant. Among agricultural labour households, however, the situation was different. Female work participation was *higher* than male; for all agricultural labour households the female work participation rate was 64 per cent against a male rate of 61 per cent and, among landless labour households, the female work participation rate was 66 per cent against a male participation rate of 60 per cent (the difference is significant at the 5 per cent level). In agricultural labour households, freedom from an operational holding made a marked difference to the work status of women—female participation rates jumped when there was no family cultivation and were highest among landless agricultural labourers and agricultural labour households that owned but did not cultivate land. Female work participation was also high among peasant households (except the highest section of peasants) and was very high among the poorest section of the peasantry.

The survey data also confirmed certain other trends in the Census village-level data: first, the very high proportion (57 per cent) of workers from agricultural labour households in the work-force of the village; and, secondly, that this high proportion was particularly marked in the case of female workers (see Table 3.4). The proportion of female workers in agricultural labour households as a proportion of all female workers in the village was 64 per cent.

[14] Sinha (1978) suggests that participation rates in the female labour force decline in states where male earnings are higher.

Table 3.4 *Workers of each class as a percentage of all workers, Gokilapuram, May 1977*

Class	Males	Females	Total
Landlords: major and medium	0.3	0.0	0.1
Non-labouring small landowners	0.3	0.0	0.2
Peasants			
Total	20.0	16.6	18.4
Upper	6.8	2.7	5.8
Middle	1.1	1.2	1.1
Lower	12.2	10.7	11.5
Agricultural labourers			
Total	51.6	63.7	57.2
Landless	39.4	52.0	45.2
Non-agricultural labourers	3.4	3.8	3.6
Main income from business activities other than direct crop production	13.1	7.8	10.6
Main income from artisan activity or from traditional caste callings	6.3	5.3	5.8
Main income from salaried persons, professionals	3.8	2.3	3.1
Main or sole income from remittances, small rents, old age pensions, etc.	0:4	0.1	0.3
Unclassified	0.7	0.4	0.6
TOTAL	100.0	100.0	100.0

Source: Survey data.

The number of workers per household in the village and the Cumbum Valley was higher (particularly in the village) than the average in the district and State (see Table 3.5 for Census data). The overall figure for the village from survey data were slightly lower than the Census results for the village (although higher than for the taluk). The sources of this discrepancy are likely to be, first, differences in the definition of 'worker' at the survey and at the Census and, secondly, the fact that the Census of India, for which enumeration takes place in March, is likely to have covered a number of households that regularly migrate to the plantations from April–May to January–February and were not covered by the survey. These households probably have a relatively large number of workers per household.

Survey data show an average of 2.88 workers per agricultural labour household, against an average of 2.06 workers per household for all households other than agricultural labourers (see Table 3.6).[15] Interestingly, the only category with more workers per household than agricultural labourers were poor peasants, whose work participation rate was as high as that of agricultural labourers and among whom the average size of household was larger.

[15] By comparison, the average earning strength of agricultural labour households for the State was 2.37 persons in 1974–5 (RLE 1974–75 (1978), table 2.6(*a*)).

Table 3.5 *Number of workers per household, by region, 1971, 1981*

Region	No. of workers per household		
	1971	1981 (excluding marginal workers)	1981 (including marginal workers)
Tamil Nadu (rural)	1.80	1.99	2.14
Madurai district (rural)	1.90	2.22	2.32
Uthamapalayam taluk	2.15	2.42	2.53
Gokilapuram village	2.64	2.52	3.06

Source: Census of India.

Table 3.6 *Number of workers per household, by class, Gokilapuram, May 1977*

Class	No. of workers per household
Peasants: upper	2.21
lower	3.44
Agricultural labourers: all	2.88
landless	2.81
All households	2.46
All households except agricultural labourers	2.06

Source: Survey data.

Caste Structure: Index of Access and Mobility[16]
(see Tables 3.7 to 3.10)

The index of access of any caste to an occupation (socio-economic category) is defined as follows:

$$\lambda_{ij} = \frac{\text{percentage contribution of caste } j \text{ to class } i}{\text{percentage contribution of caste } j \text{ to total households}}$$

$$= \frac{N_{ij}/N_i}{N_j/N}$$

$$= \frac{N_{ij}/N_j}{N_i/N}$$

$$= \frac{\text{percentage contribution of class } i \text{ to caste } j}{\text{percentage contribution of class } i \text{ to total households}}$$

[16] This index is due to K. Nagaraj.

where

N_i = total no. of households in class $i(= \Sigma_j N_{ij})$

N_j = total no. of households in caste $j(=\Sigma_i N_{ij})$

and N = total no. of households.

Thus, if i is landless agricultural labour and j stands for scheduled castes,

then
$$\lambda_{ij} = \frac{38.99}{25.23} = 1.55.$$

Now, the overall access index to any class i is defined as

$$A_i = \frac{1}{\Sigma_j|1-\lambda_{ij}|} \times 100.$$

Similarly, the mobility index of any caste j is defined as

$$M_j = \frac{1}{\Sigma_i|1-\lambda_{ij}|} \times 100.$$

The results are clear. Given its representation in the whole population, the categories to which the access of a caste is disproportionately high (or low) is given by the coefficients shown in Table 3.10. The two highest coefficients in each row are italicized.

Table 3.7 *Households in each caste as a percentage of all households, Gokilapuram, 1977*

Caste	%
Maravar	28.9
Scheduled castes	25.2
Pallar	15.2
Parayar	5.7
Chakkiliyar and unspecified	4.3
Artisan and service castes	13.4
Achari	6.8
Potmaker	4.9
Vannar	1.7
Kallar	11.7
Telugu Chettiar	10.6
Nadar	2.1
Komati	1.4
Karkatha Vellalar	1.2
Kodikkal Vellalar	0.8
Thuluva Vellalar	0.8
Others and unspecified	3.9
All castes	100.0

Source: Survey data.

Table 3.8 *Caste and economic status of households, distribution of castes among classes, Gokilapuram, 1977*

Caste	Distribution of each caste in each class (no. and % of households)												
	Major, medium land-lords	Non-lab-ouring small land-owners	Peasants (upper)	Peasants (middle)	Peasants (lower)	Agri-cultural labour-ers	Non-agri-cultural labour-ers	Business persons	Artisans etc.	Salaried persons, profess-ionals	Small rents, remit-tances, etc.	Unclassi-fied	All occupations
Maravar	13	6	23	5	18	83	7	18	0	10	3	2	188
	(6.9)	(3.2)	(12.3)	(2.7)	(9.6)	(44.1)	(3.7)	(9.6)	(0.0)	(5.3)	(1.6)	(1.1)	(100)
Scheduled castes[a]	0	0	2	0	17	124	10	3	4	3	1	0	164
	(0.0)	(0.0)	(1.2)	(0.0)	(10.4)	(75.6)	(6.1)	(1.8)	(2.4)	(1.8)	(0.6)	(0.0)	(100)
Kallar	0	0	9	1	16	34	2	7	0	5	1	1	76
	(0.0)	(0.0)	(11.8)	(1.3)	(21.1)	(44.7)	(2.6)	(9.2)	(0.0)	(6.6)	(1.3)	(1.3)	(100)
Telugu Chettiar	8	2	1	1	1	23	1	25	0	2	4	1	69
	(11.5)	(2.9)	(1.5)	(1.5)	(1.5)	(33.3)	(1.5)	(36.2)	(0.0)	(2.9)	(5.8)	(1.5)	(100)
Artisan, service castes[b]	0	2	6	0	0	36	4	7	27	4	1	0	87
	(0.0)	(2.3)	(6.9)	(0.0)	(0.0)	(41.4)	(4.6)	(8.1)	(31.0)	(4.6)	(1.1)	(0.0)	(100)
Others	0	1	1	1	2	18	5	20	5	9	1	3	66
	(0.0)	(1.5)	(1.5)	(1.5)	(3.0)	(27.3)	(7.6)	(30.3)	(7.6)	(13.6)	(1.5)	(4.6)	(100)
All castes	21	11	42	8	54	318	29	80	36	33	11	7	650
	(3.2)	(1.7)	(6.5)	(1.2)	(8.3)	(48.9)	(4.5)	(12.3)	(5.5)	(5.1)	(1.7)	(1.1)	(100)

Note: Numbers in brackets indicate households in class as a % of all households in the caste.
[a] Pallar, Parayar, Chakkiliyar.
[b] Achari, Potmaker, Vannar.

Source: Survey data.

Table 3.9 *Caste and economic status of households, distribution of classes among castes, Gokilapuram, 1977*

Caste	Distribution of each caste in each class (%)												
	Major, medium land-lords	Non-labouring small land-owners	Peasants (upper)	Peasants (middle)	Peasants (lower)	Agri-cultural labourers	Non-agri-cultural labourers	Business persons	Artisans etc.	Salaried persons, profess-ionals	Small rents, remit-tances, etc.	Unclassi-fied	All occupations
Maravar	61.9	54.5	54.8	62.5	33.3	26.1	24.1	22.5	0.0	30.3	27.2	28.6	28.9
Scheduled castes[a]	0.0	0.0	4.7	0.0	31.5	39.0	34.5	3.7	11.1	9.1	9.1	0.0	25.2
Kallar	0.0	0.0	21.4	12.5	29.6	10.7	6.9	8.7	0.0	15.1	9.1	14.3	11.7
Telugu Chettiar	38.1	18.2	2.4	12.5	1.9	7.2	3.4	31.3	0.0	6.1	36.4	14.3	10.6
Artisan, service castes[b]	0.0	18.2	14.3	0.0	0.0	11.3	13.8	8.7	75.0	12.1	9.1	0.0	13.4
Others	0.0	9.1	2.4	12.5	3.7	5.7	17.2	25.0	13.9	27.3	0.1	42.8	10.2
All castes	100.0	100.0	100.0	100.0	100.0	100.0	100.0	100.0	100.0	100.0	100.0	100.0	100.0

[a] Pallar, Parayar, Chakkiliyar.
[b] Achari, Potmaker, Vannar.

Source: Survey data.

In relation to its weight in the population, Maravar caste representation is particularly high among the landlords and peasantry; they were 29 per cent of all households in 1977. Households of the Maravar caste are the most numerous in Gokilapuram; they are also represented in every occupational group other than artisans and workers at traditional caste-based tasks. The caste group as a whole has traditionally held a greater share of the land than any other caste, a position that it was steadily losing in the 1960s and 1970s, particularly in respect of surface-irrigated land, to other castes, to Udayar landlords from Rayappanpatti, Muslim landlords from Uthamapalayam, and, from among the castes resident in the village, to Chettiar households.

There were four Maravar households among major landlords, and a high proportion among medium landlords and, predictably, among households that derive from landlord households and are now, as landlords, in decline. As we have seen, the Maravar caste is also well represented among the peasantry; 12 per cent of Maravar households were among the upper sections of the peasantry and another 12 per cent among the middle and lower sections. The largest group of Maravar households are agricultural labourers, mainly landless.

The proportion of Telugu Chettiar households in the village population was 11 per cent in 1977. In relation to their weight in the population, Telugu Chettiar representation is particularly high among landlords and business people. Of the 11 major landlord households covered by the survey, seven were Telugu Chettiar. Two of these households (whose heads are first cousins) have had traditional holdings in the area; their land was, until about the mid-1960s, largely cultivated by tenants who paid the landlords a fixed rent. These two households have since acquired more land, particularly surface-irrigated land, of which they now have the largest holdings among households resident in the village. They are also the richest households in the village, deriving income from relatively new sources of business in addition to income from crop land, trees, and plantation land, from retail outlets for the sale of furniture, electrical goods, and hardware, and from financial assets. One household has a lucrative dealership for the sale of electric water-pumps.

One of the Telugu Chettiar major landlords has had unirrigated and groundwater-irrigated land holdings (by our schedules) for at least 50 years. At least two previous generations were involved in merchant activity in addition to agriculture; the father of the present head of household did business in, among other things, gold ornaments. In 1977 the household also had a plot of surface-irrigated land leased in from a big merchant, a relative resident outside the village. The household leases out a part of its groundwater-irrigated land holdings, and derives income from trees and plantation land as well as from the tamarind trade.

The other four Chettiar major landlord households descend from a common household. The common ancestor was engaged, in the 1920s and 1930s, in petty merchant activity of diverse types; the big breakthrough for the family came with its entry into the tamarind trade in the village and surrounding areas in the Valley. During this period, the family also began to invest the money gained from business activity in the purchase of unirrigated land (much of this has been converted into groundwater-irrigated land) and small plots of surface-irrigated land. By 1977 the four households that had branched out from the common household were—although later entrants into land ownership and control—among the largest and richest landowners in the village.

Chettiars are hardly represented among the peasantry, that is, among households who contribute their own family labour to farming the land that they cultivate; at the same time, a high proportion, one-third, were agricultural labourers.

In relation to their weight in the population, scheduled castes are most highly represented among manual labourers, agricultural and non-agricultural. The working scheduled caste population of the village consists, to an overwhelming degree, of agricultural labourers. There were 164 scheduled caste households in 1977, constituting more than a quarter of all households in the village; of these, 124 households, or 76 per cent, were agricultural labour households; the great majority were landless. Non-agricultural workers comprised another 6 per cent of scheduled caste households; thus manual labourers, agricultural and non-agricultural, comprised 82 per cent of all scheduled caste households. Only two scheduled caste households (1 per cent of all scheduled caste households) were among the upper sections of the peasantry; 17 (10 per cent of all scheduled caste households) were among the hard-labouring section of the peasantry.

There are three scheduled castes in the village: Pallar, Parayar, and Chakkiliyar. Pallar households are the most numerous among the scheduled castes; they are also the caste with the second highest number of households in the village, after the Maravar caste. Among the Pallar households in 1977, 78 (79 per cent of the total) were agricultural labourers, of which 69 (70 per cent of all Pallar households) were landless; in the popular perception, of the agricultural workers in the village Pallar workers are the paddy land workers (particularly weeding, transplanting, harvesting, and threshing) *par excellence*. Of Pallar households among the peasantry, one was in the highest group and 15 (15 per cent of Pallar households) were among the lower categories. This was also the highest proportion of peasants from any scheduled caste.

Parayar households are also predominantly agricultural labourers and of them the major section are landless. Only two households were among the peasantry at the time of the survey. The Parayars had a higher proportion of plantation workers from among them than any other caste; this is particularly reflected in the fact that 14 per cent of Parayar households were non-agricultural workers. Chakkiliyar households are the fewest in number among the scheduled castes and represent their most depressed section. Here again, their main concentration is among agricultural labourers; there was only one peasant Chakkiliyar household, a poor share-cropper. There were also four households from the scheduled castes among households whose major earnings were from traditional caste-based tasks.

In relation to their weight in the population (12 per cent of all households in the village in 1977), Kallar households are the most highly represented among the peasantry. The complex history of the Kallar caste in the region and the village (the Kallars of Gokilapuram were among those of Tamil Nadu's people on whom the infamous Criminal Tribes Act was imposed by British colonial rule), and the particular features of caste cohesion among people of the Kallar caste in the village (and the southern districts of Tamil Nadu) are, of course, outside the scope of this study. Elements of this history include their migration, in the last century and this one, from the dry areas of Usilampatti and other parts of Madurai north and north-west of the Valley, and years of protest, along with other sections of the national movement, against the Criminal Tribes Act until its repeal in the 1930s (see Blackburn 1978; Arnold 1979; Dumont 1986, pp. 26 ff.).

The survey data show that members of the Kallar caste were established as peasants on dry land (and later on groundwater-irrigated land), as tenants on groundwater-irrigated land and surface-irrigated land, as agricultural labourers, as small and medium-sized traders, and some as rich peasants who combined agriculture with money-lending at very high rates of interest. They did not establish themselves as owners of surface-irrigated land and, at our survey in 1977, not a single Kallar owned any Periyar-irrigated land. The first time a substantial plot of surface-irrigated land was even mortgaged by a Kallar family was at the end of the 1970s. Among Kallar households, there were also some that sent their children out of the village to be educated: the caste with the highest number of schoolteachers in the village was Kallar, the first women schoolteachers were from Kallar households and the first person whose family was actually resident in the village to earn a research degree (an M.Phil. in Education Studies) was a young man from a Kallar household.

At the survey, 34 per cent of Kallar households were peasants, the largest proportion of any single caste. Of these, nine households were upper peasants, and the majority were in the middle and lower sections of the peasantry. Within the Kallar caste, the largest occupational group was agricultural labourers, mostly landless. There were also five households whose major earnings were from salaried persons, mainly schoolteachers.

In relation to their weight in the population, Achari, Potmaker, and Vannar households are the most highly represented among those engaged in artisan activity and traditional tasks. Achari households (traditionally blacksmith, woodworker, and goldsmith families, as well as owners and cultivators of land) are losing their position as cultivators; by 1977 the largest segment of the caste, 43 per cent, were agricultural labourers. While for many families traditional artisan activity is a dwindling source of remuneration, 32 per cent of Achari households still derived their main income from woodwork and blacksmith work. The potter's trade has fallen on even harder times and only three households from this caste derived their main income from making earthen pots; the majority of households in this caste were also agricultural labourers.

The other castes in the wide range represented in the village together constituted 9 per cent of households; their individual and relative occupational positions are clear from Tables 3.7 to 3.9. Looking at the overall access of persons of different castes to different occupations (Table 3.10, last row), the results are clear:

- the occupations that were least caste-heterogeneous were artisan and other traditional activities (there are long-standing social barriers to entering these sunset activities);
- the next was the category of landlords (there are, *inter alia*, formidable economic barriers to entry here); and
- the most caste-heterogeneous class of all was that of agricultural labourers: agricultural labour is the great leveller of castes in the village.

In summary, while there is no congruence or one-to-one relationship between occupational structure and caste structure, and while caste barriers to entry into occupations are breaking down, particularly among manual workers, certain class–caste associations persist in the village.

Scheduled castes are concentrated among the hard-labouring sections of the population, particularly among agricultural labourers. Kallars are also concentrated in the working population and among agricultural labourers, although to a lesser degree than scheduled castes and with a better representation among the peasantry, including its upper sections.

While a large number of the artisan and other service castes (that is, the Achari, Potmaker, and Vannar castes), more than two-thirds, have moved out (or have been pushed out) of their traditional activity (the largest proportion work as agricultural labourers in the present period), artisan activity continues to remain their preserve, with other castes not making any inroads into their traditional activity. This reflects the skills of the original artisan castes, as well as the fact that in the contemporary period, these tasks have taken on the characteristics of sunset activities (see the case study of Karuppayya Achari in Appendix 3) and do not attract new sections of people to learn the skills that they require.

There is a concentration of Telugu Chettiar households among households engaged in business activity, although this is obviously not their exclusive preserve. (It would also be interesting, although outside the scope of this study, to look at the qualitative differences between business ventures run by Chettiar households and by households of other, say, Maravar or scheduled castes.)

The major landlords are drawn exclusively from two castes, Maravar and Telugu Chettiar; at the other end of the occupational spectrum, agricultural

Table 3.10 *Indices of access and mobility, Gokilapuram, 1977*[a]

Caste	Major, medium land-lords	Non-lab-ouring small land-owners	Peasants (upper)	Peasants (middle)	Peasants (lower)	Agri-cultural labour-ers	Non-agri-cultural labour-ers	Business persons	Artisans etc.	Salaried persons, profess-ionals	Small rents, remit-tances, etc.	Unclassi-fied	M_j
Maravar	*2.14*	1.89	1.89	*2.16*	1.15	0.90	0.83	0.78	0.00	1.05	0.94	0.99	0.23
Scheduled castes	0.00	0.00	0.19	0.00	1.25	*1.55*	*1.37*	0.15	0.44	0.36	0.36	0.00	0.13
Kallar	0.00	0.00	*1.83*	1.07	*2.53*	0.91	0.59	0.75	0.00	1.30	0.78	1.22	0.15
Telugu Chettiar	*3.59*	1.71	0.22	1.18	0.17	0.68	0.32	*2.94*	0.00	0.57	3.42	1.35	0.08
Artisan, service castes	0.00	*1.36*	1.07	0.00	0.00	0.85	1.03	0.65	*5.61*	0.91	0.68	0.00	0.11
Others	0.00	0.89	0.23	1.23	0.36	0.56	1.70	2.46	1.37	*2.68*	0.89	*4.22*	0.14
A_j	13	25	24	27	23	61	42	20	12	31	27		

Note: The two highest coefficients in each row are italicized.

[a] The indices of access and mobility, denoted respectively by the symbols A_j and M_j, are defined on pp. 64–5.

Source: Survey data.

labour is the one that encompasses the widest range of castes.[17] An important feature of the contemporary agrarian economy of the village is, then, that the burgeoning class of manual labourers that agrarian change has generated is caste-heterogeneous.

Literacy

Literacy data mirror class, caste, and gender distinctions in the village. The survey data on literacy show a fair degree of coincidence with Census data (although both are likely to have overstated literacy levels in the population). Despite possibilities of overstatement of literacy levels, the Census data show that more than 30 years after Independence, more than half of the village was illiterate. The survey data, which can be disaggregated, further show that literacy among agricultural labourers (36 per cent) was considerably lower than for other classes; that among agricultural labourers, scheduled caste agricultural labourers had an even lower level of literacy (31 per cent), and that the literacy level of the most depressed of the scheduled castes—the Chakkiliyar caste—was still lower (see Table 3.11). Literacy among females is systematically lower than among males; and, even by our count, only 11 per cent of Chakkiliyar women were literate.

Table 3.11 *Literate persons as a proportion of total population, sex-wise, Gokilapuram, 1977* (%)

Source/category	Males	Females	Total
Census of India (1981)[a]	61.5	34.3	48.2
Survey population	60.3	38.3	49.6
Major and medium landlords	80.5	82.2	81.4
All manual labourers[b]	48.3	25.8	37.0
Agricultural labourers	47.4	25.3	36.3
Landless agricultural labourers	45.5	21.5	33.3
Scheduled caste manual labourers	41.2	24.9	33.1
Scheduled caste agricultural labourers	38.9	23.6	31.2
Scheduled caste landless agricultural labourers	35.2	19.3	27.2
Chakkiliyar manual labourers	26.3	11.4	19.2

[a] Derived from Census of India; all other figures derived from survey data.
[b] Manual labourers comprise agricultural and non-agricultural labourers.

[17] The fact that agricultural labour is a multi-caste class has been noted for different parts of India. Drèze and Mukherjee, while noting that there are still barriers to entry into caste-based occupations, make the point that 'casual labour is really the only serious "free entry" occupation available to the less well endowed households. The operation of casual labour as a "last resort option" underlies the clear link between casual labour and poverty' (Drèze and Mukherjee 1987, p. 7). See also Nakamura (1982, pp. 40 ff.); C. R. Reddy (1985, pp. 13–14); Gough (1987, p. 288).

Property Inequality among Households

The distribution of land and other productive resources in the village is marked by staggering levels of disparity, with the landlords established in their soaring Himalaya above, the poor on the flat plains below, and rich peasants and others scattered over the foothills in between. In this section we look at the relative position of the major socio-economic groups in the structure of ownership of land and other assets.[18]

Land Holdings of Households (see Tables 3.12 and 3.13)

The first major feature of land holdings and their distribution in the village is that the majority of households are landless: at the time of the survey, 69 per cent had no ownership holdings, 65 per cent had no operational holdings, and about 60 per cent neither owned nor operated land.

A second major feature of the distribution relates to the other end of the scale, to a highly unequal distribution and the concentration of ownership and management of land in the hands of major and medium landlords. These 21 households (about 3 per cent of all households in the village) owned 33 per cent of the total extent of crop land and cultivated 35 per cent. They owned and cultivated a disproportionately large share of the best land in the village, surface-irrigated land (of which they owned 67 per cent and operated 58 per cent) and groundwater-irrigated land (of which they owned 50 per cent and operated 36 per cent).

The concentration of ownership and operational holdings in the hands of a few, then, is particularly marked in the more productive tracts, where modern techniques of production are gaining ground; it is less marked on dry land, an asset of lower value and vastly inferior productivity and whose distribution among a wide section of the peasantry is relatively more even. Broadly speaking, the distribution of surface-irrigated land reflects in part the continuation into the present of the traditional land monopoly, and the distribution of dry land reflects the pattern of distribution of a backward, coarse-grain-dominated, peasant economy. Lift-irrigated land, in many ways the sector where cultivation is most modern, shows concentration of ownership and operational holdings in the hands of a few landlord households as well as greater access to operational holdings for rich peasants than on surface-irrigated land. The limitations of using the physical extent of land holding as a measure of the economic size of the farming unit is likely to be the most marked in the case of groundwater-irrigated land, where a more finely balanced combination of indicators—for instance, the investment per unit of land, the particular crop pattern selected, the costs of cultivation per unit of land, the

[18] More detailed analysis, along with additional classification by size classes of ownership and operational holdings, is in Ramachandran (1983).

Table 3.12 *Ownership holdings of households, by class, Gokilapuram, 1977*

Class	Households		All crop land			*Nanjai*			*Thottam*			*Punjai*		
	No.	%	Owners[a]	Extent[b]	%[c]	Owners[a]	Extent[b]	%[c]	Owners[a]	Extent[b]	%[c]	Owners[a]	Extent[b]	%[c]
Major landlords	11	1.7	11	231.88	29.4	10	105.29	60.2	11	49.19	33.0	9	77.40	16.7
Medium landlords	10	1.5	10	68.43	3.7	8	12.70	7.3	8	25.91	17.4	10	29.82	6.4
Small non-labouring landowners	11	1.7	11	40.13	5.1	10	17.91	10.2	3	4.52	3.0	8	17.70	3.8
Peasants														
Upper	42	6.5	37	159.76	20.3	18	20.45	11.7	19	37.59	25.3	31	101.72	21.9
Middle	8	1.2	5	8.49	1.1	3	1.04	0.6	1	0.45	0.3	3	7.00	1.5
Lower	54	8.3	30	61.48	7.8	7	2.42	1.4	8	9.49	6.4	23	49.57	10.7
Agricultural labourers	318	48.9	44	100.03	12.7	4	0.82	0.5	4	6.56	4.4	43	92.65	20.0
All others	196	30.2	55	118.08	14.9	17	14.36	8.2	10	15.17	10.2	38	88.55	19.1
TOTAL	650	100.0	203	788.28	100.0	77	174.99	100.0	64	148.88	100.0	165	464.41	100.0

Note: No. of households with no ownership holdings: *nanjai:* 573 (88.2% of all households); *thottam:* 586 (90.2% of all households); *punjai:* 485 (74.2% of all households); total: 447 (68.8% of all households).

[a] No. of households with ownership holdings in the particular category.
[b] Acres.
[c] % of total acreage.

Source: Survey data.

Table 3.13 *Operational holdings of households, by class, Gokilapuram, 1977*

Class	Households		All crop land			*Nanjai*			*Thottam*			*Punjai*		
	No.	%	No.[a]	Extent[b]	%[c]	No.[a]	Extent[b]	%[c]	No.[a]	Extent[b]	%[c]	No.[a]	Extent[b]	%[c]
Major landlords	11	1.7	11	229.95	27.1	11	107.04	52.9	10	45.51	24.0	9	77.40	17.0
Medium landlords	10	1.5	10	62.40	7.4	8	10.44	5.2	8	23.91	12.6	8	28.05	6.2
Non-labouring small landowners	11	1.7	11	32.49	3.8	10	12.27	6.1	3	4.52	2.4	7	15.70	3.4
Peasants														
Upper	42	6.5	42	187.08	22.1	27	37.51	18.5	22	47.40	25.1	29	102.17	22.4
Middle	8	1.2	8	16.39	1.9	4	2.24	1.1	5	9.05	4.8	3	5.10	1.1
Lower	54	8.3	54	127.95	15.1	29	18.81	9.3	27	46.52	24.6	29	62.62	13.7
Agricultural labourers	318	48.9	47	104.70	12.4	2	0.64	0.3	3	3.76	2.0	46	100.30	22.0
All others	196	30.2	45	86.22	10.2	14	13.26	6.6	6	8.47	4.5	32	64.49	14.2
TOTAL	650	100.0	228	847.18	100.0	105	202.21	100.0	84	189.14	100.0	163	455.83	100.0

Note: No. of households with no operational holdings: *nanjai:* 545 (83.9% of all households); *thottam:* 566 (87.1% of all households); *punjai:* 487 (74.9% of all households); total: 422 (64.9% of all households).

[a] No. of households with operational holdings in the particular category.
[b] Acres.
[c] % of total acreage.

Source: Survey data.

methods of cultivation, the rate of water utilization, gross and net profits per acre, and specific features of the socio-economic status of households that cultivate groundwater-irrigated land—are likely to provide a richer assessment of inequality in groundwater-irrigated cultivation than the physical extent of land alone.

Each subgroup of the peasantry—distinguished by definition from the landlords by the fact that they participate in the major manual operations on the land that they operate—cultivates a much smaller proportion of land than the landlords, on holdings that are, on the average, a fraction of the size of a landlord's holdings. Among the peasantry, the upper sections are, of course, better endowed with land, although here too the total holdings and holdings per household are far smaller than that of landlords. The peasantry own relatively inferior land, the proportion of dry land in their total ownership holdings being much greater than for landlords or the village as a whole.

The great majority of agricultural labourers, as we have seen, neither own nor operate land; for those who do, the holdings are overwhelmingly unirrigated.

The Gini concentration coefficients for the distribution of ownership and operational holdings strongly reflect the two tendencies: landlessness at one end and concentration at the other. For all households, the Gini ratio is 0.86 for the distribution of ownership holdings and 0.84 for the distribution of operational holdings. The Gini ratios for the distribution of holdings excluding the landless (that is, for owner and operator households alone) are still high: 0.54 for ownership holdings and 0.53 for operational holdings.

A third significant feature of the distribution of land holdings in the village is the persistence of a large sector of small and medium holdings, continuing amid landlessness on the one hand and concentration of land control in the hands of a few on the other. The vast majority of households actually engaged in cultivation, particularly the middle and lower sections of the peasantry and agricultural labourers with operational holdings, carried on production on very small plots of land (Table 3.14). The operational holdings of the poor were particularly small on surface-irrigated land: the average size of holding of the middle and lower sections of the peasantry and cultivating agricultural labourers combined was 0.62 acres. The minimum size of holding on groundwater-irrigated land, reflected in a higher average size of holding for the same three groups, was 1.69 acres.

The period of the 1960s and 1970s was, by all accounts, an active period in the land market in the Cumbum Valley. While any assessment of the land market and of the determinants of land values requires detailed and separate study (the wide range of factors that contributed to the very high prices of surface-irrigated land are, for instance, discussed in Appendix 4), it does appear that the inflation of the period, combined with high incomes among a small section of the population, irrigation water shortages in 1974–5, and the production conditions facing poor landholders played a notable role in

Table 3.14 *Average size of landholdings of households, by class, Gokilapuram, 1977* (acres)

Class	All crop land		*Nanjai*		*Thottam*		*Punjai*	
	Ownership	Operational	Ownership	Operational	Ownership	Operational	Ownership	Operational
Major landlords	21.08	20.90	10.53	9.73	4.47	4.55	8.60	8.60
Medium landlords	6.84	6.24	1.59	1.31	3.24	2.99	2.98	3.51
Non-labouring small landowners	3.65	2.95	1.79	1.23	1.51	1.51	2.21	2.24
Peasants								
Upper	4.32	4.45	1.14	1.39	1.98	2.15	3.28	3.52
Middle	1.70	2.05	0.35	0.56	0.45	1.81	2.33	1.70
Lower	2.05	2.37	0.35	0.65	1.19	1.72	2.16	2.16
Agricultural labourers	2.27	2.23	0.21	0.32	1.64	1.25	2.15	2.18
All others	2.15	1.92	0.84	0.95	1.52	1.41	2.33	2.02
All households	3.88	3.72	2.27	1.93	2.33	2.25	2.81	2.80

Source: Survey data.

redistributing land in favour of the rich. Many households that belonged to the poorer sections of the peasantry, faced with an erosion of household income from the land and confronted by their creditors, sold or mortgaged their land when the price, particularly of surface-irrigated land, rose to levels that were much higher than the cumulative returns that the households could expect from the land over a reasonable time period. The continued eviction of tenant cultivators also contributed to changes in the distribution of land holdings over this period.[19]

Despite this, the economy of the village has not been transformed into one of only two polar classes: landlord and worker. In the midst of an overall process of polarization, there has been a certain tenacity, in spite of its economic unviability, about small production in the countryside. Utsa Patnaik points out that the attachment to land is rooted in the value that a peasant household attaches to being its own master and that this attachment has a rational basis where reliance on the labour market is fraught with risk. These are, however, only relative barriers to the transition from peasant to labourer (Patnaik 1987*a*, pp. 177–8). The sweep and intensity of development of the market for hired labour may, paradoxically, have contributed to the persistence of small production, as members of peasant families attempt to stave off the spectre of having to sell their land by going out to work on the land of others.[20]

A fourth feature of land holdings is that most of the land is cultivated directly, that is, with family and/or hired help and, among major landlords (the 11 richest households in the village), who were mainly capitalist landlords employing hired labour to cultivate their fields, 98 per cent of ownership holdings were cultivated directly.

Fifthly, tenant cultivation is still a significant—though secondary and, over a long period, declining—feature of land tenure (see Table 3.15).[21] The overall index of tenancy[22] was 17 per cent, with variations by irrigation type of land, and forms of payment of rent are still complex. There is fixed-rent tenancy as well as share-cropping in Gokilapuram, and each of these takes different forms. Fixed-rent systems predominate on surface-irrigated land and share-cropping on unirrigated land and groundwater-irrigated land.[23]

[19] There is a discussion of the pattern of land-loss among households that are today agricultural labour households in ch. 4.

[20] This point is due to Amit Bhaduri; the issue has been dealt with in the context of an area of more backward cultivation, Noakhali district in Bangladesh, in Bhaduri *et al.* (1986).

[21] Declining rates of tenancy have been discussed in B. Harriss (1987, p. 35); see also K. Bardhan (1983) on the impact of tenancy legislation on landlessness and the agricultural labour force, and Parthasarathy and Pothana (1983, p. 593) on the 'rapid demise' of pure tenants and the growth of a rural proletariat in Andhra Pradesh. For evidence on Tamil Nadu see Haswell (1967), Gough (1987), and the resurveys of the Slater villages by S. Guhan and others of the Madras Institute of Development Studies.

[22] Area leased in as a proportion of the extent of all operational holdings.

[23] For studies of different types of tenancy on different irrigation types of land in other parts of Tamil Nadu, see Nakamura (1976) and Gough (1981). See also Binswanger and Rosenzweig (1984, p. 23) on the widespread coexistence of share and fixed-rate tenancies.

Table 3.15 *Land leased-in as a percentage of all operational holdings, by irrigation type of land, Gokilapuram, 1977*

Type of land	Degree of tenancy
Nanjai	21.9
Thottam	25.2
Punjai	12.2
All operational holdings	17.4

Source: Survey data.

Tenancy is higher on surface-irrigated land than on all types of land combined: 22 per cent of operational holdings were held on lease. While 40 per cent of leased-in land was leased by the lower and middle sections of the peasantry, the section with the highest proportion of leased-in surface-irrigated land was the highest group of the peasantry. This indicates that some landlords prefer to lease out their land to peasant households that also own land and are able to meet the costs of seed, fertilizer, and pesticides. They also generally have a sturdy pair of plough animals.[24] Tenancy takes the form of fixed-rent payments on surface-irrigated land. The rent generally works out to about 40 to 45 per cent of the produce—owners do not contribute to input costs—and, in a bad year, the fixed rent expressed in share terms may climb to over 50 per cent of the crop. There were still some tenant households—some 3 or 4—that were subject to other forms of rack-rent: after paying the fixed rent, they paid the landlord a half share of the remaining produce (although, in such cases, the landlord generally participated in meeting input costs).

Different types of share-cropping prevail on dry land and on groundwater-irrigated land. Share-cropping leases are generally structured around an equal division of physical output or of the value of output. There are, however, differences in the net share accruing to individual landowners and tenants. These net shares vary with the extent to which tenants meet input costs in each lease contract. On dry land, where yields are low, it is generally a simple share, with the tenant meeting (low) input costs. The two years preceding the survey were years of poor rainfall in the eastern dry-land tracts and the returns from spending a season in cultivating single-cropped unirrigated land most unrewarding—there are many cases of a tenant getting a bag or less of cereal as his family's share from the harvest. Tenancy was low on dry land—12 per cent of operational holdings—and was spread among the peasantry; the largest share of dry land on lease was leased in by agricultural labourers.

On land irrigated by wells, returns are greater, and leased groundwater-irrigated land is perceived as a source of income by lessor and lessee. Share-

[24] In Rudra and Bardhan's (1983) study, the tenant was expected to provide his own labour and bullock and plough (p. 289).

cropping on groundwater-irrigated land is generally accompanied by different forms of input-sharing arrangements. These arrangements between owner and tenant on groundwater-irrigated land vary with the crop that is planted and the investment required. If evidence were needed that modernization does not necessarily lead to a dissolution of land-leasing, groundwater-irrigated land in Gokilapuram provides an example: while it is a commercializing sector of cultivation, groundwater-irrigated land still had the highest degree of tenancy among the different types of land—25 per cent of operational holdings. Tenants on groundwater-irrigated land were completely located among the peasantry and, among the middle and lower sections of the peasantry, the leased-in portion of operational holdings predominated over the other components.

Pranab Bardhan and Ashok Rudra make the point that when the landlord does not share in inputs, the share of the tenant is higher. This was true of Gokilapuram, where, as the input share and supervision of the landowner became greater, the share of the tenant became lower, and the boundary between tenant and agricultural labourer became increasingly blurred.[25] Take the case of a landlord who employed a worker-tenant for a crop (the example is from case-study material collected in 1981–2). The landlord decided the crop that was to be grown and the timing of crop operations. He decided how many workers to employ for each operation and also met the cost of inputs. The worker-tenant played a supervisory role and worked in individual operations as well. He received one-ninth of the crop as payment. Such a worker-tenant had it in common with a share-cropper that his earnings were uncertain, being dependent on the state of the harvest: bad crop, low earnings. In terms of his actual control over the choice of crop and the cultivation process, however, he was more akin to a farm servant whose remuneration happened to be calculated in terms of a share of the crop.

Rudra and Bardhan also point out that there 'is a considerable amount of evidence that the institution [of share-cropping tenancy] has been adapting itself more and more to the needs of increasing production and profit by enterprising farmers, both owners and tenants' (Rudra and Bardhan 1983, p. 290). While this is true in certain circumstances, it is not incompatible, as the Gokilapuram material shows, with the institution of tenancy drawing within it systems of dependence and labour service and keeping them alive.

Whichever the type of land or system of rent, rents were far above the statutory limit of 25 per cent of the gross produce.

Tenancy played an important role in the household economies of the peasantry, particularly its middle and lower sections (see Table 3.16). It was not, however, confined to these sections: the upper sections accounted for slightly more than a quarter of net area leased in (Table 3.17). The nature of tenancy also differed: the upper sections leased in only surface-irrigated and

[25] Mencher's (1978) studies of villages in Chengalpattu district in the 1960s and 1970s show a movement of workers between the categories of share-cropper and landless labourer (p. 161).

Table 3.16 *Land leased in as a percentage of all operational holdings, by class and by irrigation type of land, Gokilapuram, 1977*

Class	Degree of tenancy			
	All crop land	*Nanjai*	*Thottam*	*Punjai*
Major and medium landlords	2.0	2.4	0.0	2.8
Non-labouring small landowners	4.2	11.0	0.0	0.0
Peasants				
Upper	16.9	46.5	21.0	4.1
Middle	48.8	53.6	55.3	35.3
Lower	49.9	82.6	70.1	25.2
Agricultural labourers	25.7	10.0	0.0	26.
Main income from business activities other than direct crop production	22.6	82.1	0.0	11.2
Others	2.5	0.0	0.0	3.5
All classes/all operational holdings	17.4	21.9	25.2	12.2

Note: The degree of tenancy is the leased-in portion of operational holdings as a percentage of total operational holdings.

Source: Survey data.

groundwater-irrigated land, while the poor, who must survive, leased in dry land as well. Land was leased in mainly from households resident outside Gokilapuram.

Asset Holdings of Households

The asset holdings of households of different classes have been disaggregated into seven groups: crop land, other agricultural assets, draught animals, other livestock, trees and groves (*thope*), plantation land, and non-agricultural assets (see Table 3.18). Asset categories are defined in Appendix 4.

The distribution of the *value* of assets is even more concentrated in the hands of landlords than the distribution of the *extent* of land owned. At the survey, in every category of assets (except the two livestock categories), classes whose share in the total value of assets was second to that of major landlords were very poor seconds and the assets that they owned were a fraction of those held by landlords. In every category of assets, without exception, landlords had higher average asset holdings per household than households in any other category (Table 3.19).

The Gini indices of concentration for the distribution of different asset categories between households speak for themselves. The coefficient for crop land was 0.93; 0.97 for agricultural assets other than crop land; 0.97 for trees and groves; 0.98 for plantation land; 0.89 for draught animals; 0.87 for other livestock; 0.78 for non-agricultural assets; and 0.87 for all assets combined.

The grip of the major landlord households, which constituted 2 per cent of all households in the village, was particularly vicelike on the most productive and income-bearing assets. In respect of crop land and trees, they have built on traditional holdings and in some cases have enhanced these holdings over the last 20 years. At the survey, they owned 53 per cent of the value of crop land against an ownership of 29 per cent of the extent of crop land; in other words, they owned the best land. They owned 67 per cent of trees and groves held by households in the village and 60 per cent of the value of plantation land.

Of particular significance was the extremely high concentration of ownership of 'other agricultural assets' (this category consists mainly of agricultural implements and machinery). The 11 major landlord households owned about 72 per cent of the total value of these assets; together, they have a comprehensive hold over the most modern means of production in agriculture. At the other end, the middle and lower sections of the peasantry and agricultural and other manual labourers owned a share of the total value of assets that was marginal relative to the share of major landlords and in relation to the weight of their own numbers in the population. The data give us quite incredible comparative figures in this respect: to take an example, the *combined* assets of 257 agricultural labour households (40 per cent of all households in the village) were Rs 5.64 lakhs, or only 60 per cent of the assets of *each* major landlord household, which (even by our count) were over Rs 9.39 lakhs. The assets of landless manual labour households were concentrated almost entirely in non-agricultural assets, assets other than means of production (Table 3.20).

Table 3.17 *Share of different classes in total land leased in, by irrigation type of land, Gokilapuram, 1977* (%)

Class	Degree of tenancy			
	All types of land	*Nanjai*	*Thottam*	*Punjai*
Major and medium landlords	3.9	6.4	0.0	5.4
Non-labouring small landowners	0.9	3.1	0.0	0.0
Peasants				
Upper	21.5	39.5	20.9	7.5
Middle	5.4	2.7	10.5	3.2
Lower	43.4	35.2	68.6	28.3
Agricultural labourers	18.3	1.5	0.0	47.3
Main income from business activities other than direct crop production	5.8	11.6	0.0	6.1
Others	0.8	0.0	0.0	2.2
All classes	100.0	100.0	100.0	100.0

Source: Survey data.

Table 3.18 *Asset holdings of households, by class, Gokilapuram, 1977*

Class	Households		Crop land		Other agricultural assets		Draught animals		Other livestock	
	No.	%	Rs	%	Rs	%	Rs	%	Rs	%
Major landlords	11	1.7	5,959,750	53	321,134	72	27,500	21	32,460	19
Medium landlords	10	1.5	1,034,590	9	29,712	7	14,550	11	12,685	7
Non-labouring small landowners	11	1.7	683,800	6	6,370	1	2,200	2	4,845	3
Peasants										
Upper	42	6.5	1,672,767	15	51,912	12	20,400	16	19,613	11
Middle	8	1.2	74,900	1	3,983	1	2,500	2	1,069	1
Lower	54	8.3	374,067	3	16,131	4	29,933	23	20,294	12
Agricultural labourers	318	48.9	290,552	3	1,217	n	27,000	21	55,404	32
who own and/or operate land	61	9.4	290,552	0	1,129	n	20,500	16	9,319	5
who neither own nor operate land	257	39.5	0	0	88	n	6,500	5	46,085	27
All other classes	196	30.1	1,165,797	10	16,551	4	5,950	5	27,022	16
All households	650	100.0	11,256,223	100	447,010	100	130,033	100	173,392	100

Class	Trees and groves		Plantation land		Non-agricultural assets		All assets	
	Rs	%	Rs	%	Rs	%	Rs	%
Major landlords	1,028,075	67	787,500	60	2,179,480	36	10,335,899	50
Medium landlords	218,725	14	54,000	4	451,240	8	1,815,502	9
Non-labouring small landowners	33,370	2	25,000	2	394,725	7	1,150,310	6
Peasants								
Upper	116,529	8	23,000	2	503,595	8	2,407,816	12
Middle	3,000	n	0	0	31,880	1	117,332	1
Lower	50,350	3	9,500	1	338,075	6	838,350	4
Agricultural labourers	24,538	2	4,500	0	706,894	12	1,110,105	5
who own and/or operate land	17,750	1	0	0	206,629	3	545,879	3
who neither own nor operate land	6,788	n	4,500	0	500,265	8	564,226	3
All other classes	69,139	4	400,500	31	1,401,930	23	3,086,889	15
All households	1,543,726	100	1,304,000	100	6,007,819	100	20,862,203	100

Note: n = negligible.

Source: Survey data.

Table 3.19 *Average asset holdings of households, by class, Gokilapuram, 1977* (Rs)

Class	Crop land	Other agricultural assets	Draught animals	Other livestock	Trees and groves	Plantation land	Non-agricultural assets	All assets
Major landlords	541,795	29,194	250	2,951	93,461	71,591	198,135	939,627
Medium landlords	103,459	2,971	1,455	1,269	21,873	5,400	45,124	181,550
Non-labouring small landowners	62,164	579	200	440	3,034	2,273	35,884	104,573
Peasants								
Upper	39,827	1,236	486	467	2,775	548	11,990	57,327
Middle	9,363	498	313	134	375	0	3,985	14,667
Lower	6,927	299	544	376	454	176	6,261	15,525
Agricultural labourers	914	4	85	174	77	14	2,223	3,491
who own and/or operate land	4,763	19	336	153	291	0	3,387	8,949
who neither own nor operate land	0	0	25	179	26	18	1,947	2,195
All other classes	5,947	84	30	138	353	2,043	7,153	15,749
All households	17,317	688	200	267	2,375	2,006	9,243	32,096

Source: Survey data.

Table 3.20 *Assets in each category as a percentage of all assets, Gokilapuram, 1977*

Class	Crop land	Other agricultural assets	Draught animals	Other livestock	Trees and groves	Plantation land	Non-agricultural assets	All assets
Major landlords	57.7	3.1	0.3	0.3	9.9	7.6	21.1	100
Medium landlords	57.0	1.6	0.8	0.7	12.1	3.0	24.8	100
Non-labouring small landowners	59.5	0.6	0.1	0.4	2.9	2.2	34.3	100
Peasants								
Upper	69.5	2.2	0.8	0.8	4.8	1.0	20.9	100
Middle	63.8	3.4	2.1	0.9	2.6	0.0	27.2	100
Lower	44.6	1.9	3.6	2.4	6.0	1.1	40.3	100
Agricultural labourers	26.2	0.1	2.4	5.0	2.2	0.4	63.7	100
who own and/or operate land	53.2	0.2	3.8	1.7	3.2	0.0	37.9	100
who neither own nor operate land	0.0	n	1.1	8.2	1.2	0.8	88.7	100
All other classes	37.8	0.5	0.2	0.9	2.2	13.0	45.4	100
All households	54.0	2.1	0.6	0.8	7.4	6.3	28.8	100

Note: negligible
Source: Survey data.

CONCLUSIONS

This chapter has dealt with socio-economic structure and property inequality in Gokilapuram and with aspects of change in the characteristics of the classes involved in production.

There has been change in the nature of traditional landlord economy: it is no longer the case, as it was in the early part of the century, that Maravar landlords with large land holdings, mostly leased out in small plots and operated on a small scale, dominate the village economy. There has been change in the personnel of landlords. A section of Maravar landlords has now been pushed out of positions of dominance, and Chettiar landlords and landlords from other villages have emerged as an important economic force. Changes in the scale and type of cultivation have not, however, meant a loosening of the grip of landlordism *per se* over the village economy; on the contrary, the sway of the major landlords over socio-economic life in the village—their control of land and wealth—has, if anything, been strengthened in the recent period.

The economy of landlord households resident in the village is now characterized mainly by direct cultivation with hired labour (although tenancy remains a significant secondary tenurial form) and incomes that derive from diverse sources. Landlords are the major employers of hired labour in the village (the peasantry, considered *en bloc*, are next, but come far behind the landlords).

At the other end of the scale, there is a large and expanding mass of the relatively propertyless, dominated by manual workers, particularly agricultural labourers. And in between there is a heterogeneous, discernibly differentiated peasantry, whose socio-economic characteristics have been shaped by the development of the market for hired labour, by its sweep and relative intensity. Polarization represents the overall trend; at the same time, small production continues to characterize farming among the majority of households that operate land. Work participation rates are high among manual workers and peasants, particularly among agricultural labourers. In Gokilapuram in 1977, female work participation rates (by the definition adopted here) were significantly lower than male participation rates in the population as a whole, but, among agricultural labour households, female participation rates were higher than male participation rates. The number of workers per household among agricultural labourers was higher than the State average. Another important feature of the occupational structure of the village is its diversification: more than a quarter of all households were engaged mainly in tasks outside direct crop production (although many of these tasks are related to agriculture).

While there is no direct congruence between caste structure and occupational structure, certain caste–class associations persist and continue to be an important feature of village life. (This study attempts to construct an index that uses survey data on caste and socio-economic status to measure caste mobility between occupations and the access of households to different occupations.)

A striking feature of the caste–class situation is that agricultural labour, the occupation in which the largest single section of the village population is concentrated, is the occupation that is the most caste-heterogeneous.

Literacy is still very low, particularly among women, scheduled castes, and manual labourers.

The data show that, while inequality in the ownership of land and wealth in the village has taken new and more complex forms in the present period, wide disparity remains the predominant feature of land and asset distribution. There is a concentration of land and other productive assets in the hands of a small minority of rich households, in particular the major landlords. This exists along with a relative deprivation of the poor and middle peasants and agricultural labourers, who constitute the majority of households. In the case of agricultural labourers, the dominant feature of their asset holdings is propertylessness in terms of ownership of the means of production. The poor are generally asset-poor, and their asset-poverty is an important factor in their income-poverty.[26] And it is apparent that unless the present system of ownership and production is broken, the sources of future accumulation of wealth lie primarily in the hands of those among whom wealth is already concentrated today.

These are the major characteristics of the actors on the socio-economic scene; we now turn to more detailed study of the characteristics of agricultural labourers and the relationship between this class and its employers.

[26] B. Harriss (1987) has a review of the Indian material on this; see also Mencher (1978, table 7.3); Patnaik (1987*a*, p. 165).

4

The Labour Force: Sources of Expansion

Preceding chapters described the growth and development of a large force of agricultural labourers in the Cumbum Valley and in Gokilapuram, and showed that agricultural labourers, who are essentially propertyless, constitute the largest section of the working population. This chapter attempts to examine the sources of expansion of the labour force in the village, using material on land holdings and occupational histories of households that are today in agricultural labour. The chapter also deals briefly with the size of agricultural labour households. The data show that a dramatic process of differentiation occurred over a generation, during a period of growth and technological transformation in agriculture.

Manual labour is the only alternative occupation open to the propertyless and the uprooted in the village, to those who lose their land, and to those for whose traditional trade, craft, or service the demand is dead or dying.[1] The aggregate expansion of the labour force in the village is the result of the convergence of many streams: among others, the dispossession of the peasantry; eviction of tenants; the erosion of demand for the services of village artisans and the loss of traditional rural non-agricultural occupations; the existence, over an extended historical period, of a large force of traditionally landless—often scheduled caste—agricultural labourers; and the opening up of new agricultural areas and of opportunities for labour migration. In addition, the persistent increase in the pressure of population on land and on the existing sources of employment—land hunger—adds a particularly painful twist to the process of expansion of the labour force and the differentiation of the peasantry in general. To identify the factors contributing to the expansion of the labour force and to quantify their relative importance in different regions and over different periods of time is a major task and is still a field open for data-based research, since the existing sources of macro-level data are generally inadequate in this respect.[2]

In the first section of this chapter, an attempt is made to put together data regarding

- land that was held by ascendant generations of households that are today agricultural labour households;

[1] Gough (1987) describes the category of landless labourers as a 'cul-de-sac into which many of the dispossessed villagers had fallen' (p. 288), and Drèze and Mukherjee (1987) call it the 'last resort option' for workers in the village (p. 7).

[2] There are useful general descriptions of the sources of the contemporary labour force in Dasgupta (1977, p. 15); Rudra and Bardhan (1983, p. 11); Gough (1981, 1987); Bardhan (1983, pp. 7 ff.); and Patnaik (1987*a*, pp. 159–60).

- occupations of the fathers of the heads of agricultural labour households; and
- permanent first-generation immigration among agricultural labour households.

These, taken together, provide an account of the sources of the present agricultural labour force.

Past Ownership Holdings

We asked the head of each household: 'How much land do you own today/did you own 5/10/25/30/75 years ago?', working through the male linear ascent of the head of household.

Of the 257 landless agricultural labour households, 44 per cent (114 households) reported having owned land in the current or previous generations (Table 4.1).[3] There is evidence that the pace of land-loss increased during the past 25 years; of the 114 households which had owned land in the past, 52 households (46 per cent) had lost holdings in the 25 years preceding the survey (i.e. approximately 1952 to 1977). Of the households that had owned land after 1952

- 7 households (3 per cent of all 257 households) had lost their land between 1952 and 1960;

Table 4.1 *Loss of ownership of land, landless agricultural labour households, Gokilapuram, 1977*

Status	Agricultural labour households		
	No.	% of those who had owned land	% of total
Never owned land[a]	142	n/a	55.3
Had owned land	114	100.0	44.3
Before but not after 1952	45	39.5	17.5
Between 1952 and 1977	52	45.6	20.2
Inadequate data	17	14.9	6.6
Unspecified	1	n/a	0.4
TOTAL	257		100.0

Note: n/a = not applicable.
[a] i.e. had no memory of owning land.

Source: Survey data.

[3] These 257 households include 60 households whose heads migrated to the village in the current generation. Even if these households are excluded, no qualitative difference is made to the conclusions in this section.

- 42 households (16 per cent of all 257 households) had lost their land after 1960; and
- for 3 households the data were inadequate.

Of the agricultural labour households with operational holdings but no ownership holdings, 4 households out of 18 had sold ownership holdings over the past 50 years.

Activists of the peasant movement in the Valley point to the 1950s as the period when tenants were evicted on the largest scale. Our data and interviews in the village indicate that evictions continued after the new technology in agriculture began to spread (see Appendices 2, 5, and 6). Of the 257 landless agricultural labour households, 60 reported having had leased-in holdings in either the present or the immediately preceding generation.[4]

Putting this data together with the data on ownership holdings, *at least*[5] 154 households had ownership and/or operational holdings in the past. These households constituted 60 per cent of all landless agricultural labour households.

It is likely that the extent of land resumed for owner cultivation will be higher in the case of households that are presently peasant households which have lost the leased-in portions of their operational holdings, since in many cases landlords (as mentioned earlier) prefer to lease land out to households that are already involved in household cultivation and have plough animals. An indicator of this is the fact that among the 29 agricultural labour households that owned *and* operated land, 11 (38 per cent) lost leased-in land in the present and immediately preceding generation. Data for 2 of these households were inadequate; for the remaining 9 households, 10 separate leases were withdrawn from the cultivating household between 1967 and 1977. The agricultural labour households with operational but no ownership holdings also show fairly high levels of loss of tenancy; 8 had lost tenancy rights, these evictions being mainly concentrated in the 1967–77 period.[6]

Occupations of the Fathers of Heads of Agricultural Labour Households

Agricultural labour is the largest single occupational category among the first-generation ascendants of agricultural labour households, and this indicates a trend of continuity in the existence of a force of agricultural labourers in the

[4] Of these 60, 5 were households that had holdings on tank-bed (encroachment) land.

[5] We say 'at least' because records of leasing-in that we have extracted from schedule data relate only to the present and immediately preceding generation; the number would rise if we had sufficiently accurate information to include all leasing-in over the last 75 years.

[6] On the prior occupations of households currently involved in agricultural labour, Rudra and Bardhan (1983) make the point that, from their data, agricultural labourers with land were mainly ex-tenants, and landless agricultural labourers were mainly former attached labourers and ex-tenants (p. 11).

village. Data from this study showed that the main occupation of 42 per cent of fathers of the head of each agricultural labour household in the village, as reported by the head of household, was agricultural labour (see Table 4.2). Another, and very striking, feature of the table follows from this: it is that 58 per cent of the households derive from households that were *not* agricultural labour households in the previous generation. Those whose fathers were either peasants or landholders other than peasants or artisans and workers at traditional village tasks (pot-makers, woodworkers, blacksmiths, temple priests) constitute no less than 47 per cent of agricultural labour households. Although these data constitute only a partial indicator, they do point to a remarkable process of differentiation over a single generation.[7]

The following are the major conclusions from the data on the castes of landless labour households and the occupation of the fathers of the heads of these households (see Table 4.3):

Table 4.2 *Occupation of fathers of heads of all agricultural labour households, Gokilapuram, 1977*

Occupations of fathers of heads of households	Agricultural labour households					
	Owning and/or operating land		Landless		Total	
	No.	%	No.	%	No.	%
Peasants	31	50.8	77	30.0	108	34.0
Landowners other than peasants	2	3.3	2	0.8	4	1.3
Artisans, workers at traditional caste callings	6	9.8	33	12.8	39	12.3
Agricultural workers	17	27.9	118	45.9	135	42.4
Other manual workers	0	0.0	7	2.7	7	2.2
Main income from business activities other than direct crop production	1	1.6	12	4.7	13	4.1
Others	1	1.6	1	0.4	2	0.6
Unspecified	3	4.9	7	2.7	10	3.1
TOTAL	61	100.0	257	100.0	318	100.0

Source: Survey data.

[7] The RLE data for Tamil Nadu suggest that the increase in agricultural labour households with land over the decade 1964–5 to 1974–5 (which, by the RLE, was 59%) was even greater than the increase in landless agricultural labour households (29% by RLE data); this does not appear to have been the case in Gokilapuram (RLE 1974–75 (1978), table 2.1, pp. 22–5).

Table 4.3 *Occupation of fathers of heads of landless agricultural labour households, by caste, Gokilapuram, 1977 (no. of households)*

Caste	Occupations of fathers of heads of landless agricultural labour households									
	Peasants	Landowners other than peasants	Artisan activity, and traditional caste calling	Agricultural labourers	Other manual labourers	Business activities except crop production	Others	Unspecified	Total	% of all households
Maravar	30	2	1	26	2	3	1	1	66	25.7
Telugu Chettiar	2	—	1	10	—	3	—	1	17	6.6
Kallar	13	—	1	6	—	2	—	2	24	9.3
Achari	1	—	11	—	1	1	—	—	14	5.4
Potmaker	1	—	6	1	1	—	—	2	11	4.3
Nadar	4	—	—	—	—	—	—	—	4	1.6
Other	2	—	2	6	—	1	—	—	11	4.3
Non-scheduled castes: subtotal	53	2	22	49	4	10	1	6	147	57.2
Pallar	20	—	1	43	3	1	—	1	69	26.9
Parayar	5	—	—	16	2	—	—	—	23	8.9
Chakkiliyar	1	—	6	4	—	—	—	—	11	4.3
Unspecified scheduled castes	0	—	1	6	—	—	—	—	7	2.7
Scheduled castes: subtotal	26	0	8	69	5	1	—	1	110	42.8
TOTAL	79	2	30	118	9	11	1	7	257	100.0

Source: Survey data.

Of the households whose first-generation ascendants were peasants, 67 per cent were caste Hindus. Among the scheduled castes whose first-generation ascendants were peasants, most were Pallar, among whose numbers owners of dry land and fixed-rent tenants on surface-irrigated land were more common in an earlier period.

Among households whose first-generation ascendants were agricultural labourers, the situation is almost reversed: 59 per cent were scheduled caste.

Among scheduled caste households, 63 per cent had first-generation ascendants who were already agricultural labourers (and 67 per cent whose members were agricultural and other manual labourers) and 24 per cent that were still peasant households. Depeasantization over a single generation was higher among caste Hindus, and of them among Maravars and Kallars (the first, the dominant caste of the village and the other a major peasant caste), than among scheduled castes (among caste Hindus, 36 per cent of first-generation ascendants were peasants).

However, it must also be noted that the process of 'agricultural labourization' even among caste Hindus started earlier than this generation: a significant section of caste Hindu landless agricultural labour households had been in agricultural and other manual labour for more than a single generation—in fact, the proportion whose fathers were agricultural labourers or other manual labourers was the same as the proportion whose fathers were peasants, 36 per cent.

Change in occupational status over a single generation appears to be particularly drastic among the two major artisan castes, Achari and Potmaker. Of the 25 landless agricultural labour households from these two castes, 17 were mainly occupied at their traditional caste calling until the present generation, and only a single household was an agricultural labour household in the previous generation. 'When castes lose the tools and techniques of their traditional occupations,' Nakamura writes, 'they tend to become agricultural labourers' (1982, p. 40; see also S. V. Rao 1984, p. 769). And when the demand for their services shrinks rapidly, this process accelerates (see Appendix 3).

Immigration

The mobility of the rural labour force in Tamil Nadu in the contemporary period and over longer historical time is a relatively under-researched area. Gokilapuram, as we saw in an earlier chapter, has seen substantial emigration in recent years, particularly to the plantations. There has also been the opposite process, discussed in this section, of immigration to the village.

The peasant and labour work-force of the Cumbum Valley was increased through significant waves of immigration in an earlier period, particularly during periods of large-scale colonization of land for cultivation. All of the major castes in the village trace their origins in the village and Valley to nineteenth-century settlers, and almost all of these were settlers from the dry eastern district of Ramanathapuram. Immigration to the Valley is likely to have played a large role in the rapid growth of population in Periyakulam taluk (of which the Cumbum Valley was then a part) in the decade 1891–1901. The opening of the Kannan Devan tea estate in Travancore State (now in Kerala) and the continuous development of the cardamom plantation sector across the border in Travancore stimulated a strong movement of landless labour and pauperized peasants into the Cumbum Valley, particularly after 1896, when the sale of cardamom ceased to be a Travancore government monopoly and leading 'ryots' (mainly landlords) from the Cumbum Valley were granted occupancy deeds in the Cardamom Hills.

The data here have to do with immigration to the village in the present generation. Of the 257 landless agricultural labour households, 60, or 23 per cent, had heads of households whose native villages were other than Gokilapuram and during whose lifetimes immigration to Gokilapuram took place. The 56 agricultural labour households that owned or operated land represented a more settled section, and only 7 households among them had heads of households who migrated to Gokilapuram in their lifetimes.[8]

Agricultural labourers come to the village with their families, to a place where opportunities for work are greater than in the village of their birth, generally following the trail of persons of their own caste who have established settlements in the village over a longer period of time.[9] Many migrate to the village after marriage, when one partner is from Gokilapuram. Here it is not only the case that a woman who marries a man from Gokilapuram comes to settle in the village of her husband: in addition, the advanced agriculture of her village and the opportunities for wage labour are part of the dower, as it were, of the labouring woman of Gokilapuram, and among agricultural labourers it is common for men who come from drier tracts and marry Gokilapuram women to leave their villages and come to live and work in the village of their wives.[10]

Of the 257 landless agricultural labour households in the village, the heads of 60 were born in other villages (see Table 4.4). There is clear evidence of increased permanent immigration recently. Of the 60 heads of households, 25 per cent moved to Gokilapuram in each of the following periods: 1927–42, 1943–61, 1962–70, and 1971–7. In other words, while the first 50 per cent of

[8] This section deals with 'permanent' migrants who are now residents of the village. Seasonal migration of workers is dealt with in ch. 9.

[9] K. Bardhan (1977) sums up the literature in this area neatly: 'Even with a strong push factor (unless it is extreme as in an epidemic or an acute drought), emigration from rural areas is selectively determined by the main available system of information and contact cumulated through past migration from kinship groups.'

[10] It is noteworthy, in this context, that a recent study of settlement patterns in the Cumbum Valley has shown that, in the long term, the rates of growth of population in villages with relatively high rates of irrigation have been higher than in villages with low rates of irrigation (see Rukmani 1989).

Table 4.4 *Location of native village of first-generation migrants, landless agricultural labourers, Gokilapuram, 1977*

Location of native village of head of household	Households	
	No.	%
Uthamapalayam taluk (same taluk)	36	60.0
Periyakulam taluk (neighbouring taluk)	8	13.3
Other taluks of Madurai district	7	11.7
Other districts	7	11.7
Unclear	2	3.3
TOTAL	60	100.0

Source: Survey data.

migrants came to Gokilapuram over a 34-year period (1927–61), the next 50 per cent came in less than half that time (1962–77).

Household Size

Agricultural labourers as a class live in distinctly smaller households than others in the population—in the survey the average size of household among all agricultural labour households was 4.6 persons and among landless agricultural labourers, 4.48 persons, while for all other classes in the population the average size of household was 5.22 persons.[11] The Rural Labour Enquiry results for Tamil Nadu have it that the average size of agricultural labour household was even smaller, 4.21 persons in 1974–5, and that the size increased from an average of 4.01 persons in 1964–5 (RLE 1974–75 (1978), table 2.5(*a*)). Among the other classes in the village, the peasantry as a whole and individual classes among the peasantry (average size of household among rich peasants: 5 persons; among poor peasants: 5.44 persons) and households with major income from business activity other than crop production lived in larger households than agricultural labourers; major landlords, among whom the average size of household was 8.73, had the highest average among classes in the village (see Tables 4.5 and 4.6).

There are at least four immediate contributory factors to household size: marriage and fertility, mortality, migration of individual members, and the formation of nuclear families, and each of these is likely to have contributed actively to smaller household size among agricultural labourers. There was

[11] In a review of the literature on the interaction between household occupational structures and demographic characteristics among the assetless poor, Barbara Harriss shows the general conclusion from the literature to be that 'although not all small households are poor, poor households tend to be small' (B. Harriss 1987, pp. 8–9).

Table 4.5 *Average number of persons per household, by region, 1961, 1971, 1981*

Region	1961	1971	1981
Tamil Nadu	4.64	4.74	4.71
Madurai district	4.68	4.81	4.73
Uthamapalayam taluk	4.96	5.00	4.67
Gokilapuram village	5.38	5.08	5.23

Source: Census of India.

Table 4.6 *Average number of persons per household, by class, Gokilapuram, May 1977*

Class	Average no. of persons
Major landlords	8.73
Upper peasants	5.00
Lower peasants	5.44
All peasants	5.18
Agricultural labourers who neither own nor operate land	
All agricultural labourers	4.60
Main income from business activities other than direct crop production	5.20
All households except agricultural labourers	5.22
All households	4.92

Source: Survey data.

migration by individual members (although in the major wave of annual migration, which was to the cardamom and other plantations, migrants were increasingly moving with their families). Although we do not have data on comparative levels of child mortality in the village, the effect of various demographic factors was to bring the child–woman ratio for agricultural labourers below the ratio for all other classes combined. There was also a distinct observed tendency for agricultural labourers to establish nuclear family households with high worker to non-worker ratios.[12] The ownership and control of land and other means of production played no role, or a very small one, in the economy of agricultural labourer families, and in such circumstances the advantages of a compact and independent household asserted themselves. There is also pressure on residential space and living facilities among agricultural labourers, and in most cases marriage entails setting up life

[12] For work participation rates, see chs. 2 and 3.

Table 4.7 *Growth of population among agricultural classes, Tamil Nadu, 1963–1964* (per 1000)

Household occupational category	Crude birth-rate	Crude death-rate	Natural rate of growth of population
Farmer	31	10	21
Cultivator	35	10	25
Agricultural labourer	34	13	20
Other agricultural work	36	8	28
All households	33	11	21

Source: NSS (n.d., table 1.3, pp. 20–3 and table 1.7, pp. 29–32).

in a separate hut. *Per contra*, at the other end of the scale, although much old landlord property has been partitioned, broken up (often into more manageable economic units), and sold in recent years, 9 out of 11 major landlord households were still in varying degrees joint families—there are still obvious advantages in not partitioning the property of households that have a large and varied asset base—and generally combined successful commercial agriculture with other forms of lucrative business activity. The results here are clearly consistent with Krishnaji's findings on family size among different socio-economic groups, summed up in his observation that 'a low average size of family among agricultural labourers arises mainly from high rates of mortality and family formation; and a high average size of family among big land-owners from low rates of mortality and a tendency for families to remain undivided' (Krishnaji 1983, p. 869).[13]

Concluding Notes

This chapter attempted to identify the sources of increase in the agricultural labour force in the village. It is a striking feature of development in Gokilapuram that—despite the fact that the agricultural labour force was assembled in an earlier period and a settled landless labour force is not a new phenomenon in

[13] The qualification should be added that it is useful to distinguish between agricultural labourers and poor peasants in this regard, since household size patterns are likely to differ between the two. (See also Krishnaji (1980) on family size variations and property partitioning.) It is of note here that the results for rural Tamil Nadu from an early NSS report (NSS n.d.) strongly support the hypothesis that the natural rate of growth of population among agricultural labour households is lower than for other classes, and that the crude death rate is higher (see Table 4.7). See also Nagaraj (1986) for an important discussion of the evidence on differential rates of infant mortality in Tamil Nadu.

the village—there has been an ongoing process of differentiation (involving households of different castes) over a single generation. Of the heads of landless agricultural labour households in the survey, the fathers of 58 per cent were *not* agricultural labourers; they were, in the main, peasants and artisans. Recent changes in the agrarian economy of the village and region accelerated differentiation and the expansion of the labour force through dispossession of the peasantry, erosion of traditional occupations, immigration, and other processes.

The data show that the number and proportion of manual labour and landless households in the village increased sharply over the last generation. It is reasonably clear that this process and the process of establishing the new technology and new farming practices were related. Some examples follow of the ways in which the new technology induced increased landlessness and additions to the manual labour force.

First, the demand for the services of artisans was directly affected by the new technology. Where mechanization is a component of new farming practices, this can be expected. The introduction of tractor-driven ploughs, tillers, cage-wheels, and trailers strikes at the demand for making carts and ploughs and repairing them. The installation of motor pumps has wiped out, almost completely, the demand for cattle-drawn leather water-bags. The literature on the backward and forward linkages associated with agricultural growth, in so far as it deals with the demand for rural non-agricultural products, has focused on the new demands created in the course of modernization and growth. (See, for instance, Mellor and Lele 1973; Mellor and Johnstone 1984; Hossain 1988.) A balance-sheet on growth linkages needs also to take into account the decline in the demand for the services that were associated with old farming practices.[14] (See, in this context, the subsection on ploughing in Chapter 9 and Appendix 3.)

Secondly, with the introduction of the new technology, many landlords found it less profitable to lease their land to tenants (and, indeed, often found that tenancy was an obstacle to the consistent implementation of new methods of cultivation). This was particularly true of their holdings of surface-irrigated land, which are located close to the village's residential settlement and where supervising cultivation is relatively easy. In the Cumbum Valley in recent years, tenants have not been organized, tenancies have not been registered, and tenants have been tenants-at-will. Those who were evicted were most often evicted with no compensation at all. The number of ploughs in the village is directly related to the number of families that actually cultivate operational holdings, however small the individual holding may be, and for that reason the eviction of tenants also had the effect of lowering the demand for the work of village artisans (see Appendices 2 and 3).

[14] In addition, the income elasticities of demand for some consumer goods produced by village artisans and craftsmen are negative: in the village, certain kinds of earthen cooking pots and vessels are conspicuous examples.

Thirdly, the poor have lost land through sale and mortgage. For a poor cultivator household, the current income from a plot of dry land and the potential income from that plot are similar and, in the vicinity of Gokilapuram, very low. (As we have seen, in a year of low rainfall, the gross output from an acre was often less than a bag or two—a bag contains about 58 kilograms—of coarse grain per crop.) For a rich household, the potential income from the same plot of land was vastly higher than current income. The household had, after all, the funds to convert dry land into groundwater-irrigated land (or to deepen a pre-existing well and install an electric motor) and plant crops that need high levels of investment and are exposed to risk. The difference between the current and the potential income from dry land had the effect of pushing up the price of dry land and of inducing smallholders to sell their dry land. Of course, the price of land has not been uniform for all plots of dry land. It has been particularly high for dry land with a high potential for conversion into irrigated land, such as level land adjacent to groundwater-irrigated land. Many rich farmers bought parcels of dry land, consolidating them into larger plots of groundwater-irrigated land.

Another reason for the sale of land by the poor in the recent period is the high price of failure when a farmer invests in new crop varieties and seeds and new methods of cultivation. This is true of groundwater-irrigated crops (such as banana, grape, coconut, or cotton) as well as rice cultivation on surface-irrigated land.

The price of land, particularly surface-irrigated land, rose steeply in the Cumbum Valley in the late 1970s. (This rise is discussed in Chapter 3 and in Appendix 4.) The rise in land prices was associated with new levels of productivity per unit of land, with the price of cardamom on the world market, which soared in the second half of the 1970s, and with the limited supply of Periyar-irrigated rice-growing land in the Valley. The price of an acre of surface-irrigated land varied between Rs 50,000 and Rs 60,000 in 1977 and rose to Rs 100,000 by 1985. The price of an acre of surface-irrigated land in 1976–7 was the equivalent of about 25 to 30 years' net income from paddy cultivation on an acre of surface-irrigated land in that year. In 1974–5, there was a shortage of irrigation water in the Periyar tracts. Cultivators were permitted to dig tube wells fitted with diesel pumps to help tide over the crisis that year, and it was, inevitably, only some landlords who could afford to do so. There were distress sales of surface-irrigated land during this period, necessitated, at least in part, by the new costs of failure that are associated with the cultivation of high-yielding varieties.

By contrast, the new price of land may also have the effect of helping an owner-peasant stave off depeasantization. When high-priced land can be sold in small units, as is true of surface-irrigated land in Gokilapuram, a peasant household can meet its debt and distress needs by selling a smaller parcel of land than it would have had to before the new technology contributed to raising land values. (See Osmani 1987, p. 22 n. 12 for a statement of this argument.)

Turning to household size, agricultural labourers live in smaller households than do other classes in the population, as well as in households that have a relatively high worker-density.

5

The Demand for Labour and Sources of Employment

This chapter deals with another aspect of the intensity and sweep of activity in the labour market: sources of employment in the village and changing patterns of labour absorption in different types of farming.

Paddy Cultivation on Surface-Irrigated Land

Description of Tasks

On paddy land, the demand for labour peaks sharply. Men are employed for field wall preparation, bund-cutting, and for ploughing. Tractor ploughing has replaced male labour and bullock power substantially on many fields. Men and women are employed at basal manuring: men, often helped along by young boys, cart the manure to the fields and unload it, dumping piles of farmyard manure at different places in the field, and women scatter it over the topsoil. The demand for female labour rises sharply at the time of transplantation, a female-dominated operation; it rises again when women are called for the first and second weedings. Only a few rich farmers hire labour for a third weeding, and only on some fields. The high point of the demand for hired labour on paddy-fields is at harvest time, when male and female labour power is employed. A few, generally scheduled caste, households send their women and sometimes children to glean the paddy-fields, and the women of some scheduled caste families sweep sections of the roads between paddy-field and threshing-floor to recover what they can of grain that has been shed. The first threshing of grain is done manually by men and women (the operation requires intense physical effort); the second threshing, formerly done by men driving bullocks across the grain, is now largely done by tractor. A few women, generally of the Chakkiliyar caste, sweep the threshing-floor.

Irrigation and guarding tasks at the paddy-fields are done by a group of irrigation workers (*neerpaichi*) and crop-guards (*kavalkarar*), agricultural labourers belonging to the Maravar caste, who are chosen annually for the job and are paid in paddy at the end of each crop. They also scatter chemical fertilizer over the paddy-field. Private operators with diesel-powered crop-sprayers are paid by the acre to spray pesticides. An important feature of the system of employment of workers on paddy land is the contract (*kothu*) system during the transplantation and harvest seasons, in which groups of contract labourers, each under a head labourer (*kothukkarar*), transplant or harvest paddy-fields, and are paid at piece-rates calculated by the extent of the plot (irrigation and the *kothu* are dealt with in detail in Chapter 9).

Labour Absorption

Table 5.1 shows the pattern of labour absorption in paddy; the calculations are for a single *kuzhi* (0.6 acres). On a plot where all operations other than the second threshing were not mechanized, the total number of person-days required in the period 1976–81 was of the order of 126 eight-hour person-days, or 519 person-days per hectare. Even if we assumed that the farm had tractor ploughing, transported farmyard manure by tractor trailer, and employed fewer women than was normal for weeding, and if we also leave out the days

Table 5.1 *Requirements of hired labour per* kuzhi *(0.6 acres) for paddy cultivation, Gokilapuram* (8-hour days)

Crop operation	Caste	Sex	Approximate employment per *kuzhi*		
			Males	Females	Total
Construction and maintenance of bunds	n/a	Male	2–3	n/a	2–3
Basal manuring					
Dumping farmyard manure	n/a	Male	20	n/a	20
Scattering farmyard manure	n/a	Female	n/a	3	3
Ploughing[a]	n/a	Male	15	n/a	15
Transplanting					
Carrying bundles	n/a	Male	1	n/a	1
Uprooting seedlings, bundling, transplanting	n/a	Female	n/a	15	15
Distributing seedlings	n/a	Female	n/a	1	1
Weeding[b]	n/a	Female	2	30	32
Irrigation[cd]	Maravar	Male	7	n/a	7
Guarding[d]	Maravar	Male	6	n/a	6
Harvesting[e]	n/a	n/a	n/a	n/a	15
Threshing[f]	n/a	n/a	3	4	7

Notes: These are the average requirements of a landlord household where all work is done by hired labour. Time taken in spraying pesticides is not included.

n/a = not applicable.

[a] It is assumed that all ploughing is done by men with draught animals.

[b] It is assumed that there are two weedings.

[c] The irrigation operation includes time taken in applying chemical fertilizers.

[d] Irrigation and guarding tasks are done in common; the figures here are a recalculation of the total labour time deployed by irrigation workers and guards (see ch. 9).

[e] It is assumed that the threshing-floor is next to the field; that the sheaves do not have to be transported a long distance; and that harvesting is not broken by rain or damp weather.

[f] It is assumed that the second threshing is done by tractor.

Sources: Survey data, case studies.

spent guarding the crop,[1] the number of person-days per hectare per paddy crop was still about 346.

In a farm where mechanization has not taken place (except in the second threshing), the ratio between male and female labour requirements (excluding the harvest operation, where both are employed) is marginally favourable to men. It is, however, to be noted that women's employment is concentrated in operations where relatively large numbers are required, and that when ploughing is mechanized and tractor trailers used to transport farmyard manure, two important sources of male employment, ploughing and basal manuring, are immediately hit.

Changes

Four factors influencing rates of labour absorption have been pointed out by Kalpana Bardhan: cropping patterns, intensity of irrigation and material inputs, mechanization of agricultural operations, and concentration in the distribution of land holdings (K. Bardhan 1983, pp. 11 ff.).[2] The first three factors are discussed here in the context of different irrigation types of land in Gokilapuram; aspects of the fourth, the distribution of land holdings, have been discussed in Chapter 3. The literature distinguishes between the impact of inputs of water, of seed, and fertilizer (these generally increase employment), of tractorization (whose overall impact, when the interaction of tractorization with other aspects of the production process are taken into account, need not be to displace labour),[3] and of other forms of mechanization such as the introduction of harvesters and threshers (which do unequivocally displace labour). It is a safe generalization from the available data that in the case of paddy cultivation in most parts of India, where the introduction of labour-displacing technology has been very limited, yield-raising techniques have increased labour absorption per unit of land. This is true of other rice-growing regions of Asia as well.[4] Some field studies have also shown that while new and high-yielding varieties absorb more labour per unit of land than old varieties,

[1] That is, if we change the assumptions of Table 5.1.

[2] The literature on the effect of the new technology on agricultural employment has also been reviewed in Basant (1985); see also Ishikawa (1976); P. K. Bardhan *et al.* (1978); ILO-ARTEP (1979). A detailed analysis of changing trends in labour absorption in Indian agriculture for the period 1971–2 to 1983–4, a time period subsequent to the one covered in the early ILO-ARTEP literature, is in Bhalla (1987).

[3] Swenson (1973), for instance, suggests that tractorization may have been a major factor in increases in the double-cropping of paddy in the two Thanjavur villages that he studied.

[4] On increased labour absorption as a consequence of increased irrigation and the introduction of new varieties, see Haswell (1967); Clay (1976); Bautista (1977); Dasgupta (1977); Patel (1981, p. 21); Smith *et al.* (1981); S. V. Rao (1984); Bray (1986). Analyses of changes in labour absorption in rice cultivation in the 1970s—after the new varieties had taken root, so to speak, in different parts of the country—are more scanty. A recent analysis by N. J. Kurian of official data covering the 10 major rice-growing states in India over the period 1971–4 to 1978–81 indicate that, across these states, there was an increase in labour absorption. Behind the overall figure, however, is considerable variation between states (Kurian 1986, table 5).

their labour requirement is less per unit of output, because output gains are more than proportionately higher (see, for instance, Bartsch 1977, p. 40).[5]

In Gokilapuram, surface-irrigated land has been paddy land throughout the period that we are discussing. The increase in labour absorption—the predominance of 'labour-augmenting' factors in the new technology over 'labour-displacing' ones, in the terminology of the literature—has been part of a process that began in earlier decades. Even before the new seed, fertilizer, and pesticide technology together with the associated infrastructural developments had taken over paddy cultivation, some cultivators had begun to supervise hired labour closely and ensure that individual tasks were performed with greater care. Bunds were cut more carefully and built stronger, weeding was more thorough, and transplanting done in rows, and this made for increase in employment on surface-irrigated land. In the case study of changes in farming practices on rice cultivation in Appendix 2, SNR describes this process as having begun in the 1940s on his surface-irrigated land.

Looking back today on changes over the past three decades or so, the operations where increases in the demand for labour have made the major contributions to the overall increase in labour absorption on paddy-fields are transplanting, weeding, harvesting, and threshing. There has also been an increase in the number of hired labourers employed at operations which employ workers on a smaller scale. These operations are crucial to the cultivation of the crop and some of them are spread out longer over the crop season. Construction and maintenance of field walls and bunds; irrigation and field maintenance; basal manuring on fields that are well supplied with farmyard manure; and pesticide-spraying and applying chemical fertilizers are examples of such operations. These operations were hardly known in the village in the 1950s and not undertaken over large tracts till the middle and late 1960s and—particularly in the case of pesticide-spraying—the early 1970s.

In paddy cultivation, as the foregoing suggests, the increase in employment at operations where women find work has been greater than at tasks done by men. John Harriss's data on the labour calendar for men and women and the relative demand for male and female labour in paddy cultivation in 'Randam' village in North Arcot district point in the same direction. His conclusion is

[5] Swenson (1973) attempted one of the few micro-level assessments of changes in labour productivity in Tamil Nadu agriculture in recent years; his data base has, however, been subject to criticism (Connell and Lipton 1977). His data cover the time span 1965–6 to 1970–1 in two villages in Thanjavur district, and show an increase in the productivity of labour: while labour employed in paddy cultivation increased by 7% over the period, paddy production increased by 22%. Swenson does qualify this by saying that it may be an understatement of the increase in employment over the district as a whole. An alternative index of labour productivity that is used in the literature is the elasticity of employment with respect to the output of new varieties, which can be taken as a measure of potential changes in labour productivity. The literature on this has been reviewed in Basant (1985). This figure shows considerable variation: the data in Mellor (1976) show an elasticity of 0.3; in C. H. H. Rao (1976), the corresponding figure is 0.7, and, in the revised draft of the sixth five-year plan, the figure used was 0.79 (GOI 1979).

that the 'pumpset revolution'—there are revolutions and revolutions—and the introduction of high-yielding varieties 'has led to a greater increase in the availability of work for women than for men, because the most labour intensive [Harriss means, of course, 'employment intensive'][6] operations in paddy cultivation (transplanting, weeding, and harvesting) are carried out . . . manually by women, who also do as much manuring work as men and contribute to threshing' (J. Harriss 1982, p. 97).[7]

Paddy cultivation in Asia is characterized by limited levels of mechanization in comparison with other crops (such as wheat), even in regions where modern methods of farming are practised. In most of East Asia, water control and the improvement of field conditions laid the basis for mechanization. In some regions, the first operation to be mechanized was threshing (as in Japan and even parts of central Luzon); in others, the first was land preparation, by means of tractors or power tillers. Mechanization has been more difficult with regard to other operations; Francesca Bray, for instance, points to the slow progress in the mechanization of transplanting and to barriers to the use of heavy machinery because of the nature of paddy soils in Asia (Bray 1986, pp. 56, 60).[8]

While the Cumbum Valley has it in common with regions in East Asia that water control and land improvement preceded mechanization (as did the careful supervision of manual operations), mechanization in the Valley has been restricted to the introduction of tractors and power-driven pesticide sprayers. And unlike some parts of East Asia, herbicides have not displaced manual labour at weeding. Consequently, labour absorption has declined only for bullock-supported operations, now partly done by tractor.[9] There are three such operations:

Carting farmyard manure to the field. Today those who have farmyard manure, have no carts, and own tractors or can afford tractor time, use tractors to haul it in trailers to the field.

[6] If the term 'labour-intensive' is taken to refer to the low organic composition of capital in the farming process, then operations such as the construction and maintenance of bunds or irrigation are as labour-intensive as, say, harvesting, although harvesting requires that many more persons be employed per unit of land.

[7] The research on the differential impact of new farming techniques on the demand for male and female labour has been reviewed in B. Harriss *et al.* (1987) and Basant (1985, pp. 59 ff.). Mies (1986, p. 60) and Hart (1986, p. 96) have useful discussions of the gender division of labour in rice cultivation. Bina Agarwal's evidence on the displacement of female labour in wheat cultivation as a result of the introduction of harvesting machines is particularly striking, since about 88% of female labour time in wheat cultivation is absorbed by the harvest; see Agarwal (1981).

[8] Barker *et al.* (1985, ch. 8), has a useful description of patterns of mechanization of paddy cultivation in Asia. It is also of note that even in two countries with very different agrarian regimes, Britain and Japan, increased farm outputs were initially achieved by means of biological innovations (see Collins 1969).

[9] See Kurian (1986, p. 68) on the impact of mechanization on bullock-supported operations in rice cultivation in India.

Ploughing. Major landlords own tractors, and most others, even relatively poor peasants, hire tractor time for ploughing. They do so except where plots are inaccessible or the line of contour makes it impossible to plough the field with a tractor, or where the cultivator is very poor or cultivating just a patch of, say, encroachment land.[10] Fields are ploughed with the nine-pronged tiller and the cage-wheel before and after the release of Periyar water and also after fields are flooded. (Nevertheless, it remains a feature of paddy-farming practices in the village that, despite relatively extensive tractorization, nursery ploughing and one wet-land ploughing are still done by bullock-drawn ploughs.)

Second threshing. Until recent years, the second round of threshing harvested grain was done by cattle that were made to trample grain for a day or more; today threshing by trampling has almost disappeared in Gokilapuram (although it still continues in certain parts of the Valley, notably Uthamapalayam).[11]

Mechanization in paddy cultivation—unlike, for instance, in wheat cultivation in advanced regions—has not taken over any high-employment operation; in all such operations the effect of the new technology has been to increase labour absorption per unit of land.

A useful measure of labour productivity is the physical output per person-day of labour input in the crop. The evidence relating to this indicator in respect of rice cultivation in Asia has been reviewed by Barker *et al.* (1985, ch. 9). A conclusion that emerges very strongly from comparing the Gokilapuram data with the data from East and South-East Asia used by Barker *et al.* is that cultivation in Gokilapuram was characterized by high yields per hectare (which are comparable to levels in Japan and parts of central Taiwan in the 1970s), side by side with levels of physical output per person-day that were very low by international standards (see Table 2.8). Production per hectare of paddy during the first (June to October) crop in the village varied between 6 and 8.1 tonnes per hectare of paddy (or 4 to 5.4 tonnes of rice per hectare) in the 1970s and early 1980s. At the same time, while many parts of East Asia employed less than 150 days of hired labour per hectare, in Gokilapuram, as we have seen, employment was in the range of 346 days to 519 days per hectare per crop. So while productivity measured in kilograms per person-day varied between 36 kilograms and 62 kilograms in Japan, South Korea, and Taiwan, it varied between 10.4 kilograms and 15.6 kilograms (at an output level of 8.1 tonnes of paddy or 5.4 tonnes of rice per hectare) in Gokilapuram.

This was a very low level of productivity compared with East Asia; the differences here appear to be related to differences in the technological level of farming, and to greater mechanization and the use of herbicides in East Asia. In relation to productivity levels in Java, the Philippines, and Thailand, the

[10] See Bray (1986, p. 56) on the relationship between terrain and water control in paddy cultivation, and the restriction of field size in order to ensure that fields are flat.

[11] Guhan (1983) describes a roughly similar process of tractorization in Palakurichi (pp. 44 ff.); in East Thanjavur, mechanization has also involved the use of motor pumps for field drainage.

Cumbum Valley had a higher level of labour absorption and much lower levels of productivity per person-day of labour, *as well as* higher production per hectare. The comparative situation here tends to support the conclusion of Ishikawa and other contributors to the ARTEP literature on labour absorption that at a certain level of technology, increased labour utilization goes hand in hand with increased production per hectare. The conclusion that Barker *et al.* draw, that 'with the technological alternatives available today, labour use in rice will not reach the level of 200 or more [days] per hectare prevalent in Japan and Java for a decade or more after World War II' (1985, p. 127), is interesting in that it poses that which is *possible*; it is not, however, true of the high-yielding paddy tracts of the Cumbum Valley.

Farming on Groundwater-Irrigated Land

Groundwater-irrigated land is in use throughout the year. The pattern and intensity of labour absorption depend on the cropping pattern and methods of cultivation, and crucial determinants of peaks in the demand for labour are the number and type of crops planted in an agricultural year. Employment on groundwater-irrigated land is more evenly spread over the year than on paddy-fields. The important sources of employment for men are field-bund and field preparation, including ploughing and basal manuring; preparing irrigation basins with hoes; irrigating fields; clearing fields of plants after harvest; guarding standing crops; and, in addition, on banana fields, chopping side-shoots and dead leaves from the tree. For women, the main tasks are spreading basal manure, weeding, cotton-picking, harvesting vegetable and cereal crops, and threshing and winnowing coarse grain. Grape cultivation draws on a group of specialized workers, who work most of the year only on grape fields. (This is not a very important source of employment in Gokilapuram; vineyard workers live mainly in and around villages where grape is a more important crop.) On banana fields, harvesting is done by workers employed by merchants who buy up the crop; in Gokilapuram, these workers are brought in from nearby Chinnamanur.

Changes in Labour Absorption

Let us take, by way of example, a plot of unirrigated land that has been converted to groundwater-irrigated land.

First, there is the hired labour involved in such conversion and in maintaining groundwater-irrigated land, which includes land improvement (levelling the land, clearing the field of scrub and rocks and stones, and, when the field is near the foothills, clearing it of scree and building protective contour walls), digging a well and installing a pump, deepening and maintaining the well, and keeping the well, pump, and coping-wall in good repair. The extension of groundwater-irrigated cultivation has also led to a certain proliferation of electricians' skills—many villages now have an electrician and assistant to repair electric pumps.

Next, there is the sea-change in labour absorption in field crops that takes place after the conversion to groundwater irrigation. Assume that our plot of unirrigated land was of one acre and planted with a coarse-grain crop. On such a plot in the vicinity of Gokilapuram in the period 1977 to 1980, the employment of hired labour for cultivation would have been about 23 woman-days and 5 man-days as well as 5 man-days of ploughing with animals (this excludes labour time in guarding the crop; see Table 5.2). Assume also that this acre, converted to groundwater irrigation, was given over to irrigated *cholam*/tomato/banana/cotton. (An affluent cultivator-investor would generally not, of course, stop at converting a single acre to groundwater-irrigated land; he would want to use the benefits of well irrigation over a wider extent of land and would attempt to acquire contiguous plots.) Excluding, in each case, the labour time taken on guarding and plant protection and tractor time, the main operations in

- irrigated *cholam* would have absorbed 75–90 days of female labour time and 55–60 days of male labour time plus ploughing time (see Table 5.3);
- tomato would have absorbed about 307–72 days of female labour time and 42–5 days of male labour time plus ploughing time (see Table 5.4);
- banana, which provides men with relatively more employment per unit of farmed land than other groundwater-irrigated crops, would have absorbed about 73–83 days of female labour time and 199–252 days of male labour time plus the time taken in banana harvesting (which was done by men from outside the village, directly employed by merchants who bought up the crop) plus ploughing time (see Table 5.5); and

Table 5.2 *Labour absorption for one cultivated acre, unirrigated coarse grain* (8-hour days)

Crop operation	Men	Women
Construction and maintenance of bunds	2	—
Sowing	—	1
Weeding	—	15
Harvesting (including threshing)	3	7
Total	5	23
Ploughing	5	—

Note: The data in Tables 5.2 to 5.5 were collected from selected households in Gokilapuram in the period 1977–80; they refer to plots where all operations are undertaken by hired labour. They exclude time taken in guarding the crop and plant protection and also tractor time.

Table 5.3 *Labour absorption for one cultivated acre, irrigated* cholam (8-hour days)

Crop operation	Men	Women
Construction and maintenance of bunds	3–5	—
Basal manuring	3	6
Forming irrigation basins and planting	8	8–13
Fertilizer application	2	1
Weeding	—	40
Irrigation	36	—
Harvesting	3	15
Threshing	3–6	5–9
Total	58–63	75–84
Ploughing	8	—

Note: See Table 5.2.

Table 5.4 *Labour absorption for one cultivated acre, tomato* (8-hour days)

Crop operation	Men	Women
Construction and maintenance of bunds	4	12
Basal manuring	6	—
Forming irrigation basins	6–7	—
Sowing/nursery planting	—	15
Transplanting	—	—
Weeding	—	55–70
Irrigation	18	—
Picking and packing into baskets/ crates and uprooting plants after harvest	8–10	225–275
Total	42–45	307–372
Ploughing	13	—

Note: See Table 5.2.

- cotton would have absorbed about 324–64 days of female time and 79 days of male labour time plus ploughing time (see Table 5.6).

When a plot of unirrigated land of the type that exists in the vicinity of Gokilapuram was converted to groundwater-irrigated land in the 1960s and 1970s, the old levels of employment were simply not comparable with the new.

On land that has traditionally been irrigated by wells, the major change is that traditional lift irrigation—in which water is drawn in leather bags by bullocks driven by a worker, or lifted by *ettram*, a water lift operated wholly by human labour—represents a technology that has now been driven out. In terms of cropping patterns, there have been shifts of different types:

- from irrigated food-grain and traditional non-cereal crops such as chilli to vegetables, banana, cotton, grape, and so on; and

Table 5.5 *Labour absorption for one cultivated acre, banana* (8-hour days)

Crop operation	Men	Women
Construction and maintenance of bunds	2	—
Basal manuring	6	8
Forming irrigation basins	9–11	—
Digging pits and planting saplings	12–18	—
Irrigation, general crop maintenance	125–150	—
Spreading earth, packing the soil	15–25	—
Fertilizer application	—	15–25
Weeding; cutting dead leaves	—	50
Harvesting[a]	—	—
Uprooting the trees	30–40	—
Total	199–252	73–83
Ploughing	14	—

Note: See Table 5.2.

[a] Done by workers from outside the village.

Table 5.6 *Labour absorption for one cultivated acre, cotton* (8-hour days)

Crop operation	Men	Women
Construction and maintenance of bunds	6	—
Basal manuring	4	6
Forming irrigation basins	11	—
Planting	4	8
Irrigation	33	—
Weeding	12	65–75
Fertilizer application	—	10–15
Picking, cleaning	—	260–275
Uprooting	9	—
Total	79	354–379
Ploughing	16	—

Note: See Table 5.2.

- from old varieties (older cotton and vegetable varieties, for example) to new varieties (such as MCU-5 and Varalakshmi cotton and PKM tomato varieties).

In groundwater-irrigated cultivation, as in surface-irrigated cultivation, bullock-supported operations—here, irrigation, ploughing, and carting—have declined.

One contemporary development in groundwater-irrigated cultivation that is likely to affect future trends in the demand for hired labour is the large-scale planting of coconut on groundwater-irrigated land. This is a trend that gained momentum in the second half of the 1970s and continued through the 1980s (the data for the 1970s are in Chapter 2) and its impact on employment is marked; where fields are turned from, say, cotton, vegetables, or banana to coconut, the long-term effect on the demand for hired labour in field production is to reduce it substantially (see Chapter 10).

Unirrigated Farming

Unirrigated land in the vicinity of Gokilapuram village and in most parts of the Cumbum Valley is, as may be expected, the least labour-absorbent and the poorest source of employment for workers. The fertile unirrigated groundnut fields in the south of the Valley, however, are an exception, and during the groundnut harvest season workers leave the village in groups under a contractor, pitch camp in and around Gudalur and Lower Camp near where the Periyar enters the plains, and remain there till the end of the harvest, for about a month in a good year. In the village, the standing crop on unirrigated land is guarded by a group of agricultural labourers chosen from among the Kallar caste, who are employed for about five months in a year and are paid in grain at a rate per unit of land that they guard.

Dry-land farming in the area is relatively unchanging, and the demand for hired labour varies with soil and topography and, most important, with the quantum and timing of rainfall. No peasant household, however, can live off a small plot of dry land, and no agricultural labourer family in this area can depend on dry-land farming to provide the employment that will support its members.

Tree Crops and Tasks Other than Crop Production

Over the past 25 years, Gokilapuram has emerged as the major centre of the tamarind economy of the Cumbum Valley. A few of the major landlord families, especially those of the Kali Vilas (the name of their family house) group, are the Valley's major tamarind contractors, who bid for and obtain the right to harvest tamarind trees from the Highways Department, the Public

Works Department, local bodies and other departments of government, and from private persons.

In earlier days, people from the villages and towns broke fruit from the tree for household use, part of the annual tamarind crop went to waste, and the ripe fruit were not intensively, carefully harvested. Tamarind was a commodity, bought and sold on the market, but commercialization was not yet comprehensive and the market did not hold complete sway over the usufruct of tamarind trees everywhere in the Valley. Today, virtually every tree (belonging, as we noted, to public organizations, other institutions, and individuals) has been contracted for, and a substantial labour force has been brought together by the contractors to pick every ripe fruit off them. From January each year, they employ men and women from Gokilapuram and other villages to harvest the tamarind; men climb the trees and beat the fruit down and women gather the fruit and carry it in baskets to storage places or to carts and trailers for transporting it back to the village.

After the crop is brought back to the village, women work at processing the tamarind, and this is their major source of employment outside direct crop production in the year. Women dry the tamarind fruit and shell it; then, working in large numbers in sheds or courtyards belonging to tamarind contractors and merchants, they hammer down the fruit, removing its seed and fibre. Often their children help them. They are paid by the 10-kilogram weight, the baskets of decorticated and de-fibred fruit being notoriously short-weighed. Tamarind-processing work lasts, with varying intensities of employment, for 80 to 90 days.[12]

Plantations on the eastern (Tamil Nadu) and western (Kerala) hills draw off a significant number—for the Valley as a whole, the number runs into thousands—of labouring families every year, particularly from harijan households. In 1977, workers were still employed through the *kankani* (overseer) system. An estate owner gave a *kankani* an advance of money and told him how many workers he was to recruit. The *kankani* then came down to the Valley and recruited from his own and neighbouring villages. He recruited workers whom he knew or whose families he knew, generally recruiting them in pairs, mainly married couples. He had the workers give him photographs of themselves and sign a bond and he paid them an advance on their wages. The *kankani* system was a method of control: any worker who showed signs of militancy or protest could be told to leave the plantation and go back to the village. By 1977, the employment situation for an agricultural worker in the village was grim and plantation work, particularly in Kerala, was coveted employment, and the *kankani* had a large pool of workers to choose from. Within the agricultural

[12] For purposes of analysis, in subsequent chapters we have taken tamarind-*harvesting* to be an agricultural operation, since it is directly part of crop production. The task of tamarind-*processing*, removing seeds and fibre, done by women employed by tamarind contractors in sheds and courtyards in the village, is considered a non-agricultural task, i.e. a task outside direct crop production.

labourers' sections of the village, and particularly within its harijan sections, the *kankani* households were better off than the others; they lived in bigger, more substantial houses, their women wore jewellery, and, more often than not, a *kankani* was also a petty money-lender.

The trade union movement in Kerala fought for decades for increased benefits for plantation workers and for bringing the *kankani* system to an end and, during the late 1970s and the 1980s, the system was brought to an end in the Kerala plantations. It was replaced by the establishment of a settled labour force, in which Tamil workers and particularly workers from the Valley are a major component. Although workers are still employed at a daily rate, they are on the muster rolls of the *estate* and are not employees of the *kankani*, and must be re-employed by the estate when cultivation resumes after the slack season. Workers, most of whom travel with their entire households, work on the cardamom estates for nine months of the year, returning to the village for three months, and sometimes in between for major festivals. In the east and south-east, where the development of cardamom and coffee plantations is more recent and the trade union movement still weak, workers are employed for shorter periods, with little protection of their rights under law.

There is also a cooperative society in the village for hand-pounding paddy into rice, where about 25 to 30 women are employed; employment here is subject to seasonal fluctuations. Some men work at well-digging and at public-works sites; men and women also go to the eastern hills, particularly during the lean season, to bring down and sometimes sell firewood. Men and women, particularly women, from among the poor work at a wide variety of domestic tasks for the rich—chopping wood, sweeping house and compound floors, clearing dung, taking paddy to the mill, cooking, running errands, and other such tasks.

Artisan activity in the village, which is caste-based, consists of woodwork (mainly the construction of carts and ploughs), pot-making, and blacksmith work. The skill of Gokilapuram woodworkers at making carts is well known locally. There are two small blacksmiths' forges in the village: they are mainly engaged in replacing and repairing plough-tips, making small hand implements, and forging the metalwork for carts.

Exchange Labour

Exchange labour works like this: members of a household join members of another household in working without payment at certain tasks on the land of the second household; in exchange for this, members of the second household join members of the first household in labouring on the latter's land. Labour is generally exchanged for work at the same task (weeding for weeding, ploughing for ploughing, and so on). While exchange labour is not strictly within the purview of this study, it is of note that the decline of this form of labour in cultivation, its reduction to insignificant levels relative to the labour market as a whole, is a process that has run parallel to the development of the demand for hired labour in the village. There is an extensive literature on exchange labour in different parts of Asia, whose major conclusion is that it is an institution in decline, replaced, under conditions of agricultural commercialization, by different forms of wage labour.[13]

On surface-irrigated land, exchange labour survives to a very limited extent in ploughing (particularly wet-land ploughing) and has almost disappeared in weeding, except on very small plots of land. It survives in some dry-land operations: ploughing, weeding, and reaping scanty harvests. Exchange labour exists generally between families that have plots of roughly similar extent in the same vicinity; such families are generally well known to each other and often belong to the same caste. In most cases, exchange labour represents mutual aid between peasants and poor working families that operate at low levels of economic efficiency. On occasion, however, even small and medium landlords may send out their animals with a farm servant or a rich peasant may send his animals with a family member to plough the land of another in exchange for a similar loan when their own land has to be ploughed.

Conclusions

There has been a sharp increase in the demand for hired labour at different tasks in the region; the literature indicates that this is true for other parts of the State as well. There has been a gradual increase in labour absorption at different operations in paddy production, which is most marked in the employment-intensive operations. Farming practices in paddy cultivation have become characteristic of what Francesca Bray describes as 'a highly productive system' of rice cultivation.[14]

The sharpest increase in labour absorption has been on groundwater-irrigated land. When a plot of unirrigated land on which, say, unirrigated coarse grain is grown is converted to groundwater-irrigated land, there is a multifold increase in labour absorption, whichever the new crop grown, and particularly if the crop is an employment-intensive irrigated non-cereal crop. For example, if the change is from an unirrigated coarse cereal to irrigated

[13] See Yeshwanth (1966); Gunasinghe (1975); Nakamura (1976, 1982); Dasgupta (1977, p. 14); Fredericks (1977); Kitahara (1977); Takahashi (1977); Wickramasekara (1977); Chung (1981); Hara (1981); Rahman and Das (1982); Drèze *et al.* (1986, p. 26); van Schendel and Faraizi (1984, pp. 57 ff.). An exception to this phenomenon of decline is in the work of Sudha Rao (1984), who noted a process of resurgence of this institution in a village in Karnataka.

[14] See ch. 2 and App. 2. Two studies of farming practices and land management from village studies conducted in other parts of Tamil Nadu in an earlier period also provide points of contrast with current practices in Gokilapuram; see Nakamura (1976) on Abinnimangalam in Tiruchi district and Étienne (1968) on Kila Ullur in Thanjavur district. For qualitative accounts of the agricultural cycle and the gender division of labour in Thanjavur district in the 1950s, see Sivertsen (1963) and Gough (1981).

cotton, the increase is by a factor of about 16. The conversion of poor-quality land with a well from which water is drawn by men working with draught animals to land with a well that is fitted with an electric motor pump also involves a rise in labour absorption, particularly if the plot is converted from one on which a poorly irrigated coarse grain or oil-seed crop is grown to one that grows cotton or vegetables or banana.

On unirrigated land where cultivation practices have not changed, labour absorption remains low and stagnant. Areas where commercial cultivation has developed on unirrigated land (as at Gudalur in the south of the Valley) provide a secondary source of employment to hired labourers in their vicinity and, at the time of the groundnut harvest, attract migrant workers from other villages and from outside the Valley.

The operations where labour absorption has been cut down are, in the first place, those where men work with draught animals. In paddy, this has meant a loss of jobs for men at ploughing, carting, and the second threshing. However, even on fields on which tractors are used, nursery ploughing and one wet-land ploughing are still done with bullocks. On groundwater-irrigated land, lifting water with bullocks has almost disappeared, and ploughing is also increasingly done by tractor. A consequence of this is a decline in the number of landless agricultural labour households that maintain plough-and-cart animals. An interesting related point is that exchange labour in the village is on the decline.

An important feature of the employment of hired labour in the village is that although the gender division of labour is not unchanging, almost every specific operation, with the prominent exceptions of paddy and groundnut harvesting, is not performed by men and women together. In the case of Gokilapuram, it is clear that female work participation rates are high and that the changes in labour absorption described in this chapter led to a greater increase in the demand for female labour than the demand for labour at tasks at which men were employed. However, since changes in cropping patterns and the precise tasks created and displaced by different aspects of technical change are crucial determinants of the absorption of female labour, the situation is not static. As we shall see in Chapter 10, by 1986 it had swung against women.

While paddy yields in Gokilapuram compare favourably with, for example, cultivation in Japan, South Korea, and Taiwan, labour absorption per hectare in paddy in this high-yielding region is exceptionally high, and the productivity of labour in paddy cultivation, measured in output per person-day of labour employed, is exceptionally low in comparison with East Asian agriculture and low in comparison with cultivation in South-East Asia (although output per hectare in Gokilapuram is higher than in South-East Asia).

There has been a diversification of the sources of employment. Workers seek jobs at different field crops on different irrigation types of land, at tree crops, and at different types of non-agricultural work, tamarind-processing, plantation work, hand-pounding paddy, road-building, and other non-agricultural manual tasks. Diversification and the introduction of new crops has been accompanied by a casualization of the labour force—the great majority of workers look for work wherever they can get it—along with increased specialization among a smaller group of workers at certain specific tasks, such as grape cultivation, harvesting bananas, operating power sprayers, and driving tractors.

An interesting example of accelerated commercialization in the product market serving as an impetus for the assembling of a progressively larger labour force is provided by developments in the tamarind business in the Valley. This has seen considerable centralization and concentration in recent years, with Gokilapuram emerging as its focal point: landlord households of the village function together, bidding for harvesting rights from individuals, private institutions, and government bodies, and they employ men and women to harvest tamarind in different parts of the Valley and women to process tamarind in the village.

While the work-force is settled in that it is resident in the village, it is also notably mobile. Workers look for, and walk to, daily employment on fields that are up to 6 and 7 kilometres away. They migrate to Gudalur, south of the village, for the groundnut harvest for about a month in August–September and to villages in different parts of the Valley to harvest tamarind from January to early March. At paddy transplanting and harvest time, their ranks are increased by workers who come from surrounding dry villages. There is large-scale migration to plantations in the hills that border the Cumbum Valley. Thousands of workers from different villages—pulled by the development of cardamom cultivation in Kerala and its extension into Tamil Nadu, and by the higher wages of the estates, and pushed out of the village by underemployment and poverty—seek employment for varying periods of time on the estates. (The most fortunate of these migrants are employed for nine months a year at estates in Kerala, where trade unions are active.)

Migration takes a worker out of a village and into labour markets where the traditional ties of the village do not, or seldom, extend, and where (in the case of estate workers in Kerala) he or she is protected by a trade union and by labour legislation. Both for the changes it brings to the daily life of the worker, and for the new ideas and consciousness that are disseminated by migrants to those still village-bound, migration is a great catalyst of proletarian qualities in the labour force.

With changes in the level and operation-wise distribution of demand for labour over the 1960s and 1970s, employers have had to hire large numbers of workers for specific operations at different crops on different types of land at different times of the year. The new levels of demand for labour are an important factor in the impersonalisation of the relationship between employer and employee at production tasks. This is an issue whose different aspects are dealt with in the chapters that follow.

6

The Deployment of Labour Time

This chapter deals with the deployment of the labour time of landless agricultural labourers at agricultural and non-agricultural tasks over a working year. It raises issues related to the volume of employment available to agricultural labour households, to the pattern of this employment, and to seasonality in the deployment of labour time. While the chapter is organized around these three issues, it also deals briefly with other features of the working year, namely, diversity in the tasks that agricultural labourers are employed in, child labour, and the length of the working day.

If data on labour utilization are to achieve a measure of accuracy, it is necessary that they be collected on a crop-specific, operation-specific basis. For this reason, as has been noted in the literature, micro-level studies often reflect the extent of rural unemployment more reliably than existing macro-level surveys. Rates of unemployment generated by micro-level data have also been observed to be, in general, higher than those obtained from official sources of macro-level data (P. K. Bardhan 1984*b*; Drèze *et al.* 1986). The method of collecting the data used in this chapter involved a detailed recording of the deployment of labour, crop by crop and operation by operation, separately for men, women, and children in each household. The questionnaire-based data were supplemented by labour diaries maintained for selected agricultural labourers. (There is a more detailed description of the method of data collection in Appendix 1.)

Here as elsewhere in the book the term 'agricultural tasks' is used to indicate work in direct crop production; 'non-agricultural tasks' are tasks outside direct crop production; although they may be, as tamarind-processing and the hand-pounding of paddy to rice are, tasks that derive from agricultural production.

Employment and Unemployment among Landless Agricultural Labourers

Of all workers, slightly more than 80 per cent were employed at agricultural tasks; nearly one-fifth did not seek or did not get employment in agriculture (see Table 6.1). The latter comprised the non-agricultural-worker component of landless agricultural labour households. There is variation here: this wholly non-agricultural component was about 20 per cent among men, less than 9 per cent among women, and about 65 per cent among children.

Among non-agricultural tasks, tamarind-processing was by far the most important: about a third of the total labour force in landless agricultural labour

Table 6.1 *Number of landless agricultural labourers employed, by tasks, Gokilapuram, 1976–1977*

Category of workers	Men		Women		Children		Total	
	No.	%	No.	%	No.	%	No.	%
No. of workers in category	315		338		79		732	
No. of workers employed at:								
All agricultural tasks	252	80.0	309	91.4	28	35.4	589	80.5
Non-agricultural tasks								
Tamarind-processing	4	1.3	168	49.7	48	60.8	220	30.1
Plantation labour	40	12.7	33	9.8	10	12.7	83	11.3
Hand-pounding paddy	0	0.0	25	7.4	0	0.0	25	3.4
Other manual tasks	47	14.9	10	3.0	4	5.1	61	8.3

Note: Percentages are of no. of workers in category.

Source: Survey data.

households were employed in it. This is a task performed by women, and by women who are helped by children. Next in importance, plantation labour called on about 11 per cent of the labour force, with men, women, and children participating. The third most important non-agricultural manual task was hand-pounding paddy, again an operation done by women. Other manual tasks (building roads, construction work, digging wells, chopping wood, etc.) were dominated by men, and about 8 per cent of the labour force found employment in these tasks.

An important feature of the labour force was that the most active casual workers performed a very wide variety of tasks, within agriculture and outside it. The rate of work participation among landless agricultural labourers was consistently high, and, in the age group 20 to 59 years, only 4 per cent of the population, 10 men and 13 women, were non-workers (see Table 6.2 for age-specific work participation rates). Unemployment among casual labourers in the village is a very different phenomenon from unemployment among, for instance, educated unemployed who are looking for salaried jobs. Almost every person above the age of 15 years, and 79 children below that age, did *some* kind of manual labour during the course of a working year. Most of them obtained work for only a part of the year (and some for only a few days); the differential availability of employment is an issue dealt with later.

Table 6.2 *Work participation rates for agricultural labourers, by age and sex, Gokilapuram, 1977* (%)

Age group	Work participation rates		
	Males	Females	Total
0–4	3.7	0.0	2.2
5–9	2.4	12.4	7.5
10–14	36.8	52.1	44.7
15–19	86.0	88.9	87.6
20–24	93.5	97.6	95.4
25–29	100.0	91.4	95.4
30–39	98.4	98.6	98.5
40–49	100.0	92.2	96.1
50–59	86.0	97.3	91.3
15–59	94.2	94.0	94.1
60 and above	80.8	72.2	75.8
All ages	60.4	65.9	63.2

Source: Survey data.

Number of Days of Employment per Worker

From the data on the average number of days of work available to an adult worker in landless agricultural labour households, four major points emerge.[1]

First, the average number of days of work available per adult worker was about 140 a year, the equivalent of a little less than five months. This indicates, for the adult labour force as a whole, massive unemployment. The average adult worker in a landless agricultural labour household was unemployed for about 225 days, or the equivalent, scattered over the year, of seven months.

The phenomenon of casual labourers obtaining employment for six months or less is not uncommon in other parts of India and Tamil Nadu. Casual labourers in Gujarat studied by Jan Breman were employed for 186 days a year (Breman 1985, p. 281). Data from four villages studied by Sudipto Mundle in Palamau, Bihar, show that a *mazdoor* (manual worker) in the village that had the highest levels of employment received only 140 days of agricultural work a year, while in another village total employment per annum per worker was only 68 days (Mundle 1979, pp. 76–7, table 3.32). Village surveys conducted by the AERC, Madras, as early as the 1950s and the first half of the 1960s in

[1] See Table 6.3, which provides data for an initial assessment of the overall availability of employment per adult worker for all adult workers in agricultural labour households. The concern here is not with the number of days of employment at a task per worker who is actually employed at that task, for instance, the number of days of irrigation labour done by irrigation workers alone.

Table 6.3 *Employment among landless agricultural labourers, average number of days per adult worker in the agricultural year, Gokilapuram, 1976–1977* (days)

Caste, sex	Agricultural tasks			Non-agricultural tasks					Total
	All except guarding	Guarding	Total	Tamarind-processing	Plantation labour	Hand-pounding paddy	Other manual tasks	Total	
Scheduled castes									
Male	94	7	101	0	28	0	6	34	135
Female	116	0	116	34	22	11	1	68	184
All	105	3	108	18	25	6	4	53	161
Non-scheduled castes									
Male	47	27	74	2	12	0	15	29	103
Female	91	0	91	22	11	9	1	42	133
All	70	13	83	12	11	4	8	35	118
All households									
Male	68	18	86	1	20	0	11	32	118
Female	103	0	103	28	16	10	1	55	158
All	86	9	95	15	18	5	6	44	139

Note: Each cell is the quotient of the following fraction: in the numerator, the sum of all calendar days of hired employment; in the denominator, all adult agricultural and non-agricultural workers in landless agricultural labour households.

Source: Survey data.

different parts of Tamil Nadu show similar results.[2] Kathleen Gough provides material on employment that is rarely available, namely, the number of days of work as a hired labourer available to a male casual labourer in two villages—Kumbapettai and Kirippur in Thanjavur district, Tamil Nadu—at two points of time. In Kumbapettai, the number of days of employment of a male casual labourer declined from 200 days a year in 1952 to between 76 and 180 days a year in 1976–80; in Kirippur, the decline was from between 160 and 180 days a year to between 60 and 110 days a year over the same time period (Gough 1987, p. 290).[3]

Secondly, of the 140 days about 96 (or a little over three months) were days spent at agricultural labour and about 44 (about a month and a half) were spent at non-agricultural tasks.

Thirdly, it is interesting to note that women, in 1976–7, fared better, on average, than men: they received an average of 160 days of employment against the men's average of 118 days, a difference of about a month and a half. They received, on average, 19 more days of employment at agricultural tasks, and 22 more days at non-agricultural tasks.[4] Comparisons with other parts of South Asia are useful in this respect, particularly in the light of the earlier discussion of female participation rates in the rural work-force in Chapter 3. Two examples may serve to illustrate the contrasts that exist. Data from Haryana in the early 1970s analysed by Utsa Patnaik show that the number of days of employment that women obtained was only 39 per cent of the corresponding figure for men (Patnaik 1987*a*). Differences in cropping patterns and the displacement of female work as a result of the mechanization of cultivation, particularly wheat cultivation, in Haryana are likely to be important factors in explaining this contrast between the Cumbum Valley and Haryana. In the Comilla district of Bangladesh, a region vastly different from Haryana, women contributed only a third of the total time spent on 'directly productive activities', a third of the total time spent on agricultural work, and *only 1.5 per cent* of the time spent on pre-harvest operations. These time-budget data were collected in 1976, in a village characterized by a rice-based agricultural economy.[5] Socio-cultural and demographic factors play an important role in explaining the contrast between this region and the Cumbum Valley.

Fourthly, behind the average of 140 days of employment was a very uneven distribution (see Table 6.4). In 27 per cent of households, each working

[2] See the AERC surveys, particularly AERC (undated *a*, *b*, *c*, 1961, 1962, 1964, 1969*a*), and Yeshwanth (n.d.).

[3] Gough notes that the days of employment available to women workers had been similarly curtailed. At the all-India level, a major study of the 1972–3 employment data of the NSS by Pravin Visaria showed that unemployment was 'markedly higher' among casual labourers, that the incidence of unemployment in 17 states of India showed a statistically significant, positive association with the percentage of casual labourers in the labour force, and that the importance of casual labour in the labour force seemed to explain a good part of the interstate variation in unemployment rates (Visaria 1980).

[4] There were important changes, in this respect, between 1976–7 and 1985–6 (see ch. 10).

[5] Barkat-e-Khuda (1980); the study is of Barkait village in Comilla.

Table 6.4 *Days of agricultural employment per agricultural worker in landless agricultural labour households, frequency distribution of households, Gokilapuram, 1976–1977*

No. of days of employment per worker employed at agricultural tasks	No. of households	% of all landless agricultural labour households	% of days of employment at agricultural tasks	Column (4)/ column (3)
1	2	3	4	5
0–30	30	11.7	1.8	0.15
31–60	39	15.2	6.5	0.43
61–100	63	24.5	18.0	0.73
101–120	28	10.9	10.5	0.97
121–150	29	11.3	14.2	1.26
⩾ 151	68	26.4	49.0	1.85
TOTAL	257	100.0	100.0	

Source: Survey data.

member received less than 60 days of employment. Households whose working members each received more than 151 days of employment constituted a similar proportion of households: 26 per cent. The Gini coefficient for the distribution is 0.33, relatively high when we consider the fact that *only* landless labour households are being dealt with here.

This inequality in the number of days of employment can be traced to a range of factors. First are events such as pregnancy or illness or accidents, which can push workers out of the labour force for periods of varying length. A second factor is the age structure and the demographic composition of the individual household—the number of able-bodied earning members, the need to withdraw labour on occasion for child care, and so on. Migration by able-bodied members of households to other areas is also an important related aspect of the demographic factor. Thirdly, there are factors to do with the 'contacts' of workers with employers, their ability to seek and gain employment with certain landlords or other employers, and lastly, factors to do with the skills of workers. Some of these skills are long-standing traditional skills, such as the knowledge of the irrigation system of a Maravar irrigation worker and the traditional association of Chakkiliyar women with threshing coarse grain. Others are recently acquired special skills, such as the skills of a worker employed exclusively in a vineyard and, at another level, the skills of a tractor driver. These are some of the issues involved in differential access to employment in the village; while some of them are dealt with in Chapters 8 and 9, a more detailed examination must await further research.[6]

[6] In an interesting discussion of their data from the Kosi region of Bihar, Rodgers and Rodgers write that changes in employment levels can be explained in terms of 'life cycle effects, illness,

The major conclusion here is that despite increases in the demand for labour in the recent period, a huge problem of unemployment persists among agricultural labourers in the village. I do not have similar data on the number of days of employment available to a landless agricultural labourer for points in time before 1976–7, although there is evidence of the increasing pressure of an expanded work-force on the existing sources of employment in the village, and less employment for a worker in specific tasks such as transplanting and harvesting. The available evidence is presented in Chapter 9.[7]

The Deployment of Labour Time

Agricultural Tasks

Paddy is the key factor in the year's employment at agricultural tasks, and at the time of the survey occupied about 53 per cent of the total labour time of all workers, 57 per cent of the total labour time of women, and 45 per cent of the total labour time of men. Within paddy, male labour was divided almost equally between operations that are concentrated at the beginning of a crop season (hoeing at bunds, basal manuring, ploughing) and at the end of the crop season (harvesting and threshing), each set of tasks occupying about 22 per cent of male labour time in all field operations. Female labour in paddy farming was, of course, spread out less unevenly over the working year. The most important sources of employment in paddy farming for women were transplanting, weeding, harvesting, and threshing and to a lesser extent, field preparation. (See Tables 6.5 to 6.7.) For the rest, crops irrigated from groundwater sources—non-cereal crops and the irrigated component of coarse grain and groundnut—were the next most important sources of employment of hired labour, with tamarind trees, from which workers harvest the fruit, next in importance.

Briefly stated, the pattern of labouring out reflects the fact that the location of Gokilapuram makes it essentially a village with surface-irrigated farming, with an important subsidiary sector of groundwater irrigation in contiguous villages, and with the additional feature—owing to its leading position in the

random factors in job access, availability of irregular work . . . and so on' (1984, pp. 20–1). See also Gough (1981) for data that show wide variations in the number of days of work available to casual labourers; B. Harriss (1987) on the effect of age on access to employment; and Breman (1985, pp. 282–3). See Hart (1986) for a discussion of unequal access to wage labour located in the larger context of social and political change in Indonesian society.

[7] From a comparison of all-India rural employment data from the 38th round (1982–3) of the NSS with earlier rounds (1972–3 and 1977–8), Utsa Patnaik has found that of 'all workers in rural areas, the percentage reporting unemployment has slowly risen under every head of employment status (usual status, weekly status and daily status)' (Patnaik 1987*a*, p. 16). See also Nayyar (1977) for data on unemployment per male worker per annum in Uttar Pradesh.

Table 6.5 *Deployment of agricultural labour among adult workers, landless agricultural labour households, Gokilapuram, 1976–1977*

Crop/s	Scheduled castes (SC)		Non-scheduled castes		All workers		Percentage share of aggregate employment		
	No. of days	% of total days	No. of days	% of total days	No. of days	% of total days	SC	Non-SC	All workers
Paddy	16,021	47.6	13,784	48.7	29,805	48.1	53.7	46.3	100.0
Irrigated non-cereal crops	6,476	19.2	3,396	12.0	9,872	15.9	65.6	34.4	100.0
Coarse grain	3,857	11.4	3,155	11.2	7,012	11.3	55.0	45.0	100.0
Groundnut	958	2.8	608	2.1	1,566	2.5	61.2	38.8	100.0
Tamarind	5,284	15.7	1,829	6.5	7,113	11.5	74.3	25.7	100.0
Others	59	0.2	952	3.4	1,011	1.6	5.8	94.2	100.0
Guarding	1,038	3.1	4,560	16.1	5,598	9.0	18.5	81.5	100.0
TOTAL	33,693	100.0	28,284	100.0	61,977	100.0	54.4	45.6	100.0

Source: Survey data.

Table 6.6 *Deployment of agricultural labour among men workers, landless agricultural labour households, Gokilapuram, 1976–1977*

Crop/s	Scheduled castes (SC)		Non-scheduled castes		All workers		Percentage share of aggregate employment		
	No. of days	% of total days	No. of days	% of total days	No. of days	% of total days	SC	Non-SC	All workers
Paddy	5,911	40.5	3,894	31.0	9,805	36.1	60.3	39.7	100.0
Irrigated non-cereal crops	2,514	17.2	1,107	8.8	3,621	13.3	69.4	30.6	100.0
Coarse grain	1,468	10.1	589	4.7	2,057	7.6	71.4	28.6	100.0
Groundnut	509	3.5	216	1.7	725	2.7	70.2	29.8	100.0
Tamarind	3,114	21.3	1,270	10.1	4,384	16.1	71.0	29.0	100.0
Others	42	0.3	915	7.3	957	3.5	4.4	95.6	100.0
Guarding	1,038	7.1	4,560	36.3	5,598	20.6	18.5	81.5	100.0
TOTAL	14,596	100.0	12,551	100.0	27,147	100.0	53.8	46.2	100.0

Source: Survey data.

Table 6.7 *Deployment of agricultural labour among women workers, landless agricultural labour households, Gokilapuram, 1976–1977*

Crop/s	Scheduled castes (SC)		Non-scheduled castes		All workers		Percentage share of aggregate employment		
	No. of days	% of column total	No. of days	% of column total	No. of days	% of column total	SC	Non-SC	All workers
Paddy	10,110	52.9	9,890	62.9	20,000	57.4	50.5	49.5	100.0
Irrigated non-cereal crops	3,962	20.7	2,289	14.5	6,251	17.9	63.4	36.6	100.0
Coarse grain	2,389	12.5	2,566	16.3	4,955	14.2	48.2	51.8	100.0
Groundnut	449	2.4	392	2.5	841	2.4	53.4	46.6	100.0
Tamarind	2,170	11.4	559	3.6	2,729	7.8	79.5	20.5	100.0
Others	17	0.1	37	0.2	54	0.2	31.5	68.5	100.0
Guarding	0	0.0	0	0.0	0	0.0	0.0	0.0	100.0
TOTAL	19,097	100.0	15,733	100.0	34,830	100.0	54.8	45.2	100.0

Source: Survey data.

Valley's tamarind economy—that Gokilapuram-based tamarind contractors hire workers who are normally resident in the village to harvest tamarind in different parts of the Valley.

The number of adult female workers in agriculture (309) was higher than the number of adult male workers (252).[8] The absolute number of days of employment obtained by women in hired labour was also higher than the number of days obtained by men, with women working for 56 per cent of all days of hired labour in agriculture (including crop-guarding). The employment of women relative to men was particularly high in day-to-day manual tasks, that is, tasks excluding guarding. A woman agricultural worker was employed for an average of 113 calendar days in such tasks, while the average for adult male workers was 86 calendar days. The disparity was wider among workers who are other than scheduled caste. Among them, women worked, on the average, for 102 days in the year at all tasks excluding guarding, while men worked for less than 70 days at all tasks excluding guarding. On aggregate, the deployment of adult female labour in tasks other than guarding was 66 per cent of total labour deployment at such tasks.

Men made up a large part of the difference through employment at guarding (*kaval*) and irrigation-cum-guarding tasks. Men obtained, on average, 108 days of labour at *all* tasks (including guarding), while women obtained 113 days of employment at all tasks. (Women are not employed, as a rule, to guard fields or irrigate them.)

There are different aspects to the question of the relationship between caste and the pattern of employment. The most important one is that over time, caste barriers to employment have broken down. There is qualitative material on this, some of which appears in the case studies of specific forms of employment in Chapter 9. I shall cite only one example here, which relates to the lifting of a traditional barrier on the employment of scheduled caste workers. Persons of the Parayar and Chakkiliyar castes in the village did not, by tradition, work in transplanting and harvesting operations in paddy cultivation. At present, however, transplanting and harvesting are important sources of employment for men and women of both castes, and it is noteworthy that people of these castes join *multi-caste* contract groups at the time of harvesting and transplanting, groups in which they work side by side with caste Hindu workers. In general, when an operation requires the employment of large numbers of workers, caste-specific employment is not the practice.

Some differences, however, persisted in the labour schedules of scheduled caste agricultural labourers and of non-scheduled caste agricultural labourers. From the survey data, it is evident that scheduled caste agricultural labourers obtained more days of employment in the year at agriculture, in aggregate terms and in terms of days of employment per worker, than non-scheduled

[8] For the calculations in this section, only adult agricultural workers in landless agricultural labour households are considered. The non-agricultural component of these households has been excluded.

caste agricultural labourers. Adult male scheduled caste workers gained, on average, 119 days of employment at agricultural tasks, against 97 days obtained by adult male non-scheduled caste labourers. Adult female scheduled caste workers obtained 124 days of employment at agricultural tasks, against an average of 102 days obtained by adult female non-scheduled caste workers.

If we consider only the days spent at day-to-day manual tasks (that is, if the days spent guarding the standing crop on unirrigated land and groundwater-irrigated land and the days spent guarding and irrigating the paddy crop are eliminated from the calculation), the difference between the annual labour schedules of male scheduled caste workers and male non-scheduled caste workers is thrown into sharp relief. Male scheduled caste workers worked, on the average, for 110 days in such tasks, while male non-scheduled caste workers worked for less than 70 days: a highly significant difference.

By custom, however, irrigation and guarding tasks on surface-irrigated land and guarding tasks on dry land were given to adult male workers from specific (non-scheduled) castes. A group of Maravar workers was employed for irrigation and guarding on surface-irrigated land, and a group of Kallar workers was employed to guard the standing crop on unirrigated land. These workers were employed for relatively long periods of time in the year (the duration of their employment is discussed in Chapter 9), and these extreme values in the labour patterns of non-scheduled caste workers pushed up the average level of employment among male non-scheduled caste workers. Caste-specific guarding and irrigation tasks, then, served to narrow the gap between the labour patterns of scheduled caste workers and non-scheduled caste workers.

The labour of scheduled caste agricultural labourers appears to have been more varied and to an extent less dependent on paddy than on other crops. One reason for this is that scheduled caste agricultural labourers are more mobile. A higher proportion of a scheduled caste agricultural labourer's working year was spent on groundwater-irrigated fields and groundnut fields than that of the working year of a non-scheduled caste worker. These are tasks that are likely to take the worker out of Gokilapuram and to neighbouring villages. Their willingness to move has also helped agricultural labourers of the scheduled castes to gain a pre-eminent place in the employment of workers in harvesting tamarind. The Chettiar landlords-cum-tamarind merchants of Gokilapuram obtain contracts to harvest tamarind trees all over the Cumbum Valley. The first and largest component of the labour force that they assemble for this purpose is drawn from the scheduled castes of the village.

Non-Agricultural Tasks

Non-agricultural tasks, as has been shown, are an important source of employment: work at such tasks occupied about a third of the total labour time deployed in the year (see Table 6.8).

Table 6.8 *Deployment of labour time at non-agricultural tasks among adult workers, landless agricultural labour households, Gokilapuram, 1976–1977*

Non-agricultural tasks	Scheduled castes (SC)		Non-scheduled castes		All workers		Percentage share of aggregate employment		
	No. of days	% of total days	No. of days	% of total days	No. of days	% of total days	SC	Non-SC	All workers
Tamarind-processing	5,609	34.4	4,158	34.0	9,767	34.2	57.4	42.6	100.0
Plantation labour	7,711	47.3	3,899	31.9	11,610	40.7	66.4	33.6	100.0
Hand-pounding paddy	1,910	11.7	1,474	12.1	3,384	11.9	56.4	43.6	100.0
Other manual tasks	1,067	6.5	2,697	22.1	3,764	13.2	28.3	71.7	100.0
Total	16,297	100.0	12,228	100.0	28,525	100.0	57.1	42.9	100.0
Total non-agricultural plus agricultural tasks	49,990		40,512		90,502		55.2	44.8	100.0
Non-agricultural tasks as % of all tasks		32.6		30.2		31.5			

Source: Survey data.

The major source of non-agricultural employment for women in the village is tamarind-processing, at which, in 1977, 168 women and 48 children from 128 landless labour households were involved and which provided workers with a little over 55 days of employment each. Plantation work was also important (in fact, in terms of the number of days of employment, the most important); this was often not supplementary manual labour, but was the primary source of employment for those who obtained regular employment on the estates, among whom scheduled caste workers were predominant.

For men, a major source of non-agricultural employment was at tasks such as construction work, digging wells, chopping wood, carting (referred to in the table as 'other' tasks), which provided about 74 days of employment each to about 47 men—a small amount, in relation to need—from landless agricultural labour households. Hand-pounding paddy provided about 25 women from these households with substantial employment, about 135 days in the year each. Apart from these occupations, some persons from these households sought work as domestic servants (mostly women and children) and tending cattle for wages. From among landless agricultural labour households, 17 had earnings from artisan work and other traditional callings; 12 had members who were salary-earning; 24 had incomes from dairying; and 40 had members who brought in earnings from petty business activities.[9]

Table 6.9 *Wage-paid employment of agricultural labourers from landless households, Tamil Nadu, 1974–1975* (days)

	Agricultural employment	Non-agricultural employment	Total
All men	153	20	173
Men (scheduled caste)	129	20	149
All women	116	8	124
Women (scheduled caste)	92	8	100
All children	164	17	181
Children (scheduled caste)	176	7	183

Source: RLE 1974–75 (1981), pt. 2, tables 3.3 (i)(a) .1M, .2M, .1W, .1C, .2C.

[9] There are problems here with using comparative data from the RLE, and the four results from Table 6.9, which is derived from RLE reports, are at variance with our results: men obtain more employment per person than women; scheduled-caste agricultural labourers (men and women, although not children) obtain less employment per person than the average for all agricultural labourers; children obtain more days of employment per person than adult men or women; and scheduled caste agricultural labourers *with* land (men and women) obtain more days of wage-paid employment per person than landless scheduled caste agricultural labourers. The first two results will vary between regions; the third and the fourth are unlikely to be correct and throw doubt on the quality of the data base of the RLE.

Seasonality: Distribution of Labour Time over a Working Year

An important feature of agricultural production is the dependence of the production cycle on the season; seasonality is a very important determinant of the distribution of labour time over the working year. Table 6.10 charts the months in which different operations occur. This chart omits a calendar for crop operations for groundwater-irrigated crops, since good groundwater-irrigated fields are in use all the year round and are not characterized by the relatively rigid seasonal patterns of other crops.

Seasonality has certain general features. First, there is a discernible three-month lean season, from mid-February to mid-May. Secondly, since paddy is the dominant crop in the village, the paddy cultivation calendar sets the pace for seasonality in employment in Gokilapuram. Consequently, and this is the third point, October to early December is the high season for agricultural employment. Fourthly, the seasonality factor imposes different rhythms on male and female employment; absence of employment in the early part of the lean season was partly mitigated for women because of employment in

Table 6.10 *Calendar of major tasks done by agricultural labourers, Gokilapuram, 1977–1982*

Month	Crop operations					
	Nanjai paddy	*Punjai* crops	Groundnut at Gudalur	Tamarind	Tamarind	Plantations: cardamom
June	Field preparation	Field preparation				
July						
	Transplanting					
August		Weeding, if any				
	Weeding		Harvest			
September			↕			
	Weeding		Harvest			
October						
	Harvest	Harvest				
November	Field preparation					
	Transplanting					Harvest
December	Weeding					↕
	Weeding					
January	Harvest			Harvest		Harvest
February				↕		
March				Harvest	Processing	
April					↕	
May					Processing	

Note: Arrows indicate the duration of the task.

tamarind-processing, while men's employment was, on the whole, more sporadic than women's employment.

There are two factors, however, that work against seasonality and tend to even out peaks in the employment pattern. One is non-agricultural labour (such as tamarind-processing in the village) and the second is the existence of the sector of groundwater irrigation, in which fields are in use the year round (although this is not the dominant aspect of Gokilapuram's farming pattern). In Gokilapuram in 1977, both these sources of employment provided more days of work to women than to men.

The specific features of employment patterns in the village made for important differences in the way that seasonality affected women's employment in Gokilapuram compared with other parts of the country. To take two examples, data from the semi-arid tropical regions of central India showed that women had less chance of employment in the lean season than men, and a study of female labour in West Bengal in the 1970s concluded that seasonal fluctuations were greater for women than for men (K. Bardhan 1984, p. 202; Ryan and Ghodake 1984, p. 183).

There is an important policy-related conclusion from the data on unemployment and seasonality: the pattern of deployment of labour indicates that even in an area of relatively advanced agriculture such as Gokilapuram, state-run employment schemes, timed for the lean season and providing work at public works and other projects, have a potentially large role to play in filling the long periods of joblessness in a working person's year.[10]

Child Labour

Two points made by Michael Moerman in a study of a village in Thailand are of relevance to research on village-level child labour in different parts of the world (Moerman 1968, pp. 48–9, cited and discussed in Connell and Lipton 1977, p. 48). The first is a methodological point: Moerman documents how and at what age children begin to learn particular kinds of agricultural work. Secondly, Moerman calls attention to the fact that child labour gradually merges with adult labour on the peasant farm, and that 'there is no single cut-off point' (Connell and Lipton 1977, p. 48) between child and adult labour. This point is certainly valid for Gokilapuram (although for calculating child labour later in this study, the norm that has been used to separate children from the rest of the population is the widely accepted one: children are persons below the age of 15 years). In Gokilapuram, some tasks can be characterized as generally being done by children; an example is cattle-grazing, which is usually done by young

[10] I do not have data on the employment generated by the food-for-work programme introduced in 1979–80 in the Cumbum Valley; data from the Tamil Nadu budget for 1981, however, indicate that an average of only six days of employment per person were created in that year for beneficiaries of the scheme (Ramachandran and Balaji 1980).

boys. There are other tasks, such as hoeing work in field preparation, where a boy works with men at what is a man's task, but does so at a woman's wage. During the paddy harvest, when some men and women work at the threshing-floor and the numbers of workers in contract labour gangs are depleted, girls and boys between the ages of 11 and 15 years join these groups to harvest the fields. At such times, they are paid wages that are the same as for adults in the group to which they belong. A common form of child participation is as helpers in piece-rated tasks, where children help other members of their family to boost household earnings. One example of this comes from groundnut harvesting, when wages are paid on the basis of the total amount harvested, cleaned, and winnowed by a household. Another example is tamarind-processing, when children help their mothers to clean the fruit. Lastly, there are tasks for which children are not employed as hired labourers at all: prominent examples from the village are threshing paddy and managing the surface-irrigation network.

In 1976–7, child workers constituted about 10 per cent of the work-force in Gokilapuram, although the work they did was a small proportion, only about 2 per cent, of total labour *time* deployed in agricultural and non-agricultural tasks.[11] At agricultural tasks, less than 1.5 per cent of total labour time (this excludes guarding) was child labour time. (Only 17 children from the 257 households considered in Table 6.1 were agricultural labourers.) Child labour time took a larger share of guarding work, particularly on groundwater-irrigated land and unirrigated land, although the share in aggregate guarding time was still low: about 5 per cent.

Children who work at guarding standing crops on groundwater-irrigated land and dry land generally help their fathers, or other older men of the family, and also stand in while the older ones take a break. On grape fields, the practice of employing young boys to keep away birds and guard the crop is common. Here there are tin cans positioned at different parts of the *pandal* (bower), which are attached to a system of strings that converge on a nest-like shelter at a corner of the *pandal* where the guard-boy sits, pulling the lead string, and making a racket when a bird appears. He is often armed with a wood and rubber slingshot.

As for tasks outside agriculture, child labour accounted for 7 per cent of total labour time. Children help their mothers at tamarind-processing; sometimes children under 10 years sit all day with their mothers, doing their little bit to push up the volume of seeded, de-fibred tamarind in the basket at the end of the day. Some children go to the plantations with their parents and work there. They also take cattle out to graze on the outskirts of the village and sometimes

[11] Summarizing the all-India data on child labour, Kalpana Bardhan writes: 'According to the 1977–78 NSS, 13 per cent of 5–14 year-old boys and 9 per cent of girls in rural India are workers by "usual status", adding up to nearly 15 million. Total hours of rural child employment was about a tenth of total hours of adult employment . . . Over a third of usually working children are labourers, girls mostly casual labourers and boys mainly regular labourers' (1983, p. 61).

further away, grazing family cattle and minding others' for a small payment. Children, particularly girls, work as domestic servants, mainly in the houses of landlords and other village rich, doing every kind of manual task in and around the house. Many of them have to live in at their employers' houses, and the great urge of these young Cinderellas to run home to their parents' meagre huts is kept firmly in check by the threat of dismissal by the landlord, or, more likely, by the women of his house.

In sum, the role of child labour in direct agricultural production was small. At non-agricultural tasks, children often played a supportive role, and sometimes an independent one, and in some cases were important little cogs in wheels.

The Length of the Working Day

In general, a single day is clearly demarcated into two component parts for the agricultural labourer: work time and the rest of the day. By contrast, family labour on the *peasant* farm is characterized by an ill-defined working day, with the time spent on field labour not strictly separate from the rest of the day. The working day of the farm servant, who is on call at all hours of the day (and often at night as well), is also not clear-cut.[12] A more regular working day at production tasks, with the working day more clearly defined, emerges from systems that employ hired casual labour.

Across all operations, the average working day was 8 hours 48 minutes for all workers. The duration of the actual working day, however, varied over different operations. The eight-hour day is emerging, or has emerged, as the standard working day in daily-rated, cash-paid operations. Examples of such operations are basal manuring, hoeing and weeding on paddy-fields, and a variety of operations in groundwater-irrigated farming. In other operations, there are deviations from the eight-hour day. This applies especially to piece-work, in which large groups work under labour contractors, and the length of the working day fluctuates according to the amount of work the head labourer is able to contract. Examples of such operations are harvesting and transplanting paddy and harvesting groundnut. Smaller groups of men are often employed at piece-rates paid in cash to build irrigation basins on groundwater-irrigated land; the groups generally cut down on an extra person and work longer than eight hours in order to be able to split the piece-rates into shares that are larger than the wage for a regular eight-hour day's hoeing. Women who are employed in tamarind-processing, which is piece-rated, also set their own working day.

Ploughing is a task that does not generally go on for more than five hours a day; it is exhausting work, for the ploughman and the draught animals.

[12] See Kandaswamy (1964), M. A. Reddy (1983), Vidyasagar (1985) for examples of the long working hours of farm servants and attached labourers.

The longest working day is when paddy is threshed and workers are paid in kind, at a daily rate. Agricultural workers start work very early in the morning and it is not uncommon for men and women to work for 12 to 13 hours, with two or three breaks of half an hour to an hour each in between. Operations that are not performed by large groups of workers but by one or two workers, such as irrigation and plant protection, are also not generally eight-hour jobs.

The emerging eight-hour day is not a clear-cut rigorously enforced practice, as in factories or even on the cardamom plantations in Kerala. But, since a kind of eight-hour working day has become almost standard for daily-rated, cash-paid operations, and with the added influence of those who work on the plantations, the eight-hour day has become a norm that all workers recognize. Whether or not the eight-hour day is actually put into practice, casual daily labourers can no longer be robbed of their consciousness of time or of their knowledge of the eight-hour day as a norm. No longer guided only by the position of the sun, workers now have other measures of time: the buses that pass through the village, for instance, and even the aeroplane on the Madurai–Trivandrum run when it flies overhead. Most important, there is a siren at the town panchayat office at Uthamapalayam that blows at 8.30 a.m. and at 5 p.m. and can be heard in the village. This signals the start and end of the working day for many workers, and more than one landlord told me that the siren has 'spoiled' the workers. One said, 'If you ask me, just installing the siren at Palayam was the first big mistake.'[13]

Concluding Notes

Despite increases in the demand for labour, huge unemployment persists: the average number of days of employment at hired tasks for landless agricultural labourers was 140 days. The distribution of employment over the year is markedly seasonal: there are three consecutive lean months and scattered periods of unemployment over the year. Although features of the labour market that tend to counteract seasonality (non-agricultural labour and year-round cultivation on groundwater-irrigated land) exist and play a role, they do not yet significantly affect this pattern.

The study of the deployment of labour time has also confirmed and amplified certain conclusions from previous chapters: the diversity of tasks performed by the great majority of agricultural labourers and the mobility of the labour force (in particular, of its scheduled caste component). In a

[13] There are interesting comparisons to be made here with E. P. Thompson's study of time and work discipline during the transition to early industrial capitalism: 'It was exactly in those industries . . . where the new time-discipline was most rigorously imposed that the contest over time became most intense. At first some of the worst masters attempted to expropriate the workers of all knowledge of time' (1974, p. 60). There is a remarkable similarity between the 'time-signals' used by the workers in Gujarat who are described in Breman (1985, p. 69), and those used by workers in Gokilapuram.

previous chapter, it was shown that the increased demand for hired labour at a wide variety of tasks has been an important factor in creating a labour force of casual labourers and in depersonalizing the relationship between employer and hired labour in production tasks. At the same time, the unemployment and seasonality that have been described in this chapter are among the principal determinants of the very high level of poverty that persists among agricultural labourers and of the continuing relations of dependence of agricultural labourers on those who employ them.

7

Some Aspects of the Material Conditions of Life of Agricultural Labourers

The material conditions of life of agricultural labourers, such as they are, are crucial determinants of the persistence of different forms of socio-economic dependence of the agricultural labourer on the employer in the village. This chapter attempts, in particular, to describe and analyse agricultural wage rates and changes in wage rates; aggregate earnings and sources of earnings; the indebtedness of agricultural labour households; and overall household consumption patterns.

Wage Rates in Agriculture

Wage rates in agriculture are crop-specific, operation-specific, and gender-specific. There are, for instance, separate wages for harvesting paddy and harvesting groundnut; for harvesting paddy, for transplanting it, and for threshing it; and there are separate wages for men and women at each of these tasks. Even for harvesting paddy, when men and women join contract gangs and piece-rated wages are divided equally between all members of the group, the gross payments received by men are higher than the gross payments received by women, since traditional additions to harvest wages are allocated unequally. There is one set of operations that can be said to be characterized by a standard wage: these are operations that are daily-rated and paid in cash. In these operations, too, however, the rates paid to men and women are different. This section deals with average wage rates for men and women belonging to landless agricultural labour households across all agricultural operations during our reference year (see Table 7.1). Wage rates for individual operations are taken up in the case studies in Chapter 9.

Wage rates for men in Gokilapuram were generally higher than the statutory minimum for operations performed by men. Wage rates paid to women were below the statutory minimum. Wage rates in Gokilapuram were higher than those recorded in more backward regions around the same period.[1] When compared with wage rates in the advanced agricultural region of Haryana during the same period, wage rates in Gokilapuram were lower (Patnaik 1987*a*, pp. 193 ff.).

How do the wage rates in Gokilapuram compare with agricultural wage rates in Tamil Nadu as a whole during the same agricultural year? The reference

[1] Mundle (1979, p. 114, table 5.33); Nagaraj (1981, p. 131); J. Harriss (1982, 1986*b*); S. V. Rao (1984, p. 766); Ryan and Ghodake (1984, p. 178); Vidyasagar (1985).

Table 7.1 *Average wages for agricultural tasks, Gokilapuram, 1976–1977 (Rs per calendar day)*

Wage category	Men	Women	Children	All	Women's wage as % of men's wage
Daily rate excluding guarding					
Scheduled caste	5.73	3.02	2.69	4.11	52.7
Non-scheduled caste	6.83	3.00	3.32	4.24	43.9
All	6.13	3.01	3.16	4.17	49.1
Scheduled caste wage as % of non-scheduled caste wage	83.9	100.7	81.0	96.9	
Daily rate including guarding					
Scheduled caste	5.36	3.02	2.67	4.01	56.3
Non-scheduled caste	6.14	3.00	2.83	4.32	48.9
All	5.70	3.01	2.71	4.15	52.8
Scheduled caste wage as % of non-scheduled caste wage	87.3	100.7	94.4	92.8	

Source: Survey data.

point that we have for comparison is the government's annual *Season and Crop Report (SCR)* for 1976–7—a not altogether satisfactory source, but the only one available. *SCR* results indicate that daily wage rates for men in cash terms were higher in the Cumbum Valley for the category of ploughmen and hoe workers (who presumably come under the *SCR* category of 'other agricultural workers'), while they were lower than the *SCR* rate for 'reapers and harvesters'. For women, the daily wage rate in the Valley was lower than the *SCR* for weeding, and the same, or marginally higher, for 'reapers and harvesters'.

Wage rates in the village were also lower than for unskilled manual jobs in urban areas; the disparity between an urban unskilled worker's wage, low as it is, and an agricultural labourer's wage has nothing to do with differentials in subsistence needs. Take, for instance, the wages of the lowest-paid worker in the textile industry (a major industrial employer, employing workers at centres in different parts of Tamil Nadu) in 1976–7. The cash wage, i.e. basic wage plus cost-of-living allowance (terms which, at the present rate, are not likely to become known to agricultural workers in Gokilapuram for the rest of the twentieth century and into the next) of the lowest-paid operatives in cotton textile mills at Madras and Coimbatore was Rs 402.90 per month in 1976 and Rs 432.50 per month in 1977. The average wage rate at agricultural tasks in the village was about 27 per cent of the wage rate for the lowest-paid textile operative.[2]

Wages were low in comparison with international levels (see Table 7.2). The table gives the number of kilograms of rice that could be purchased with a day's

[2] We are not, of course, considering variation in levels of productivity here.

Table 7.2 *Amount of rice purchasable with a day's agricultural wage in selected areas, 1976–1979*

(*a*) *Selected areas of Asia*

Selected area	Kilograms of rice per day of work
East Asia	
Japan	15.6
South Korea: Suweon	7.4
Taiwan: Taichung	20.0
South-East Asia	
Burma: Rangoon	7.1
Indonesia: Central Java	6.1
Malaysia: Selangor	9.3
Philippines: Central Luzon	3.2
Thailand: Suphanburi	7.9
South Asia	
Bangladesh: Joydebpur	3.3
India: Coimbatore	3.0
Cuttack	1.5
Waltair	2.1
Nepal: Kathmandu	2.4
Pakistan: Punjab	3.3
Sri Lanka: Kurunegala	2.6

(*b*) *Gokilapuram*

Task/operation in paddy cultivation	Kilograms of rice per day of work		
	Men	Women	Men and women
Daily-rated, cash-paid tasks	3.06	1.31	—
Harvest	—	—	3.22
Threshing[a]	4.02	3.22	—

[a] The working day at this operation is about 12 hours; at other operations about 8 hours.

Sources: (*a*) A. Palacpac (1982), table 41, cited in Barker *et al.* (1985), table 9.9.
(*b*) Survey data.

agricultural wage in selected regions of Asia. These regions can be roughly divided into three: those where 15 to 20 kilograms of rice could be bought with a day's wage (Japan, Taiwan); those where the figure was between 7 and 10 kilograms (South Korea, Malaysia, Thailand, Burma); and those where the daily wage brought in less than 3.5 kilograms of rice (Indonesia, Philippines, South Asia). The data put Gokilapuram squarely in the group that was worst off; at certain operations, women earned slightly less than the equivalent of 1.5 kilograms of rice a day.

A major feature of the data on wage rates is the glaring difference between wage rates for operations performed by men and operations performed by women. The standard cash-paid, daily-rated crop operation done by women (such as weeding, cotton-picking, and gathering tamarind) was remunerated at Rs 2.50 per day; the wage rate for the standard cash-paid, daily-rated crop operation done by men (hoeing of different kinds, beating down tamarind fruit) was Rs 6 to 7 per day. Women in agriculture received less than half a man's wage (except in harvesting, for which men's wages were 10 to 15 per cent higher); this is one of the most important features of backwardness and inequity in the village today. This disparity is particularly high in agriculture and lowest in the plantation sector, in the Kerala segment of which trade unions are active.

Disparities in the wage rates paid to men and women in agriculture in India are a fact; this has been well documented in the literature (Patnaik 1985*a*, p. 22, 1987*a*; B. Harriss 1987, pp. 102–3).[3] The disparity varies across regions, and it has been suggested that the gap tends to be narrower in more developed areas (K. Bardhan 1984, p. 204). A recent review of official data on the subject, covering the period 1960–1 to 1980–1, suggests that differences in male and female wage rates tended to narrow over the period (Nayyar 1987). The Cumbum Valley presents a picture of a relatively advanced region where the disparity between male and female wage rates is very wide and, as we shall see, in the case of cash-paid, daily-rated operations, the gap actually *widened* between 1948 and 1976, and between 1976 and 1986.

Across all agricultural operations, the average daily wage rate for scheduled caste men was lower than for non-scheduled caste men. This does not mean, however, that total *earnings* were necessarily lower. Nor does it indicate wage discrimination at individual tasks, since workers at a single field worked at the same rates. What it indicates is the *variety* of tasks that scheduled caste men performed. They worked, to a greater extent than others, at tasks such as guarding groundwater-irrigated land at night and at contract tasks (discussed in greater detail in Chapter 9), where the average daily wage rate was low.

Trends in Wage Rates: Data from a Landlord Household of Gudalur

Gudalur is a large village—by Census of India classification, a small town—in the south of the Cumbum Valley, a place with perhaps the busiest and most diverse market for labour power in the area. I obtained a set of account books from Gudalur, which had been kept by a prominent Vokkaliga landlord from

[3] B. Harriss makes the point that minimum wage legislation, by laying down lower rates of remuneration for tasks that are generally performed by women, tends to reinforce these disparities (1987*a*, pp. 103–4).

1948 to 1976. Although information on annual expenditure is not complete (I do not have all the account books of each year, and no account books at all for 1968, 1971, and 1972), the account books did permit a reconstruction of trends in wage rates. While conditions in Gudalur are not precisely the same as in Gokilapuram, the villages are close enough to suggest that the trends may not be too different.

Method. Information was available on wage rates at cash-paid, daily-rated operations performed by men (hoeing) and cash-paid, daily-rated tasks performed by women (weeding, primarily, but also tasks such as scattering farmyard manure and clearing fields of grass). The wage rates for men and for women were noted monthly and the frequency of occurrence of each wage rate was recorded. There are two indices of real wage rates. The nominal wage rates for men and women were deflated by the Rural Price Index;[4] in addition, the amount of rice that could be purchased with a day's wage (male and female) was used to give an alternative index of real wages over the period (see Table 7.3).[5]

Results. The overall trend in wage rates from 1948 to 1976 was characterized by stagnation and marginal decline.[6]

The wage rate for men deflated by the Rural Price Index (1936 = 100) was Rs 0.29 in 1949 and Rs 0.34 in 1976; even the peak wage rate was only Rs 0.41. In terms of rice that could be purchased with a day's wage, it was 3.02 kilograms in 1948 and 3.06 kilograms in 1976. The data show a decline in the period after the mid-1960s—the heroic age of the 'green revolution'.

The position of women is even more dismal. Using the Rural Price Index deflator, the real wage was the same in 1949 and 1976, namely Rs 0.14, with a peak wage rate in 1961 of Rs 0.18. Using the second deflator, the amount of rice that could be purchased with a day's wage marginally declined, from 1.34 kilograms to 1.31 kilograms. The peak year was 1965, when the amount was 1.9 kilograms.

In real terms, the disparity between male and female wage rates appears to have widened. The ratio of male wage rates to female, using the first deflator, increased from 2.07 in 1949 to 2.43 in 1976. The years of the highest disparity were 1963 and 1964, when the ratios were 2.53 and 2.75 respectively. When the second deflator was applied, the ratio varied from 2.25 in 1948 to 2.34 in 1976.

Trends in Wage Rates: Some Comparisons

How does this trend of stagnation and decline in cash-paid daily wages in the Cumbum Valley compare with other parts of Tamil Nadu and the country?

[4] The Rural Price Index (1936 = 100) was collected for eight centres in Tamil Nadu; Gokilapuram was one of them.

[5] In a useful study of agricultural wages in Tamil Nadu, Sridhar Venkataraman has shown that the prices of rice and of coarse cereals moved more or less similarly between 1950–1 and 1980–1 and that the use of a single deflator does not pose problems. He also argues that cereal prices play an important role in determining money and real wage rates (Venkataraman 1987, pp. 109–12).

[6] See ch. 10 for changes in wage rates between 1977 and 1986.

The trend is clearly not an aberrant one: similar movements have been noted in studies conducted in different parts of the country.[7] For India as a whole, Utsa Patnaik cites data from the Rural Labour Enquiry to show that there was a decline in the average real earnings per rural labour family between 1964–5 and 1974–5. This was due to a combination of a decline in the real wage, a small

Table 7.3 *Real wage indices for cash-paid, daily-rated tasks, Gudalur, 1948–1976*

Year	Real daily wage (Rs per person per day) (1936 = 100)		Index of real daily wage (male rate, 1948 = 100)		Rice[a] purchasable with one day's wage (kg)		Index of columns (6) and (7) (male rate, 1948 = 100)	
	Men	Women	Men	Women	Men	Women	Men	Women
(1)	(2)	(3)	(4)	(5)	(6)	(7)	(8)	(9)
1948	n/a	n/a	–	–	3.02	1.34	100	44
1949	0.29	0.14	100	48	3.28	1.64	109	54
1950	0.23	0.12	79	41	2.36	1.18	78	39
1951	0.28	0.14	97	48	2.95	1.48	98	49
1952	0.27	0.14	93	48	2.83	1.41	94	47
1953	0.29	0.14	100	48	2.45	1.23	81	41
1954	0.23	0.15	79	52	1.77	1.10	57	36
1955	0.31	0.16	107	55	3.20	1.60	106	53
1956	0.33	0.16	114	55	2.69	1.34	89	44
1957	n/a	0.15	–	52	n/a	1.07	–	35
1958	0.35	0.15	121	52	2.96	1.22	98	40
1959	0.33	0.17	114	59	2.48	1.24	82	41
1960	0.33	0.16	114	55	2.19	1.09	73	36
1961	0.32	0.18	110	62	2.07	1.20	69	40
1962	0.41	0.18	141	62	2.67	1.16	88	38
1963	0.38	0.15	131	52	2.94	1.10	97	36
1964	0.33	0.12	114	41	2.94	1.10	97	36
1965	0.26	0.13	90	45	3.81	1.90	126	63
1966	0.35	0.15	121	52	3.93	1.64	130	54
1967	0.32	0.16	110	55	2.71	1.36	90	45
1968	n/a	n/a	–	–	n/a	n/a	–	–
1969	0.35	0.15	121	52	2.58	1.11	85	37
1970	0.27	0.14	93	48	2.44	1.22	81	40
1971	n/a	n/a	–	–	n/a	n/a	–	–
1972	n/a	n/a	–	–	n/a	n/a	–	–
1973	0.30	0.15	103	52	3.03	1.52	100	50
1974	0.35	0.17	121	59	3.37	1.68	112	56
1975	0.27	0.11	93	38	n/a	n/a	–	–
1976	0.34	0.14	117	48	3.06	1.31	101	43

Note: The base year for the Rural Price Index, Gokilapuram, is 1936.
n/a = not available.

[a] The price of second-sort rice has been used here as deflator.

Source: Household accounts from a landlord household of Gudalur.

[7] e.g. G. H. Rao (1975, cited in Parthasarathy 1977); Nakamura (1976); Nayyar (1977, pp. 79–81 and table 17, p. 99); M. A. Reddy (1978); Rodgers and Rodgers (1984); C. R. Reddy (1985); van Schendel and Faraizi (1984).

fall in the number of days of annual employment, and a rise in real indebtedness. She has also shown, from national accounts statistics, that the real value of the wage bill in the primary sector fell between 1978 and 1986 (Patnaik 1987*b*, pp. 16–18). The literature on trends in agricultural wage rates over the 1960s and early 1970s has been reviewed by Kalpana Bardhan (1977; see also Krishnaji 1971; P. K. Bardhan 1973; Jose 1974). The evidence from her review suggests that agricultural labourers' wages lag behind and are poorly adjusted to increases in productivity; that in certain regions increases in productivity have in fact had a positive impact on wage rates; that regional variations can be large and may have widened as a consequence of widening differences in agricultural productivity between regions; and that the component of wages paid in kind does not offer adequate protection to agricultural labourers' earnings, particularly in a year of harvest failure.

At the state level, C. T. Kurien's was the first careful analysis of secondary data on wage rates in Tamil Nadu in the recent period. He used the annual wage rate series constructed by the Agro-Economic Research Centre of the University of Madras. Money wage rates from 1951–2 to 1973–4 were deflated by using price series for three cereals separately—the price series for first-sort paddy, second-sort rice, and *ragi*—and the Rural Price Index. The results were dramatic:

> the most striking aspect of the wage movements in rural Tamil Nadu during the period since 1951–52 is the decline in real wages of all categories of rural workers examined irrespective of the deflator used. (Kurien 1981)

A more recent survey of wage rates in Tamil Nadu agriculture from 1950–1 to 1980–1 by Sridhar Venkataraman concludes that agricultural wages have stagnated in most districts, risen slightly in two districts, and fallen sharply in Tiruchi district (Venkataraman 1987). He has also shown that real wages, as computed from official sources, peaked in the late 1960s and early 1970s.[8]

While almost all the surveys based on analyses of official data in Tamil Nadu point to a decline or to stagnation in wage rates,[9] studies of particular villages give useful insights into interregional variations in trends. Chambers and Harriss, for instance, suggest from a comparison of wage rates in 12 villages studied by the Cambridge Project that wage rates were higher in villages where production was more continuous and more labour-intensive (Chambers and Harriss 1977). The resurveys of the 'Slater villages' organized by S. Guhan and Venkatesh Athreya also provide material on trends in wage rates. In Iruvelpattu, South Arcot district, Guhan and Mencher (1983) conclude that while 'between 1937 and 1981, harvest wages might have gone up in real terms, . . . wages for transplanting and weeding have either stagnated or declined'.[10]

[8] This was also the case in Punjab and Uttar Pradesh: see S. Bhalla (1979, pp. A-57 ff.); Nayyar (1977).

[9] Vaidyanathan (1984) reviews attempts at deriving wage rates from data on days of employment and gross earnings in the RLE.

[10] They have used post-harvest paddy prices.

In Dusi in the Cheyyar taluk of North Arcot district, near Kanchipuram, Guhan suggests that there has been a secular decline in wage rates in all operations except harvesting (where there was a decrease over the period 1937 to 1983, although there was an increase between 1957–8 and 1983) (Guhan and Bharathan 1984, p. 74, table 36). Athreya makes a similar point in regard to Vadamalaipuram (now in Kamarajar district): there was a decline in wage rates paid to men and women and to permanent and casual labour from 1916 (the year of the first survey) to 1984, although there was a rise in wage rates between 1957–8 and 1983 (Athreya 1984, pp. 98–9). In Gangaikondan, Tirunelveli district, Athreya records an increase in wage rates from 1958 to 1984 (he has data for only two points of time) (Athreya 1985).[11] In his study of two Thanjavur villages, Marudanallur and Karuppur, both in the vicinity of Kumbakonam, C. G. Swenson indicates that real wage rates for landless labourers at paddy operations increased by 11 per cent for men and declined by 5 per cent for women between 1965–6 and 1970–1 (Swenson 1973).[12]

Two village-level studies from Thanjavur district stress the influence of unionization and the process of collective bargaining on the determination of wage rates. In Palakurichi in Nagapattinam taluk, Guhan's work shows that a long series of wage struggles led by the Communist Party have led to a substantial increase in the statutory wage rate, particularly in harvesting, although at the same time, for a variety of reasons, including a very sharp increase in the number of agricultural labourers and acute problems of seasonality in labour demand, real *earnings* have shown a stagnant trend (Guhan 1983, p. 89 and elsewhere). Kathleen Gough notes that in her study villages, Kirippur and Kumbapettai, real daily wages rose between the early 1950s and the late 1970s as a result of organized action by agricultural workers' unions. At the same time, the annual incomes of agricultural labourers, in general, remained stationary or declined (Gough 1987, pp. 290–1).[13]

In sum, the secondary data for Tamil Nadu as a whole show a decline or stagnation in real wage rates from the 1950s to the 1970s. Primary data collected through village studies present a more complex picture. These show that trends in wage rates vary across regions, crops, and operations, and vary with the gender of the worker. Moreover, variations in wage rates over time are very different in different agro-economic regions. There are, however, certain features that the data have in common: wage rates are low in absolute terms, wages paid to men and women at daily-rated tasks are very unequal, and there

[11] Vadamalaipuram and Gangaikondan present special cases in terms of supply aspects of the labour market. Gangaikondan has a high proportion of non-workers in the population, and in Vadamalaipuram, there has been a large diversion of workers into non-agricultural employment.

[12] Swenson's tables are difficult to interpret, since he has not given data by operation.

[13] A village study in Kadathuchery in Kanniyakumari, which borders Kerala, suggests an interesting association between literacy and educational levels among agricultural workers, the influence of labour organization, and a rise in real wages over the 1960s and 1970s (Nair *et al.* 1984, p. 955).

has been no significant secular increase in wage rates as a consequence of the 'green revolution'.

While it is clear that the time series data from the Cumbum Valley do not show an aberrant trend, it is noteworthy that this stagnation and decline in wage rates occurred in an agrarian region that is relatively advanced, and where there was a very steep rise in the demand for labour in the 1960s and early 1970s. In Punjab, by contrast, real wage rates rose between 1961 and 1977, and it has been suggested that the major reason for this was the increased demand for labour in agriculture (S. Bhalla 1979).[14]

In a 1976 paper that is quoted frequently, Deepak Lal argues that wage rates rose as a consequence of the 'green revolution'. Based on NSS data, his paper states that 'real agricultural wages have risen in India during the period 1956–7 to 1970–71' and explicitly relates this to the gains of the new technology: 'the 1970–71 NSS data show that [the 'green revolution'] was associated, at least in its initial impact, with large rises in real wage rates' (Lal 1976, p. A-59). While Lal does not take up the changes that have occurred in the size of the labour force in agriculture over this period (factors of 'supply'), he is certainly cheerful about the net effect of changes in the *demand* for labour on wage rates: 'there is evidence to suggest that rural wages are determined within the normal supply and demand framework, and, as this framework would suggest, the demand-inducing effects of agricultural growth do have a positive effect on real wage rates' (Lal 1976, p. A-59). The specific reference period of Lal's study is likely to have made a substantial contribution to his optimism. The terminal year of Lal's study is 1970–1. Even for the 'green revolution' state of Punjab, it has been shown that wages declined after 1973. Sheila Bhalla notes that wage rates, which peaked between 1969 and 1973, 'fell in 1974 and 1975 to levels below those reached during the early 1960s, and well below those of the all-time peak. In 1976 and 1977, real wages . . . were still below those recorded from 1969 to 1973' (S. Bhalla 1979, p. A-57).

It is clear, of course, that Lal's conclusions do not apply at all to cash-paid, daily-rated wages in our area of study, where wage rates either remained stagnant or marginally declined between 1948 and 1976, where the gap between (grossly unequal) wage rates paid to men and women has actually widened, and where wage rate fluctuations have continued well into the period of the 'green revolution'. While our attempt is not to over-generalize from the data on the Valley, one conclusion that does emerge is that the 'demand-inducing effects' of the new technology do not guarantee an automatic increase in wage rates or in the earnings of agricultural labourers, particularly in a 'green revolution' region where growth in the absolute and relative size of the agricultural labour force has been steep.

To pursue this theme at the all-India level, analyses of changes in the size of the labour force during the early years of the 'green revolution' showed that

[14] Bhalla also shows that output rose by a factor of 2.5 over the period 1961–77, while the level of wages rose 15%. See also Dhaliwal and Grewal (1982).

despite the fact that new varieties require less labour per unit of output, the productivity of labour measured in terms of output *per member of the labour force* did not necessarily increase. From a study of changes from 1962–5 to 1970–3 in output per worker across 281 districts, G. S. Bhalla showed that, in general, gains in output were not necessarily accompanied by changes in output per worker. Of the districts he studied, 45 showed a compound rate of growth of output per male worker that was greater than 2.5 per cent, 84 showed between 0 and 2.5 per cent, and 152 showed a negative rate of growth. The disaggregated results showed the strong impact of increases in the supply of labour in bringing down rates of growth. The first category of districts (with a growth rate greater than 2.5 per cent) included, as may be expected, districts in Punjab, Haryana, and western Uttar Pradesh; it also included districts in dry areas of Rajasthan, Madhya Pradesh, and Gujarat. At the other end, the districts with negative compound rates of growth of output per male worker included six districts where output grew by more than 3.5 per cent per annum. This category included Ludhiana in Punjab, where, although output increased at a compound rate of growth of 6.8 per cent per annum, the male labour force grew at a rate of 7.5 per cent per annum (G. S. Bhalla 1978, pp. 14 ff.). Bhalla's results are consistent with the findings in his joint paper with Alagh and Bhaduri (1978), in which an interdistrict study covering the period 1965–6 to 1972–3 found that districts with rising yields per unit of cropped area were also characterized by corresponding increases in the supply of male agricultural labourers, keeping productivity near-stagnant. Sheila Bhalla's data from her study of real wage rates in Punjab from 1961 to 1977 show that output per male worker (i.e. cultivators and agricultural labourers together) rose in all districts other than Ludhiana, but that output per male agricultural labourer declined, mainly because of a sharp increase in the size of the agricultural labour force (S. Bhalla 1979, p. A-59).[15]

Earnings of Households

Landless agricultural labour households in Gokilapuram are grouped by per capita household income in Table 7.4. 'Per capita household income' refers to the total income and earnings of a household (the aggregate earnings of all its members) divided by the number of members of that household; the category is analogous to the NSS category 'per capita household consumption'. Table 7.5, which derives from the previous table, shows the shares of income from different sources in the total income of households. Incomes from dairying and other sources refer to net household income, after deducting costs in that particular category. Table 7.6 shows comparable data for Tamil Nadu from the Rural Labour Enquiry (RLE) 1974–75.

[15] On the increase in the supply of labour being greater than the increase in the demand for hired help in the country as a whole, see the review in K. Bardhan (1983, p. 20).

Table 7.4 *Distribution of annual income per capita of landless agricultural labour households, Gokilapuram, 1976–1977*

Per capita income of household (Rs)	Households No.	Households %	Source of earnings and income (Rs): Agricultural labour	Tamarind processing	Plantation labour	Hand pounding paddy at co-operative society	Tending cattle	Other non-agricultural manual tasks	Domestic service	Artisan activity, traditional caste calling	Salaries; professional earnings	Dairying	Interest on loans	Remittances	Petty business, rent, etc.	All sources	% of column total	Earnings from all sources per household
⩽ 150	22	8.5	9,909	1,318	240	18	0	1,160	650	0	0	0	0	0	−2,040	11,255	2.2	512
151–250	40	15.5	25,186	2,750	805	1,548	720	2,605	230	2,965	0	1,344	0	1,620	1,350	41,123	8.2	1,028
251–350	45	17.5	38,744	4,560	6,278	25	0	4,801	1,675	2,200	1,060	2,640	0	640	2,821	65,444	13.0	1,454
351–450	45	17.5	42,465	4,755	9,652	2,547	2,995	3,014	550	5,138	5,820	2,825	0	240	2,002	82,003	16.3	1,822
451–540	27	10.5	33,274	5,175	5,923	0	660	1,145	1,260	620	2,700	6,426	390	0	2,100	59,673	11.9	2,210
Poverty line, rural Tamil Nadu, 1976–7: Rs 540 per capita per year																		
541–650	30	11.7	33,131	1,671	14,309	4,644	0	4,768	0	3,720	840	3,368	720	350	4,850	72,371	14.4	2,412
651–750	13	5.1	19,016	598	3,941	5,094	0	1,226	700	1,220	5,640	1,696	0	1,800	3,088	44,019	8.8	3,386
751–850	11	4.3	19,353	1,995	1,950	0	0	2,810	2,040	1,926	0	3,005	1,140	0	410	34,629	6.9	3,148
851–950	11	4.3	24,049	2,010	8,310	840	0	2,260	0	0	1,980	4,740	0	0	0	44,189	8.8	4,017
⩾ 951	13	5.1	16,751	1,325	6,162	0	420	0	780	0	5,775	5,180	1,440	900	8,860	47,593	9.5	3,661
All households	257	100.0	261,878	26,157	57,570	14,716	4,795	23,789	7,885	17,789	23,815	31,224	3,690	5,550	23,441	502,299	100.0	1,954
% share of total income			52.1	5.2	11.4	2.9	0.9	4.7	1.6	3.5	4.7	6.2	0.7	1.1	4.7	100.0		

Source: Survey data.

Table 7.5 *Percentage share of different sources of income in total income, landless agricultural labour households, Gokilapuram, 1976–1977*

Source of income	Source of income for households		
	Below the poverty line	Above the poverty line	All
Agricultural labour	57.6	46.3	52.1
Tamarind-processing	7.2	3.1	5.2
Tending cattle	1.7	0.2	1.0
Other non-agricultural manual tasks	4.9	4.6	4.7
Domestic service	1.7	1.4	1.6
Artisan activity, traditional caste callings	4.2	2.8	3.5
Subtotal	77.3	58.4	68.2
Hand-pounding paddy	1.6	4.3	2.9
Plantation labour	8.8	14.3	11.5
Salaries, professional earnings	3.7	5.8	4.7
Dairying	5.1	7.4	6.2
Interest on loans	0.1	1.4	0.7
Petty business, rent, etc.	2.4	7.1	4.7
Remittances	1.0	1.3	1.1
Subtotal	22.7	41.6	31.8
TOTAL	100.0	100.0	100.0

Source: Survey data.

Some conclusions from the data follow:

1. The per capita household earnings of a very large number—179, or 70 per cent—of landless agricultural labour households were below the poverty line for rural Tamil Nadu for the reference year,[16] which was Rs 536.40. This is a high figure *per se* and relative to estimates of the proportion of households below the poverty line that are made from NSS data. From NSS data, 50 per cent of agricultural labour households in Tamil Nadu were below the poverty line in 1977–8 (Subramanian 1985, table 18, p. 47). However, the data for this study refer to *earnings* (they exclude money borrowed by households), and not to *consumption* levels, as NSS data do.

Estimates from the RLE give higher levels of poverty than even our data (see Table 7.6). The RLE provides, for Tamil Nadu, the percentage distribution of agricultural labour households by per capita annual income classes[17] in 1974–5. The reference year of the RLE, 1974–5, was a year of very high prices, and the Rural Price Index for Tamil Nadu was 405 (1960–1), giving a poverty line of

[16] Using the official poverty line of Rs 15 per month per capita at 1960–1 prices.
[17] i.e. the data are grouped similarly to the data in Table 7.4.

Table 7.6 *Percentage distribution of agricultural labour households by annual per capita income classes, Tamil Nadu, 1974–1975*

Annual per capita income class (Rs)	% of households in group
0–50	5.2
51–100	5.0
101–150	6.4
151–200	9.2
201–250	9.6
251–300	6.2
301–350	8.0
351–500	19.1
501–650	14.7
⩾ 651	16.5
All classes	100.0

Source: RLE 1974–75 (1983).

Rs 729 per capita per annum. By this calculation, the poverty line fell in the very highest per capita income group given in the RLE tabulation, which is Rs 651 and above; even if we exclude the households in this group, a minimum of 84 per cent of agricultural labour households had per capita household incomes that were below the poverty line. Again, if data from the RLE 1974–75 are taken as a point of comparison, average household earnings per annum for landless agricultural labour households in Gokilapuram were *higher* than the figure for Tamil Nadu as a whole. The RLE figure for Tamil Nadu was Rs 1,696 in 1974–5, which, when adjusted for 1976–7 prices,[18] is Rs 1,255. The survey data for Gokilapuram give an average household income of Rs 1,954 per annum (see Table 7.4).[19]

Let us also compare the earnings of landless labour households with a standard norm for working-class earnings—the minimum earnings recommended by the Third Pay Commission in 1972. The recommendation of the Pay Commission related, of course, to *urban* workers in major urban centres.

[18] The Rural Price Index (1960–1 = 100) was 405 in 1974–5 and 298 in 1976–7; the conversion factor is 0.74.

[19] C. G. Swenson's data indicate that in the two Thanjavur villages that he studied, the average real income per household from a sample of landless agricultural labourers employed at casual labour increased from Rs 422 in 1965–6 to Rs 473 (at constant prices) in 1970–1, an increase of 12%. To stretch Swenson's data to comparability with ours, the cash income of a landless agricultural labour household in Swenson's study was Rs 615, or about Rs 1,051 in 1976–7 prices, lower than our figures for the average landless agricultural labour household in Gokilapuram (Swenson 1973, table 5.50, p. 196).

Price and expenditure levels are different for such workers and for agricultural workers; the composition of the consumption basket (particularly expenditure on items such as residential accommodation) is also different. The Pay Commission recommended a wage of Rs 343.69 per month for a family of 3 consumption units at a consumer price index (1960–1 = 100) level of 200 points. By this reckoning, the minimum wage for Madurai in 1977 for 3 consumption units (the consumer price index using the base 1960–1 = 100 was 328 in Madurai that year) would have been Rs 563.65 per month. This amounts to Rs 6,763.80 per year, or about 3.5 times the average household earnings of landless agricultural labour households (which consist of *more* than 3 consumption units; see Chapter 4), which were Rs 1,954 a year.

2. Among landless agricultural labourers, just over half the aggregate income was from labour at agricultural tasks (tasks in direct crop production), indicating the need, even in a relatively advanced agrarian economy, for agricultural labourers to seek employment outside agriculture.

3. Agriculture and other tasks with marked seasonality (tamarind-processing, plantation labour) accounted for about two-thirds of total income; employment in off-peak seasons brought in the other one-third.

4. Earnings from agricultural labour formed a significantly higher proportion of the total earnings of households below the poverty line (a 58 per cent share) than of households above the poverty line (the corresponding share is 46 per cent). (By the RLE 1974–75 results for Tamil Nadu, 78 per cent of the income of landless agricultural labour households came from wage-paid manual labour in agriculture: RLE 1974–75 (1983), table 3.3.11.) When agricultural labourers say, as they often do, 'If we are trapped in agricultural coolie work, we cannot advance, we must remain poor,' they speak truly.

5. For households above the poverty line, what were the ways out, the escape routes, in terms of alternative sources of income? The data indicate that these were, first, regular, better-paid manual labour (employment in the plantations); secondly, regular, better-paid non-manual employment (employment at salaried occupations); and thirdly, income from sidelines (dairying and petty business; see Table 7.5).

6. To look at earnings from agricultural labour more closely, a male landless agricultural labourer, on average, earned Rs 652 per annum from agricultural tasks alone; a year's work at agricultural labour brought in, for men, an earning that was only 1.2 times that year's poverty line. The situation for women was, of course, even worse—average earnings from agriculture for a woman landless agricultural labourer was Rs 241, about Rs 200 below that year's poverty level. And landless agricultural labour households earned an average of only Rs 1,019 from agricultural labour over the year.

There is a clear conclusion here, relevant to public policy: agricultural labourers require higher wages, equal wages for men and women, more days of employment in a year (and particularly the creation of jobs during the lean season), and new and steady incomes from sideline occupations.

Poverty and Household Size

Do the poor belong to larger households? The literature in this field, and problems of macro-level data for whole populations (in India and other countries) have been assessed by Krishnaji (1984, pp. 248 ff.).[20] In this section we are dealing with family size and relative poverty levels only among agricultural labourers. While the survey data show very clearly that the most propertied section—major landlords—have bigger households than those at the other end of the spectrum—landless agricultural labourers—if we restrict ourselves to agricultural labour households, neither the results from the RLE nor from this survey provide any clear-cut, unambiguous relation between household size and poverty levels.

The RLE provides some material on this matter (see Table 7.7). The results for 1963–4 suggest that the poor (i.e. households in the lower per capita consumer expenditure classes) do in fact have more persons, on average, per household than households in the higher per capita consumer expenditure classes. The data for 1974–5 are more ambiguous. While households in the two highest per capita expenditure classes have an average number of persons per household (4.13 persons and 3.32 persons) that is lower than the average for all

Table 7.7 *Average number of persons per agricultural labour household, by annual per capita expenditure class, Tamil Nadu, 1963–1964, 1974–1975*

Annual per capita expenditure class (Rs)	Average no. of persons per household	
	1963–4	1974–5
0–50	5.00	—
51–100	5.32	—
101–150	4.66	3.00
151–200	4.46	5.23
201–250	3.91	6.27
251–300	3.82	5.47
301–350	3.02	5.42
351–500	—	4.69
⩾ 351	3.12	—
501–650	—	4.13
⩾ 651	—	3.32
All classes	3.99	4.19

Source: RLE 1974–75 (1983).

[20] See also Kuznets (1980) for conclusions from data across countries.

households (4.19 persons), the per capita expenditure class where the average size of household is the lowest, 3 persons, is also the lowest expenditure class.

Our survey data indicate that the poor belonged to households that were larger, but only marginally so (see Table 7.8). The size of landless agricultural labour households below the poverty line was 4.63 persons and above the line, 4.11 persons, and persons in households below the poverty line were 72 per cent of all persons in landless agricultural labour households (the proportion of landless agricultural labour *households* below the poverty line was 70 per cent).

Table 7.8 *Average size of landless agricultural labour households, Gokilapuram, 1977*

Category of household	No. of households	Persons	Persons per household	Percentage of persons
Below the poverty line	179	830	4.63	72.1
Above the poverty line	78	321	4.11	27.9
All households	257	1,151	4.48	100.0

Source: Survey data.

INDEBTEDNESS OF AGRICULTURAL LABOUR HOUSEHOLDS

Questions on indebtedness were part of this study's household schedule. There are, of course, problems with data on indebtedness: people are naturally inhibited about talking about the debt they are in, and the data they provide are also subject to some inaccuracy and are less easy to check. I use the data on loans taken by agricultural labour households only to draw very broad conclusions about their indebtedness status.

Some points of clarification are needed about the categories used here. 'Amount borrowed' refers to the original amount borrowed in the case of loans that were wholly or partly outstanding on the date of survey. Debts that are incurred through crop hypothecation are not included in 'amount borrowed'. 'Interest due' refers to that part of interest that was due but unpaid on the date of the survey, and does *not* include interest that would become due on the loan in future.

Sources of Credit and Terms and Conditions of Loans

The *formal* sector as a source of credit for agricultural labourers refers almost entirely to commercial banks (loans from Land Mortgage Banks, Land Development Banks and even cooperative society loans do not generally go to agricultural labourers). And for the period under consideration this meant, almost exclusively, loans given through loan-cum-subsidy schemes at differential interest rates for the purchase of milch buffalo.

Two aspects of the role of the formal sector as creditor are noteworthy. The amount borrowed by agricultural labourers from the formal sector was a small part (about 4 per cent for landless agricultural labourers) of the total amount borrowed by households in the village from the formal sector (Table 7.9). Among landless agricultural labourers, the share of the formal sector in total amount borrowed was 17 per cent, the other 83 per cent coming from the informal sector. The table shows the low access of agricultural labourers to the credit market in general and especially to the organized sector and to merchants, millers, and money-lenders; and their relatively greater access to landlords, miscellaneous part-time lenders, and others. Agricultural labourers do not have access to large-scale credit; such loans are cornered by the creditworthy, landlords and the rural rich to finance production and trade.

The RLE data for Tamil Nadu illustrate the distance that has existed between the formal credit sector and agricultural labourers. In 1974–5, the average amount of debt incurred by landless agricultural labourers in Tamil Nadu from cooperative societies and banks together was 5 per cent of all debts incurred; in 1964–5, the debt incurred by agricultural labourers as a proportion of all debts was minuscule: 0.4 per cent (see Table 7.10).[21] Although, at the survey, the share of loans from the formal sector in the total amount borrowed by agricultural labourers was only 17 per cent, it is a higher share than existed earlier; it is no longer true that commercial banks and the formal credit sector make *no* contact with agricultural labourers and the rural poor. Access to formal credit sources in the present period improved in the late 1970s and early 1980s. For all the lacunae in government-sponsored schemes based on subsidies and differential interest rates, such schemes were perceived by many agricultural labourers as initiating a regular process of borrowing, asset creation, and debt repayment.[22]

We turn now to the other, *informal*, sources of credit. There is no specific and exclusively money-lending caste or community in the village. There are a number of medium and small-time money-lenders and, more importantly, persons who combine money-lending with other economic activity. An important source of credit for agricultural labourers are landlords. Loans taken from 'shopkeepers, merchants, and others' refer, in the case of agricultural labourers, mainly to shopkeepers in the village, from whom they also make small purchases (see Table 7.9). Among the small-time money-lenders are also elderly women who lend money to obtain an income to live on in their old age.

[21] Data from an official survey of agricultural labour households conducted in 1971 in 10 villages in each of two districts, North Arcot and Salem, showed that the households surveyed received no credit at all from the organized sector (GOTN 1972).

[22] How the credit system works and the impact of these schemes on the poor are the subject of a separate study: see Swaminathan (1986, 1989).

Table 7.9 *Sources of credit, Gokilapuram, 1976–1977*

Source	Amount borrowed by						% share of landless agricultural labour households in total credit	Index of access[a]
	Landless agricultural labourers		Other than landless agricultural labourers		All households			
	Rs	% of column total	Rs	% of column total	Rs	% of column total		
Banks, government etc.	19,400	17.4	447,750	32.9	467,150	31.7	4.2	0.12
Landlords	24,970	22.4	133,205	9.8	158,175	10.8	15.8	0.44
Merchants, millers, etc.	13,710	12.3	223,937	16.5	237,647	16.2	5.8	0.16
Money-lenders	21,230	19.1	276,890	20.4	298,120	20.3	7.1	0.20
Miscellaneous part-time lenders	18,910	17.0	137,541	10.1	156,451	10.6	12.1	0.34
Others	9,912	8.9	72,455	5.3	82,367	5.6	12.0	0.33
Unspecified	3,130	2.8	68,175	5.0	71,305	4.8	4.4	0.12
TOTAL	111,262	100.0	1,359,953	100.0	1,471,215	100.0	7.6	0.21

[a] Index of access: Numerator = % of borrowing accounted for by landless agricultural labourers; denominator = % of population accounted for by landless agricultural labourers.

Source: Survey data.

Lastly, without exaggerating the phenomenon in terms of the total amount that they are able to lend, there is a significant section of lenders among the poor.[23] Even an agricultural labourer who earns a small surplus from a year of hard work may lend it to another and collect interest payments regularly rather than deposit it with a bank or other establishment in the formal sector.

These results have points of similarity and contrast with RLE 1974–75 data for Tamil Nadu on informal sources of credit for landless agricultural labour households. The contrast lies mainly in the fact that full-time money-lenders are not as distinct a section in Gokilapuram as they perhaps are elsewhere: by RLE data, 61 per cent of the total debt of landless agricultural labourers in Tamil Nadu came from money-lenders. The similarity lies in the fact that in the RLE data, employers—a rough counterpart of the category 'landlords' in our survey data—are a significant source of credit for landless agricultural labourers. (The proportion of debts from 'other' sources in the RLE is very large, and suggests that the category requires further disaggregation; see Table 7.10.)

Table 7.10 *Sources of loans to agricultural labourers, Tamil Nadu, 1964–1965, 1974–1975*

Source	1964–5	1974–5	
	All agricultural labourers	Landless agricultural labourers	All agricultural labourers
Co-operative societies, banks	0.4	5.0	8.5
Employers	12.8	10.5	6.3
Money-lenders	18.9	61.1	65.3
Shopkeepers	2.4	2.5	1.9
Others	65.5	20.9	18.0

Source: RLE 1974–75 (1978).

There is a wide variety of loans and rates of interest outside the formal sector in the village and, while these are not beyond being systematized or classified, an important feature of loans outside the formal sector is that they generally derive from specific relationships between borrowers and lenders. The credit system—borrowing, lending, enforcing repayment, and scheduling and re-scheduling repayment—does not operate impersonally, but, on the contrary, reflects a highly personalized system of lending and borrowing.

[23] Some of these come into the category of miscellaneous part-time lenders.

Loans are generally given in cash (we did not have any loans in our data where the original amount borrowed was in grain, as the oppressive grain loan that is the theme of Prem Chand's short story 'A Seer and a Quarter of Wheat'), although they are often paid back in grain. A creditor waiting to collect repayment 'at the entrance to the threshing-floor itself', when the debtor emerges after collecting wages in kind for threshing, is a common experience (and expression).

One of the most familiar methods of calculating rates of interest can be described as follows. Ask the debtor at what rate of interest he borrowed, say, Rs 100, and he may say 'on Rs 100, 6 *marakkal* of paddy per harvest', that is, as interest on the Rs 100 that he has borrowed he has to pay half a bag (a 58 kilogram bag of dry paddy consists of 12 *marakkal*) per paddy crop. For agricultural labourers, this amount is generally paid from harvest-time wages. This was a commonly specified rate of interest at our survey, but it can vary: interest can be specified at more or less *marakkal*, and can be specified as payable over two paddy crops (the agricultural year). Whether interest that is due and unpaid is carried over to the next harvest is compounded or simply added to the amount due at the next harvest is a matter that is worked out between the borrower and lender. Implicit monetary rates of interest can be worked out by using the harvest price of paddy for conversion (this is how the interest on such loans is calculated in Table 7.16). For instance, in the example, if the average harvest price of paddy was Rs 65 per 58 kilogram bag and where the rate of interest of half a bag per crop could be stated to be one bag per year (there are two paddy seasons to the year), the interest rate would be Rs 65 on Rs 100 per year, or 65 per cent.[24] The important feature of the system, of course, is that it is indexed, since it is calculated against bags of paddy.

Another form of loan is what is called a *kanthu* loan. This is characterized by high rates of interest, by very short intervals between dates of repayment (interest is often collected weekly and there are cases where it is even collected daily), and by a deduction of the first interest payment at the time the principal is borrowed. Persons who lend at *kanthu* rates are often full-time money-lenders, who have the capital, time, energy, and clout to enforce repayment. From conversations with agricultural labourers, it does appear that the most oppressive forms of *kanthu* (in particular the use of physical force in enforcing repayment) are on the decline.

Pawning is another prevalent form of borrowing and lending, with jewellery and, more commonly, metal vessels, given in security. There was no pawnshop in Gokilapuram in 1977 and pawnbrokers were mainly women who belonged to households that had other sources of income; the wives of some of the Chettiar landlords were particularly known for their pawnbroking.

There are also no-interest loans: *kaimattru*, or where money 'changes hands'. When the lender is also employer, as is often the case, the amount is

[24] This is the *minimum* rate of interest; the actual rate varies with the length of time between the loan and the harvest.

sometimes deducted from wages. For a landlord who makes such loans to an agricultural labour household—which borrows from the landlord to tide over household problem periods—these loans are often an important factor in having the workers of that household at his beck and call.

Mortgages generally involve immovable property such as land—agricultural and non-agricultural—or houses and huts. Mortgages, although generally built around the principles of a usufructuary mortgage, also vary in respect of their terms and conditions. These are typically taken on terms unfavourable to the mortgagor, and a significant proportion of the transfers of small plots of land, particularly dry land, was through foreclosures on mortgages.

How do agricultural labourers' credit profiles differ from other sections of the population in respect of sources of credit? The data suggest that the differences between landless agricultural labourers and others are particularly striking in respect of the share of loans obtained from formal-sector sources (see Table 7.9). For landless agricultural labourers the share was 17 per cent and, for all others, the share was 33 per cent. This contrast would be even more striking if we separate landlords, rich peasants, and business people, whose share in formal-sector credit was high, and in whose credit position the formal sector played a particularly important role. Further, from among other sources of credit, landlords were a more important source of credit for agricultural labourers than for others (although, here also, the share of landless agricultural labourers in the total amount borrowed from landlords was only 16 per cent).

The Purposes of Loans

An agricultural labourer's main reason for borrowing, the purpose-of-loan category that swamps all else, and accounted for two-thirds of the total amount borrowed, was the need to meet day-to-day living expenses and expenses on rituals and ceremonies (see Table 7.11). Add to this the loans taken for constructing and maintaining the huts they live in, and the loans taken to meet living expenses as a proportion of the total amount borrowed rose to about three-quarters. This was so with the State-wide results available from the RLE 1974–75 as well; loans for 'productive purposes' accounted for 11 per cent of the debts of landless agricultural labourers, while household consumption and ceremonies took up 68 per cent (see Table 7.12). The next major purpose of borrowing was for the purchase of livestock, generally under government-sponsored loan-cum-subsidy schemes.

By contrast, for other households in the village, crop loans were the largest single category of loans, loans for living expenses came next, and loans for business ventures and sidelines came a significant third. For the richer landlords, loans for land improvement (including land-levelling and conversion to groundwater-irrigated land) and the purchase of tractors were important reasons for borrowing.

Table 7.11 *Purposes of loans, by households, Gokilapuram, 1976–1977*

Purpose of loan	Amount borrowed by						Borrowings of landless agricultural labourers as % of total borrowings
	Landless agricultural labourers		Other than landless agricultural labourers		All households		
	Rs	% of column total	Rs	% of column total	Rs	% of column total	
Land purchase	0	0.0	54,550	4.0	54,550	3.7	0.00
Seeds, fertilizer, other agricultural inputs	3,700	3.3	571,925	42.1	575,625	39.1	0.6
Purchase of livestock	15,950	14.3	27,345	2.0	43,295	2.9	36.8
Industry, business	6,840	6.2	162,956	12.0	169,796	11.5	4.0
House construction, etc.	6,725	6.1	37,966	2.8	44,691	3.0	15.1
Ceremonies, living expenses	74,552	67.0	419,041	30.8	493,593	33.6	15.1
Unspecified	3,495	3.1	86,170	6.3	89,665	6.1	3.9
TOTAL	111,262	100.0	1,359,953	100.0	14,711,215	100.0	7.6

Source: Survey data.

Table 7.12 *Purposes of loans of landless agricultural labour households, Tamil Nadu, 1974–1975*

Purpose of loan	% of total debt
Productive purposes	10.5
Household consumption	49.4
Marriage and other ceremonial occasions	18.6
Other purposes	5.8
More than one purpose	15.7
TOTAL	100.0

Source: RLE 1974–75 (1978).

The Burden of Debt

It is clear that the burden of debt on agricultural labourers is very great. A large proportion of agricultural labourers are, in varying degree, in debt; in our record, of the 257 landless agricultural labour households, 163 (63 per cent) were in debt, and of the 61 agricultural labour households that owned or operated land, 44 (72 per cent) were in debt.[25]

The ratio of the amount borrowed by a household to the value of all assets owned by it—the household debt–asset ratio—is a measure of the burden of its debts. The average debt–asset ratio for landless agricultural labour households in debt was 0.28, which means that debt was equivalent to about 28 per cent of the worth of all household assets (see Table 7.13). This is an average around which there is variation: of the indebted households, 31 per cent had debts of over one-third the value of their assets, more than 25 per cent had debts of over half the value of the assets they owned, and almost 13 per cent had debts that exceeded the value of household assets. (Note that landless agricultural labourers owned only 3 per cent of all assets in the village.) If an industrial definition of sickness was imported to the village and households with liabilities exceeding 50 per cent were declared 'sick' units, then sickness here would be endemic indeed.

Households that were very deeply in debt, with debt–asset ratios that were greater than unity, were characterized by debt levels that were above the

Table 7.13 *Landless agricultural labour households, by household debt/asset ratios, Gokilapuram, 1976–1977*

Debt/asset ratio	Households			Debts		Assets	
	No.	% of all households	% of indebted households	Aggregate	Per household in group	Aggregate	Per household in group
0.00	94	36.6	—	0	0	174,711	1,859
0.01–0.10	25	9.7	15.3	5,772	231	134,345	5,374
0.11–0.20	31	12.1	19.0	18,220	587	125,009	4,033
0.21–0.33	27	10.5	16.6	11,560	428	44,292	1,640
0.34–0.50	14	5.4	8.6	12,660	904	31,860	2,276
0.51–0.75	22	8.6	13.5	17,910	814	27,824	1,265
0.76–1.00	11	4.3	6.8	12,670	1,152	14,404	1,309
> 1.00	33	12.8	20.2	29,940	907	10,861	329
All landless agricultural labour households	257	100.0	—	108,732	—	563,306	2,199
Indebted landless agricultural labour households	163	63.4	100.0	108,732	667	388,595	2,384

Source: Survey data.

[25] By RLE 1974–75 results for Tamil Nadu, the incidence of indebtedness among agricultural labour households without land was 72%; for agricultural labour households with land, 83%; and for all agricultural labour households, 76%. The extent of indebtedness was Rs 682 or, in 1976–7 prices, Rs 505 per indebted agricultural labour household (RLE 1974–75 (1978), tables 3.1*a* and 3.2*a*).

average for all households. An even more striking feature of these households was that the average value of assets owned by each of them was much lower (Rs 329) than the average for all indebted households (Rs 2,384). The average asset holding per indebted household was, in fact, higher than the average for households not in debt.

The data for agricultural labourers with ownership and/or operational holdings show similar patterns (see Table 7.14). Among indebted households in this category, the average debt–asset ratio per household was 0.16, although the absolute level of indebtedness, Rs 1,623, was much higher than among the landless.

Table 7.14 *Agricultural labour households that own and/or operate land, classified by debt/asset ratios, Gokilapuram, 1976–1977*

Debt/asset ratio	Households			Debts		Assets	
	No.	% of all households	% of indebted households	Aggregate	Per household in group	Aggregate	Per household in group
0.00	17	27.9	—	0	0	84,900	4,994
0.01–0.10	22	36.1	50.0	9,105	414	245,861	11,176
0.11–0.20	4	6.6	9.1	2,900	725	20,381	5,095
0.21–0.33	9	14.7	20.4	28,500	3,167	116,612	12,957
0.34–0.50	3	4.9	6.8	19,160	6,387	46,034	15,345
0.51–0.75	5	8.2	11.4	10,530	2,106	16,678	3,336
0.76–1.00	0	0.0	0.0	0	0	0	0
> 1.00	1	1.6	2.3	1,225	1,225	915	915
All households in the group	61	100.0	—	71,420	1,171	531,381	8,771
Indebted households in the group	44	—	100.0	71,420	1,623	4,446,481	10,147

Source: Survey data.

A closer look at the data brings out an even more significant pattern in the debt–asset ratios of agricultural labour households. At the two extremes of the debt–asset ratio scale are households with less than average asset holdings. Thus, for households with a debt–asset ratio that is greater than 0.5, the situation is one of low assets and high debts—these are poor households that get into debt for sheer survival (and this is particularly true of households with debt–asset ratios that are greater than unity). At the other end of the scale, those with no debt (debt–asset ratio = 0) were also low-asset households, reflecting the absence of 'creditworthiness' among these households. This pattern appears to hold for both landless agricultural labourers and agricultural labourers with land.[26]

[26] We have used the debt–asset ratios of households as a measure of the burden of debt because

Where lurks the debt trap that many agricultural labourers, many of the rural poor, find themselves in? It lies in the fact that most agricultural labourers have only their wages with which to repay their debts. It usually stretches their means to keep up interest payments, and it is very difficult for them to accumulate the money needed to pay off the principal. Among landless agricultural labour households, while indebted households had still to pay 81 per cent of the principal, the sum of interest payments already made and currently due amounted to 80 per cent of the sum they originally borrowed (Table 7.15). The picture is more sharply drawn if loans taken from the formal sector are excluded: when that is done, while 89 per cent of the principal was outstanding, the interest paid and interest currently due added up to 94 per cent of the amount originally borrowed.

Rates of Interest

A major feature of the rates of interest charged on loans was that the overwhelming share of loans was borrowed at rates of interest that were above the statutory limit, and a hefty share of these were rates that can only be termed usurious (see Table 7.16). More than 32 per cent of loans were taken at rates of interest greater than 40 per cent per annum and more than 56 per cent at rates greater than 18 per cent. A second feature of the loans taken by agricultural labourers was that about 13 per cent of the total amount borrowed was in the form of interest-free loans—this is consistent with the fact that landlords, who lend to workers who labour on their land, were an important source of loans. Thirdly, the share of the total amount borrowed in the interest-rate category $0 \leq r \leq 18$, which was 19 per cent, roughly coincided with the share of loans from the formal sector, which was 17 per cent.

How does this compare with the rates of interest charged on loans taken by other sections of the population? If we compare loans to landless agricultural labourers and loans to all other households, we find the rates of interest on the former are generally much higher. In the survey data, the modal rate of interest on loans to landless agricultural labourers fell into the class $40 \leq r \leq 100$, followed by $18 \leq r \leq 40$, with almost 8 per cent of the total amount borrowed being burdened with rates of interest that were greater than 100 per cent per annum. For other classes, the modal rate of interest fell into the group $0 \leq r \leq 18$ and $18 \leq r \leq 40$ was the next-to-modal class. Again, in the case of households other than those of agricultural labourers, there was a rough coincidence between the share of the amount borrowed from the formal sector and the share in the interest-rate bracket $0 \leq r \leq 18$, which also goes to show that

this is a conventional measure and one that is familiar, and also because the data are readily available. But much of the amount borrowed by agricultural labourers is not lent to them against the security of their assets: more often than not, it is a charge on their earnings, particularly harvest wage earnings.

Table 7.15 *Indebtedness of landless agricultural labour households, Gokilapuram, 1976–1977*

Source	Amount borrowed (Rs)	Principal paid (Rs)	Principal outstanding (Rs)	Interest paid (Rs)	Interest due (Rs)	(Interest paid + interest due)/ amount borrowed (%)	Principal outstanding/ amount borrowed (%)
Banks, government, etc.	19,400	10,994	8,406	1,687	560	11.6	43.3
Money-lenders	21,230	2,937	18,293	28,399	10,848	184.9	86.2
Miscellaneous part-time lenders	18,910	1,185	17,725	4,951	1,952	36.5	93.7
Landlords	24,970	2,325	22,645	15,153	4,973	80.6	90.7
Shopkeepers, etc.	13,710	1,455	12,255	7,369	3,446	78.9	89.4
Others	9,912	1,010	8,902	5,149	2,070	72.8	89.8
Unspecified	3,130	1,400	1,730	2,336	60	76.6	55.3
All sources	111,262	21,306	89,956	65,044	23,909	80.0	80.9
Informal sector total	91,862	10,312	81,550	63,357	23,349	94.4	88.8

Source: Survey data.

Table 7.16 *Amount borrowed by households, by rate of interest, Gokilapuram, 1976–1977*

Row no.	Rate of interest/ type of loan	Amount borrowed by					
		Landless agricultural labourers		Other than landless agricultural labourers		All households	
		Rs	% of row (7) total	Rs	% of row (7) total	Rs	% of row (7) total
1	$r=0$	13,340	12.7	42,503	3.4	55,843	4.2
2	$0<r<40$	19,560	18.6	429,930	35.1	449,490	33.8
3	$18<r<40$	25,020	23.9	415,690	33.9	440,710	33.1
4	$40<r<100$	25,730	24.5	87,995	7.2	113,725	8.5
5	$100<r$	8,160	7.8	7,300	0.6	15,460	1.2
6	Unspecified	13,152	12.5	242,435	19.8	255,587	19.2
7	All non-mortgage loans	104,962	100.0	1,225,853	100.0	1,330,815	100.0
8	All mortgages	6,300	—	134,100	—	140,400	—
9	TOTAL	111,262	—	1,359,953	—	1,471,215	—

Note: I have not calculated the implicit rate of interest on usufructuary mortgages.
Source: Survey data.

these rates of interest generally prevailed only in the formal sector. To step out of the formal sector is to step into the quicksand of usurious rates of interest.

Consumption Patterns

This section focuses on the *shares* of different categories of expenditure in total consumer expenditure. Absolute levels of consumption are not discussed here. Our consumption data are based on recall and not on regular monitoring of consumption, and for that reason I have put them to only limited use. I took 13 landless agricultural labour households (5 per cent of all landless agricultural labour households) and, to ensure the accuracy of the data, I chose households whose total household consumer expenditure was roughly equivalent to household earnings plus current indebtedness, and (by way of cross-check) which had reasonable levels of food-grain consumption.

In summary, consumption patterns are marked by a large share of expenditure going to buy food and a very small share going to buy consumer durables. In greater detail, for 10 out of the 13 households, all categories of food, beverages, and stimulants ate up between 64 and 88 per cent of total expenditure.[27] The three exceptions are also interesting: two had heavy

[27] By the RLE 1974–75 (1983), the share of expenditure on food and stimulants in the total household expenditure of landless agricultural labourers in Tamil Nadu was 83% (table 4.1.11, p. 160).

expenses in the reference year on rituals and ceremonies, and their food expenses were consequently lower shares of total expenditure (48 per cent and 54 per cent); the third was a household consisting of a young widow and her two small children, and there was no family expenditure on tea and coffee or stimulants, and the share of expenditure on food was consequently lower (58 per cent).

Within the category of food expenditure, food-grain takes a very major proportion (an average of 60 per cent),[28] as do tea, coffee, betel and nut, tobacco, and intoxicants (an average of 14 per cent). Relatively little is spent on vegetables—not of the best quality!—or on milk and eggs. There are households which spent nothing on milk and eggs over a year, the only milk in the diet being in the glasses of tea and coffee they bought. Meat consumption was very low—for the selected households, about 1.9 kilograms of goat's meat per annum per capita. (The section of the village population whose meat consumption is higher is the Parayar Christian community, who eat beef: beef cost Rs 3 to Rs 4 a kilogram, while goat's meat cost Rs 10.) Fresh fish was hardly eaten; dried fish—the dried fish here is of poor quality, inadequately dried, and considered an inferior food—is bought frequently, though used more for flavouring than for substance. Food, then, is frugal and not well-balanced, with a heavy component of food-grain, particularly cereal.

Durable consumption goods were hardly part of the budget. For the selected households, the average expenditure per capita under this heading during the reference year was very low, Rs 5.09. Fuel becomes an item of relatively heavy expenditure when there is no one in the household to collect dung or to go to the hills to collect twigs and bark or cut down wood and carry it back. (In the lean season, I have seen four men of a family go to the foothills, more than 15 kilometres away, and uproot a tree trunk that was a few feet in diameter and anchored by tough and difficult roots—the main tree had already been cut down—and split it up and bring it back to Gokilapuram by nightfall, to cut it down for firewood the next morning.) Only five of the selected households had any expenditure at all on education, the average expenditure per household being Rs 11.85 per annum. The figure was pushed up by the fact that one household set aside what for an agricultural labour household was a princely sum—Rs 92 in the year—on education. Special mention must be made of the head of that household, M. Raju, long-time Dravida Kazhagam supporter, who married out of his caste, does not allow any god–goddess pictures on the walls of his hut, and who had spent the money on vocational training for his *daughter*.

Every household spends on rituals and ceremonies, although the amount varies from year to year. Medical expenditure also varies, and is generally spent on children, adults receiving treatment only when really necessary. Transport is a larger component of the expenditure of households whose members work

[28] The corresponding figure for landless agricultural labour households in Tamil Nadu from the RLE 1974–75 (1983) was 76% of household expenditure on food (table 4.2.11, p. 208).

at tasks in other parts of the Valley and on estates outside it than for village-bound households. An item of expenditure that appears in just about every household schedule is the cinema (there is one, Murugan Talkies, in the village): some see more and some see less, but everyone sees *some* films. The cinema is far and away the most important form of entertainment; films are a point of reference, subjects of much discussion, and are generally an agricultural labourer's most important window on the world. ('In this place,' an elderly one drily says, 'even if they built a cinema on a high hill with only a steep and difficult path leading to it, the theatre owner still would not have to worry at all about making a profit.')

Summary

1. Absolute levels of wages in the Valley were, in general, higher than in more backward regions in Tamil Nadu. The male wage was higher than the statutory minimum in most operations. Wages were lower than in Haryana. They were less than one-third the minimum wage of the lowest-paid workers in the organized industrial sector. When a number of selected countries in East, South-East, and South Asia were ranked according to the amount of rice that could be purchased with a day's wage, the Valley (along with the entire South Asian region) fell in the lowest category.

2. In real terms, wage rates showed a trend of stagnation and marginal decline from 1948 to 1976. (I have local data on cash-paid, daily-rated operations.) This was the period of major technical change in agriculture and certainly the period during which the sharpest increase in the demand for labour took place. At present, reviews of secondary data for Tamil Nadu over the period show the same trend.[29]

3. The wage rate for women at cash-paid, daily-rated tasks was below the statutory minimum and was less than half the rate for men in 1976. The data indicate that the difference widened over the period from 1948.

4. In terms of per capita household earnings, about 70 per cent of landless agricultural labour households were below the poverty line. The share of earnings from agricultural wages was higher in the case of households below the poverty line; those above it earned incomes from more diverse sources.

5. The burden of debt on agricultural labour households was very great. At the survey, 65 per cent were in debt and the average ratio of debts to assets among indebted landless agricultural labour households was 0.28 (20 per cent of them had debt–asset ratios that were greater than unity). Three further features of the indebtedness profiles of agricultural labourers were interrelated

[29] There are, of course, variations in specific regions and at specific tasks; some of these are discussed in the chapter. See Boyce (1989) for a useful analysis for Bangladesh of the decline in agricultural wage rates between the 1950s and the late 1970s and of inter-district variations in wages.

—the major purpose of borrowing was for consumption and other household expenditure; the major source of loans were lenders in the informal sector; and, at the survey, the modal rate of interest on loans was very high: $40 \leq r \leq 100$ per annum (8 per cent of loans were taken at rates of interest that were greater than 100 per cent per annum). Overall access to credit was low: agricultural labourers received a disproportionately small share of all loans disbursed in the village, particularly from the formal sector. At the same time, the share of the formal sector in the total amount borrowed by agricultural labourers increased significantly over the second half of the 1970s and the early 1980s, mainly because of the introduction of IRDP and other government loan-cum-subsidy schemes.

8

Labour Services and Unfree Labour

> Life creates forms that unite in themselves with remarkable gradualness systems of economy whose basic features constitute opposites. It becomes impossible to say where 'labour service' ends and where 'capitalism' begins.
>
> (Lenin, *The Development of Capitalism in Russia*)

The overall impression of Gokilapuram village and the Cumbum Valley that has hitherto emerged is one of a modernizing agrarian regime, with a growing contingent of agricultural labourers. Previous chapters, while showing that the agricultural labour force is largely propertyless and is tormented by unemployment and great poverty, have stressed the factors in the economy that work towards impersonalizing the relationship between hired labourer and employer. This picture is still, however, somewhat one-sided; the account must be qualified, textured by a description of the constraints on the freedom of the agricultural labourer to sell his or her labour power. Prominent in this chapter will be a discussion of labour services performed by agricultural labourers. Labour services are tasks performed by the worker for the employer at rates (if there is payment at all) that are far below the rates of remuneration under conditions of 'free' hire. Labour service also 'always presupposes the personal dependence of the one hired upon the one who hires him, it always presupposes the greater or lesser retention of "other than economic pressure" ' (Lenin, 1972*a*, p. 197). Labour services are part of a larger and pervasive system whose basis is the absence of a general democratization of village life.

This chapter discusses labour services performed by landless agricultural labourers and members of their households. It does not take up in any detail the labour services of others, of peasants, non-agricultural workers, and domestic servants, or of village odd-job men and women, although these also constitute an important part of the system of labour service.

Defining Types of Unfree Labour

At the start, it is useful to make a distinction between the duration of an employment contract on the one hand, and the degree of socio-economic dependence ('attachment') of the labourer on the employer on the other hand. An agricultural labourer who is employed as a casual daily labourer and who is paid wages can still be an *attached* worker not completely free to sell his or her labour power to the employer of his or her choice. Conversely, a farm servant employed on a long-term contract may be free to break the contract and work for another employer and in other tasks. (Most farm servants, however, are likely to have to work off wage advances or loans from the employer.)

The labourer who is on a long-term contract, then, is not *ipso facto* an attached labourer (or, for that matter, a bonded labourer), just as a daily labourer is not necessarily a completely free wage worker, unencumbered by prior obligations to the employer. Although the distinction has been pointed out in the past, it bears re-emphasizing, since there are still contributions to the literature that treat 'attached labour' and 'long-term labour' as coterminous. Worse, conflation of the two categories (or at any rate, a failure to make a rigorous distinction between them) has distorted the collection of data on attached labour and bondage in some large-scale surveys.[1]

The empirical data are reasonably clear on some of these issues. Bonded labourers are invariably long-term workers, not daily labourers. Workers who are not bonded, but whose relationship with their employers is nevertheless characterized by varying degrees of unfreedom, are generally scattered across the duration-of-employment spectrum. Among them, for example, are daily workers, workers who are paid at piece-rates, workers who are employed for a crop season, farm servants on monthly wages, and workers who are employed for a year at a time.

The labourer in bondage and the free wage labourer stand at two ends of a continuum of degrees of unfreedom. Even in village situations where there are no bonded labourers, different aspects of the unfreedom that characterizes the relationship between bonded labourers and their employers may be discernible in the relationships between other sections of the labour force and their employers. Even where explicit bondage does not exist, unfreedom and different forms of extra-economic coercion can play a compelling role in production relations. It is also of note here that in rural areas, relations of unfreedom (including the system of bondage) in agriculture are often replicated in non-agricultural production. This is also the case in Gokilapuram.[2]

By definition, a bonded labourer is one who cannot choose freely between alternative employers, and cannot work for any person other than his or her current employer, except by physically escaping from the ambit of the coercive power of the employer. At first sight this may appear to be too inclusive a characterization, since it may appear that no worker, after all, is free to work for another employer in violation of the terms of a contract that is already in force. In fact that is not so. A worker who is not bonded can, in theory, break off a contract and quit any employment (whether it be daily employment, piece-work, or even long-term employment contracts) if he or she is willing to forgo the remuneration or pay a penalty due under the contract. A long-term farm servant who is also in debt to his or her employer is not bonded if he or she

[1] e.g. the large-scale survey reported in Marla (1981).

[2] See Nagaraj (1981) on clay-work, *beedi*-rolling, and shell-picking in Dakshina Kannada district; Chopra (1985) on brick-kiln workers in Muzaffarnagar district; Bharathan, in Guhan and Bharathan (1984) and Vidyasagar (1985) on weaving in North and South Arcot districts.

can leave the current employer and repay the debt while working for another employer. In the case of the bonded labourer, by contrast, such options are non-existent, since the ways and means of enforcing bondage exist and prevent any such free sale and withdrawal of labour power by the worker. There are many ways, which are not mutually exclusive, in which bondage is enforced. Consider some examples. Bondage may be enforced by the overt use of force by the employer. The worker may apprehend that the employer will physically prevent him or her from working for another employer, an apprehension that may arise from direct experience, or from fear of the power of the employer, from a general feeling that these are not the sort of people to be messed around with. Bondage may also be enforced by collusion between powerful employers to prevent a worker who has escaped bondage from being given a job, or by instilling in the worker the fear of such collusion. The effect of these and other methods of enforcing bondage is the annulment of the freedom of the worker in the labour market.

Bonded labour, which continues to exist in India, is formally prohibited by law, by the Bonded Labour System (Abolition) Act of 1976.[3] Although bonded labour takes many forms in India, it is characterized, empirically, by two attributes that are well-nigh invariable. First, the duration of employment of a bonded labourer—the period of time at the end of which he or she is unconditionally free to sell his or her labour power—is either uncertain or long compared to the duration of the employment contracts of other agricultural labourers. This uncertainty can be of different types. The duration of employment may be completely unspecified in terms of calendar time. The duration of employment may also be specified in terms of an event the occurrence of which is uncertain: 'until your debts are repaid', for instance. It is not possible, however, to specify a precise cut-off point that distinguishes bonded labour with regard to the length of the contract—to say, for instance, '*x* years or more represents bonded labour; less than *x* years does not'. At the same time, it is true to say that bonded labourers are not daily labourers, or labourers employed on piece-rates, or workers employed for a single season. The bondage relationship, as it is described in an official report, 'lasts for months and sometimes years, occasionally for an entire lifetime and not infrequently follows the male heir' (GOI 1974, p. 216). The current literature provides some examples of the duration of bondage in different parts of India. The average duration of bondage[4] among currently bonded labourers from Vidyasagar's data, collected in South Arcot district in the late 1970s, was eight and a half years. He has case studies of two bonded agricultural labourers; one had completed 17 years of bondage, and the other 24 years. In Nagaraj's case

[3] It should be noted that there are differences between the definition of bonded labour proposed here and the definition of bonded labour in the Act. Forced labour is formally prohibited by the Constitution of India, Part III, Article 23. For international labour conventions and recommendations of the International Labour Organization related to forced labour, see ILO (1985) and International Labour Conference (1954, 1955*a*, *b*, 1956*a*, *b*, 1957).

[4] i.e. the average number of years that currently bonded labourers had been in bondage.

studies of three bonded labourers of Dakshina Kannada district, one had been employed for 8 years, another for 30 years, and the third for 15 years (Nagaraj 1981, pp. 108 ff.; Vidyasagar 1985, pp. 146, 150–1). Of his study area in Palamau district, Sudipto Mundle writes that 'a man usually gets bonded at an early age—often the day of his marriage . . .—and his years in bondage simply increase as he grows older. Very few cases were noted where a bonded labourer had been able to free himself by paying off his "debt" ' (Mundle 1979, p. 109). A bonded labourer (*kamiya*) of the Chhatisgarh region of Madhya Pradesh had this to say to a commission appointed by the Supreme Court of India:

> I don't know how I became a *kamiya*. I am told by the creditor that I owe him a *murahi* (principal) of Rs 120 for which I have been working for thirty five years now. I have three children—two sons and a daughter. As I was unable to feed them I had to engage both my sons as *kuthiyas*. They are now *kamiyas* with the same creditor. (Kumar 1986, pp. 4–5)[5]

Table 8.1 *Duration of bondage among bonded labourers, Jaunsar-Bawar area, Dehradun district, Uttar Pradesh*

Duration of bondage	No. of bonded labourers
Less than 5 years	6
5 to 10 years	4
More than 10 years	6
'Since birth'	6
Unspecified	1

Source: Indian School of Social Sciences, Calcutta (1976, p. 17; see also p. 28).

Table 8.1 shows information on the duration of bondage from a survey of 23 bonded labourers in the Jaunsar-Bawar area of Dehradun district in Uttar Pradesh. In 1977, G. Selvam, the bonded labourer of Gokilapuram described in Appendix 6, had worked for seven years as a bonded labourer.

The second important empirical attribute of bonded labour is that the conditions of work are more onerous and oppressive than the conditions of work of other categories of workers: bondage generally implies full-time, servile labour. It is, of course, impossible to apply precise quantitative measures to such a thing as 'onerousness'. It is possible, however, to elucidate this onerousness in terms of three criteria. The first is that bonded labourers invariably work at production tasks on and off the farm and *also* perform domestic and other labour services for their employers. In the master–servant relationship between a bonded labourer and his employer, the employer's control over the labourer is not only in the process of production, but also in

[5] A *kuthiya* is a bonded child labourer.

the ordinary round of daily life. The second criterion is that bonded labourers must work for very long hours. For most of them, in fact, no fixed working hours are specified and they are on call at all times. Nagaraj points out that there is little, if any, job specification in the terms of employment of a bonded labourer; consequently, there is hardly any concept of work measure in determining the payments made to them (Nagaraj 1981). Thirdly, the position of the bonded labourer *vis-à-vis* the employer is characterized by servility greater than is characteristic of employer–labourer relations among other sections of the labouring population. There is an extensive descriptive literature on the nature and degree of this servility, which varies greatly in different parts of India. The empirical material shows that in general, a bonded labourer is denied personal freedom and is subject to utter poverty. The labourer is also, in most cases, a victim of usury and caste oppression. In addition, in many cases, verbal denigration of the worker and continual violations of his personal dignity are part of the stock-in-trade of the bonded labourer's employer.[6]

Five further empirical features of the system of bondage in agriculture are also of relevance to this discussion. First, bondage in contemporary Indian agriculture takes the form of bondage to an individual employer or household. We saw in Chapter 1 Saraswathi Menon's analysis of collective bondage in the late eighteenth and early nineteenth centuries in Thanjavur district. The class of agricultural labourers was bonded as a group to the proprietors of land, and the remuneration of the class of labourers was a share of village produce, paid to the class of labourers as a whole. Menon documents the transition, over the nineteenth century, from collective to individual bondage. By the turn of the century, bonded labourers were bonded not to the community of village proprietors, but to individual employers; in most cases, to the individual *mirasdar*.

Secondly, by most accounts there has been a transition from a system in which the family of the labourer is bonded to a system in which the individual labourer is bonded.[7] This is not to say that other members of the household of the bonded labourer are free in relation to the bonded labourer's employer. When an employer has a bonded labourer working for him, it is indeed

[6] On aspects of servility, see Indian School of Social Sciences, Calcutta (1976); Mundle (1979); Gough (1981); Nagaraj (1981); Patnaik and Dingwaney (1985). Bikramaditya Singh, in Mundle (1979, pp. 145 ff.), the case-studies in Vidyasagar (1985, p. 150), and the Indian School of Social Sciences, Calcutta (1976, p. 6) give examples, from Palamau in Bihar, South Arcot in Tamil Nadu, and other areas in the 1970s, of landlords enforcing their control over bonded labourers by means of cruel beating. The Indian School of Social Sciences, Calcutta (1976) study also has important descriptions of how the allocation of duties and roles at village rituals and festivals is used as a means to reinforce positions of inferior social status among bonded labourers. The case-studies were conducted in the Jaunsar-Bawar region of Uttar Pradesh (ch. 2, pp. 29 and *passim*). See also the description in Kumar (1986) of the conditions of life of bonded labourers in the Chhatisgarh region and of the public-interest litigation undertaken on their behalf.

[7] This process has been documented by Gough (1981) for Thanjavur district. For data from Thane district, dealing with a period when the landlord provided for a hereditary farm servant and his wife, see Saldanha (1989). Mies (1986) has case studies of the patterns of labour of the wives of bonded labourers.

generally the case that other members of the labourer's family, particularly his wife, do regular field and off-farm work for the employer and perform labour services for him as well. Nevertheless, the family is not employed or remunerated as a unit. The payment for bonded labour, as such, is made to the individual bonded labourer.

Thirdly, in practice, the great majority of bonded labourers belong either to scheduled castes or tribes or to castes that are ranked low in the caste hierarchy.

Fourthly, the bonded labourer is most often a male.[8]

A fifth feature of contemporary agricultural bondage is that in most contemporary cases the proximate cause of bondage is indebtedness. The initial loan is taken, for example, for consumption purposes, for expenditure on rituals (particularly life-cycle rituals)[9] or, as in the case of Selvam, to build a hut (see Appendix 6). Indebtedness is not necessarily the most frequent cause of all kinds of labour attachment, since there are many reasons why a labourer may not be completely free to sell his or her labour power. The most frequent cause of bondage, however, is indebtedness.

At this point it may be useful to return to an argument that was referred to in the introductory chapter of this book, Sudipto Mundle's argument that the different forms of non-contractual, extra-economic coercion that accompany bondage do not represent a survival into the present of pre-capitalist forms of oppression, and that the *kamiuti* system of bondage represents a break with the forms of forced labour that preceded it. Apart from a detailed account of the appalling poverty of bonded labourers in Palamau district and the factors in its continuance, Mundle's text and case studies also include examples of bondage over generations, often the consequence of a small initial loan; of a landlord who finds an 'absconding' worker in the market of a neighbouring village, and ties him up and beats him unconscious; as well as descriptions of extremely long working hours and the combined exploitation of the bonded labourer and his wife. Bondage of the sort described in Palamau presupposes and is predicated upon the despotic position of the landlord, the landlord's grip on socio-economic and political life, his ability to police the village (and often nearby villages as well), and his ability to control the life and labour of the bonded worker. Mundle's methodology in dealing with changes in agrarian relations that took place in the nineteenth century, while emphasizing the difference between the *kamiuti* system of bondage and the older system of *begari* or forced labour, tends to de-emphasize elements of *continuity*, aspects of oppression carried over and adapted to new forms (Mundle 1979, e.g. pp. 95–6). The question, surely, is not whether the old system was transferred *in toto* to the new period—this could hardly have been the case—but whether new

[8] There are exceptions to this; see Nagaraj (1981) for a case study of a female bonded labourer in Dakshina Kannada district.

[9] On wedding and marriage expenses as the proximate cause of debt bondage, see Indian School of Social Sciences, Calcutta (1976); Mundle (1979); Vidyasagar (1985).

forms of oppression appropriated to themselves, albeit in transmuted form, elements of the old.

To turn now to other, weaker forms of socio-economic dependence of the worker on the employer, the most pervasive kind of dependence is characterized by what may be called the 'right of first call' over a labourer. This approximates very closely to Daniel and Alice Thorner's category, in their pioneering work on the subject, of 'beck-and-call' labour (Thorner and Thorner 1962). It works like this: before going to work in the morning (it may be the evening before), a worker calls in at a particular employer's house to see if there is any work on the field. If there is, he or she works on the field for a regular day's wage; if there isn't, he or she can seek work elsewhere. Alternatively, if there is no work in the fields, the employer may instruct the worker to do another, unwaged job, and the worker does it for a small hand-out or for no payment at all. The worker refers to the employer as 'my *mudalali*' (employer, landlord) and, whatever the season, whatever the other job alternatives open to the worker, 'his' or 'her' *mudalali* has a superior right over his or her labour, a right of first call. The worker is, in other words, ready to forgo wage employment elsewhere to serve an employer who has a right of first call over him or her.

K. Nagaraj provides a useful account of the general features of the relationship between the employer and the worker over whom the employer has a right of first call. These features characterize, in substantial measure, the system as it exists in Gokilapuram as well. First, this relationship is weaker than bonded labour, in that the worker is free after he or she has served the obligations that derive from the employer's superior claim over his or her labour time. Secondly, the burden of fulfilling obligations under this form of labour are entirely on the worker: no general security is provided by the beck-and-call institution. Thirdly, in Nagaraj's data from Dakshina Kannada, all cases of beck-and-call labour were based on one relation or another *outside* the labour market (Nagaraj 1981).

Unfree Labour in Gokilapuram

Wage labour relationships are the rule in production tasks. At the same time, socio-economic life in the village is a complex web of interrelationships; there is an immense variety of bonds that link the agricultural labourer to the landlord. While it is a fairly straightforward task to make the distinction between wage labour and labour service at a conceptual level, as systems, differentiating between freedom and unfreedom in the daily life of an agricultural labourer is altogether more problematic, for the one passes imperceptibly into the other. The segmentation of labour services from the market for hired labour cannot be described as hermetic; on the contrary, in the lives of agricultural labourers, the spheres of operation of the two overlap. Labour services are not always easy to detect, to quantify, to assess, and to describe.

However, rather than avoid the issue, which is an important one, this chapter attempts at least to scratch the surface.

First, a description of the method of inquiry. One question in the schedule read, rather unsubtly, 'Are you free (to work for the employer of your choice)?' Although many households did in fact respond to the direct question, it was clearly not adequate to uncover the relationships that we were looking for. For the rest of the schedule-based interview and in subsequent conversation, we kept returning to this issue: 'Do you work for any employer more often than for others? If so, why?' 'When you go to work, is there any house at which you look in first to see if there is any work? And if there is, and the employer tells you that there is no wage work, but there *are* some other jobs—errands, odd jobs—to do, do you give up the day's work to do them?' 'Do you perform labour services, in the field or at the employer's house?' 'At such-and-such an agricultural operation, why did you work nine hours at what is usually an eight-hour task? Is it because employer B employed you for the job?' '"Oh, if there is work to be done on his field, he will surely call us"—is there any employer of whom you can say this? What do you do in return for this assurance?' 'Why do you work longer hours when you work at employer N's field than when you work at K's?' 'Is there any task that is the traditional preserve of your family or caste, that you do because you were born into your family?' 'Is there any task that you undertake only with men or women of your own caste?' 'Does your family perform any particular role in the village?' 'Employer R gave you an interest-free loan—do you have to do anything in return?' 'You get food doles from employer T's house; what does your family do for him, what is the relationship between your household and his?' 'In the lean months, when there is scarcity in your house and you and your husband have no food and the child must be fed and there is no work to be had, to whom do you turn?'—These were the kinds of questions that we asked, and we wrote down relevant answers. Even these questions cannot possibly cover all types of constraints on the sale of labour power, but this is, for the present, what we have.

Extent and Types of Labour Service

Of the 257 landless agricultural labour households in the survey, 110 households, or *42 per cent* said that they either perform labour services for certain employers or were in some way unfree in relation to them. There was one worker, Selvam, who was a bonded labourer (see Appendix 6).

The most important conclusion to be drawn from our data and interview material is that labour services at wage-paid production tasks in agriculture are diminishing; they are much more marked in domestic and other tasks. Labour services are not, in general, a substitute for wage-paid agricultural or non-agricultural work: the system of labour services plays a supplementary role in the labour time of agricultural labourers, a subordinate role in the overall system of labour hire. Furthermore, labour services are not concentrated in operations where large numbers of workers are employed—weeding, transplanting, harvesting, threshing, picking cotton, harvesting vegetables, forming irrigation basins, other hoeing work, and so on. The forms of labour that have emerged at employment-intensive operations (such as the system of contract labour at transplanting and harvesting) reinforce the tendency towards the creation of formally free wage labour at these tasks. However, in the system of farming that prevails in the village, there are a number of tasks that are vital to the crop, but for which only a single or, on occasion, two or three workers are required. These are desultory, bits-and-pieces jobs, and many of them are done by regular farm servants or through labour service. Agricultural labourers can be called in to do just about any of these tasks for little or no remuneration. In many of these tasks, although job specification is not absent, it is often ambiguous—'look after a field', 'see that the crop is doing well', and so on.

Apart from these tasks, a common form of labour service is where such labour is integrated with a regular working day. A worker works along with others on a field, say at weeding or hoeing. But she gets to the field a little earlier than the others, surveys the field, sees where the weeds are thickest or hoeing work particularly necessary and organizes the day's work accordingly. At the end of the day when the others leave, this worker clears the field bund of the uprooted weeds and throws them away or packs down the earth where hoeing has been superficial, and generally tidies up the field before leaving. For the eight hours that the others have worked, this worker will have worked more; the actual extra time varies.

Another common form of labour service is guarding a standing crop, particularly on groundwater-irrigated land and in groves. Men, and sometimes young boys from their households, guard the fields overnight, spending the night in a rough thatch lean-to erected at the corner of the field or in the motor-pump room near the well; they guard the crop and shoo away birds. They also do odd jobs on the field and report to the employer on the state of the crop in the morning.

It is in domestic and other tasks that the system of labour service comes into full play, and the image of a powerful landlord getting all his errands run and odd jobs done by workers who are at his beck and call is still, in contemporary Gokilapuram, an accurate reflection of reality. What follows is an illustrative list of domestic and other labour services.

Going to the market to shop.

Holding the shopping bags when the employer or members of his family make purchases at Uthamapalayam or at Theni, or even (when the load is bulky and heavy) at Madurai.

Bringing bricks from the kiln at Uthamapalayam.

Hand-pounding paddy and other grain; crushing coarse grain into flour at a millstone.

Soaking paddy for parboiling.

Drying damp paddy (not an ear of grain must be allowed to germinate!) or grain that has been parboiled.

Decorating the home and running errands on a festival day or when there is a ritual or ceremony at the employer's house (the landlord's daughter is to be married, and his prestige must be enhanced before those with whom his family is entering into marriage alliance).

When there are guests at the employer's house, buying tea from the tea shop.

Chopping wood into splinters.

Taking paddy to the rice-mill.

Carrying messages, running small errands.

Taking a pass-book to be updated to a bank in Uthamapalayam.

Tidying haystacks and taking care of them.

Travelling to nearby towns to buy spare parts for motor vehicles.

Driving the landlord's car.

Helping in construction work; transporting stones for masonry.

Buying milk from the milk society.

Guarding a building where tamarind is stored.

Doing all kinds of housework: sweeping the house and courtyard; washing vessels and dirty dishes; clearing cobwebs; cleaning latrines; drawing water (and hewing wood); laying out chilli and foodstuff for deep-frying in the sun to dry; clearing dung; filling the farmyard manure pit; cooking and washing clothes; and doing other household chores.

There are many tasks connected with tending the employer's cattle that are done through labour service: bringing hay to the cattle shed; mixing cattle feed and feeding the animals; milking the animals; washing and watering the cattle; and grazing them.

There is the case of Valliyammal, who does a wide variety of labour services for landlord P ('my landlord'). Valliyammal is the head of a Pallar scheduled caste agricultural labour household; her husband is unable to work and her two growing-up children also do agricultural labour. The landlord does not have a regular cattle-shed for his plough animals, so he has told Valliyammal: 'Your hut is big, keep the cattle there at night.' The hut is somewhat bigger than the ordinary harijan agricultural labourer's hut, since there is an open space—an inchoate courtyard—near the entrance. So Valliyammal goes every evening to the landlord's house, brings back the cattle to her hut, and keeps them there through the night. She brings hay from P's house with her, too. In the morning, she clears the dung (which goes to the landlord's manure heap), puts it in a basket, and, with basket in hand, leads the cattle back to P's house. At all production tasks, however, she is paid a regular day's wage, the same as a free labourer.

Duration

The duration of labour services is difficult to measure. Tasks are not specified by the clock, nor is the number of days spent at these tasks subject to a seasonal or annual schedule. Some respondents were able to say 'about 10/20/40/50/120 days a year', in other cases the response was 'a few hours a day, two or three times a week', or 'for one or two hours every day throughout the year'. The lowest was 'three days last year'. In many cases our questions on the duration of such tasks were met with a laugh and the response 'How can you put a time on such work?' The time taken at labour services is often woven into an agricultural labourer's work schedule, is a part of his or her daily life, and separate measurement of the time spent at labour services, distinct from the time spent at wage labour, becomes difficult. Being of the nature that they are, the aggregate time spent at such tasks are a fraction of the most important component of agricultural labourers' work schedules, namely, the time spent at wage-paid agricultural and non-agricultural work.

Remuneration

An important feature of remuneration under the system of labour service is the absence of standardization of work and commensurate wage norms. This lack of standardization is reflected in, among other things, diversity in forms of payment for work done. There are also tasks whose duration is not specified at all.[10] The neatest summing-up of remuneration for labour service is in the term that is used in the village for labour service itself, o.c. *velai*,[11] or unpaid work. The most frequent form of payment (there are days when there is no payment) is a small cash hand-out—what is called 'coffee expense' in the village. This was about 50 paise in 1977 (when the daily wage rate for hoeing was Rs 6). There is no regular agreement, of course, that this will be paid and often, when nothing is given, the worker hangs around, drops a hint, waits until the employer's house is clear of other people, perhaps runs another errand in the meantime, and hopes that some money will be forthcoming.

There are other forms of payment:

A glass of tea for a day's odd jobs.
1.5 measures of husked grain (for 2 days' work at hand-pounding paddy).

[10] There are interesting points of comparison here with a specific situation described by Marx, in which, in a system of capitalist hire, an employer pays wages only for the hours that he chooses to employ labour: 'The unit measure of the price of labour . . . loses all meaning as the working day ceases to contain a definite number of hours. The connection between the paid and unpaid labour is destroyed. The capitalist can now wring from the labourer a certain quantity of surplus labour without allowing him the labour-time necessary for his own subsistence. He can annihilate all regularity of employment, and according to his own convenience, caprice and the interest of the moment, make the most enormous over-work alternate with relative or absolute cessation of work' (1977, p. 511).

[11] The expression 'o.c.', or 'free' (in the sense of without payment), derives, of course, from 'on company'—a bourgeois phrase for an archaic practice.

About 18 measures of paddy (the equivalent of Rs 22.50), dried chilli, and tamarind, and occasional hand-outs (for a year's labour services, 1 to 2 hours every day).

All the paddy *husk* from two seasons and one sari, with no cash payment (for a year's work, 1 to 2 hours a day).

12 measures of paddy (worth Rs 15) and small hand-outs (for odd jobs every day).

Regular doles of cooked rice.

1.5 bags of paddy a year (worth Rs 90), plus a sari at the Pongal festival (for a year's labour service, 3 hours a day).

Permission to build and live in a hut on the employer's coconut grove.

Mariamma, a harijan woman 45 years old, works a few hours every day in the house of landlord SNR; she works a longer time when there are guests in the house and stays in on festivals and other occasions. For her efforts, she is given a sari at Deepavali, 24 measures of paddy per harvest (the equivalent of Rs 60 in the year), 3 bags of paddy husk per harvest, and she does not pay rent for her hut in the *cheri*, which has been built on SNR's land. Her son Mariappan works from 1 to 4 hours daily in the house of landlord SCC and guards his crop 40 nights a year. He is paid small hand-outs, sometimes Rs 2 when he works for 4 hours or more and about 50 paise per night for guarding work.

And the lowest recompense for labour service is when the worker is told 'Get lost, not now—I have already given you so much.'

It need hardly be further established that payment for labour service is far lower than for regular wage labour, that payment can be irregular and unspecified, and that the amount paid depends on the whim and general disposition of the employer. It still makes a difference to the day's earnings whether the landlord is a generous one—a generosity that must often be induced, cajoled out—or a miserly one.

Types of Obligation

The most important reasons for performing labour services have their basis in the poverty of the agricultural labourer. An agricultural labourer does such tasks to ensure employment at wage work—to buy, as it were, tenure with an employer. 'If we do unpaid work for an employer today, he will employ us on his fields tomorrow'; 'If we work regularly at unpaid jobs at his house, then whenever there is wage work to be had on his fields, he will call us.' Unemployment and problems of acute seasonality in employment also play their role. Labour services can be a form of insurance against lean periods for agricultural labour households; an agricultural labourer looks to an employer for odd-job employment, food doles, or money loans during the slack season. Labour service is their shield, however effective, against complete household disaster.

The fact of poverty is, of course, compounded with other features of the relationship between agricultural labourer and employer, such as traditional 'respect' and other forms of obligation.[12] These are some of them:

The worker household has to work off loans taken from the employer.[13]
Members of the labouring household are former tenants of the employer.
The employer has allotted a house site to the worker.[14]
The head of the worker household is a poor cousin of the employer.
The employer provides the worker with privileged employment in non-agricultural tasks in the cardamom estates and in tamarind-processing.
One worker in the household is a permanent farm servant or a domestic servant at the employer's house; the others in the household do labour service for the employer.
A member of the agricultural labourer family is the village sweeper, employed by the panchayat; he cleans the latrines in some employers' houses in the course of the day's work.
The employer is the full-time secretary of the co-operative society and a man of landed means and the young worker the son of a poor peasant. The boy is studying in the pre-university class at the Arts College in Uthamapalayam and, in return for being allowed to study at night by the light of the electric bulb in the society building, he runs errands and does household chores for the employer.
'Decades ago, when there were attacks on harijans by people of the Kallar caste in Gokilapuram and many surrounding villages, Maravar landlord K's late father gave shelter to our fathers' families. We still respect his family and house and work there, and we voted for K in the panchayat election.'
An irrigation worker often does extra and special tasks, tasks that can be tagged

[12] See Nagaraj (1981) on what he terms the 'bundle of prior obligations' and, on traditional loyalty, Mundle (1979).

[13] The point has been made in the literature that indebtedness *per se* is not evidence of extra-economic coercion or of bondage (Bardhan and Rudra 1978). This is correct, although, as in some of the situations described here, indebtedness can well be the proximate cause for succumbing to different forms of extra-economic coercion and, along with poverty, unemployment, and other factors, may reinforce relations of dependence in the village. Given the socio-economic context of the village, agricultural workers who are indebted to landlord employers are placed very differently *vis-à-vis* their employers in terms of what they are obliged to do in return, the labour services they are obliged to perform, than an 'office worker borrowing from his provident fund' (to take the example used by Bardhan and Rudra). Taking a loan from his provident fund is not often likely to oblige a stenographer or section clerk to clean cow-dung from the employer's courtyard or sun-dry food for the employer's kitchen, or chop wood for the fire, or send her or his little daughter to clean dishes and cooking vessels in the employer's house. Given the system that prevails in the village, taking a loan from a landlord employer often entails that the agricultural worker must do so.

[14] See S. Menon (1983) on the historical antecedents of the allotment of house sites as a source of attachment. Vyas (1964) also has a discussion of this practice. He writes that government housing schemes, which freed attached labourers from obligations that derived from the landlords' provision of house sites, were a blow to the *hali* system. See also Nagaraj (1981) on the allotment of rent-free homestead land to unfree labourers.

on to his regular work, for particular employers, particularly landlords who are on the village irrigation committee (see Chapter 9).

Now these are factors deeply rooted in the system. Among agricultural labourers there is acute recognition of the fact of poverty, of its description, its features, its detail. There is often a boldness and absence of reticence in talking about it, in speaking out aloud. There is a knowing recognition of the landlords, who eat when they do not, an assessment and criticism of the drones and parasites that is often mocking, tongue-in-cheek and uninhibited. Many times it has been our experience that when we probe the seemingly servile, the rebel appears beneath. At the same time, there are many barriers to freedom, many chains that prevent agricultural labourers from acting on the consciousness that is already, in elemental form, theirs. However sagacious the individual agricultural labourer's assessment of his or her own socio-economic predicament, he or she cannot just walk out of the system.

Who Performs Labour Service?

Men, women and children perform labour services, people of all ages and castes. The oldest man in our group was irrigation worker Rayar Karuppana Thevar, an old and distinguished man and, at the time of his death in 1980, perhaps the most experienced in the village on matters of Gokilapuram's surface-irrigated agriculture. He carried shopping bags for the landlord and brought bricks from the kiln and ran the errands that a man can run at the age of 72. The oldest woman was 65 years old. The youngest was a little girl, Sakthi, who scrubbed and washed dishes in Kali Vilas for 90 days in the year, and was all of 9 years old. Boys worked at guarding and grazing from the age of 12 upwards.

The list of castes of persons who perform labour services covers all major castes: Pallar, Parayar, Chakkiliyar, Maravar, Kallar and Telugu Chettiar. There were Achari families and one Thuluva Vellalar family as well.

Does caste still function as a system where an unalterable, pre-ordained division of labour is superimposed on the system of free wage labour? Not in the major agricultural and non-agricultural operations, particularly those that require large numbers of workers at a time, that are employment-intensive. There are, however, some tasks that are still caste-specific. All irrigation workers and guards on surface-irrigated land are Maravar and dry-land crop guards are Kallar. Menial tasks associated with rituals and temple activity have a long history in Tamil Nadu and elsewhere (see, for example, Indian School of Social Sciences, Calcutta 1976; Gough 1981; Nagaraj 1981; and S. Menon 1983); each caste has specific functions at temple festivals in Gokilapuram. Members of Parayar households, as in other parts of Tamil Nadu, beat the drums on ritual occasions. The right to sweep the roads after harvest for grain that has been shed is shared by Parayar and Chakkiliyar households. Goldsmiths, carpenters, and blacksmiths are Achari. Potters and persons who wash clothes are separate castes. Corpses are prepared for cremation by Chakkiliyar families; threshing coarse grain and sweeping threshing-floors are tasks generally done by Chakkiliyar women; and village sweepers, latrine cleaners, and those who handle dung heaps are generally Chakkiliyar. (These forms of work are discussed in greater detail in Chapter 9.)

Who Receives Labour Service?

This analysis makes it clear who the recipients of labour services are: above all, they are the landlords. The extent to which they can command traditional 'respect' and extract labour services is an important component of the overall position of socio-political dominance of the landlords in the village. Only three households other than landlord households appear in our data: one rich peasant, who is also engaged in sharp money-lending practices, and two others, both of whom are also of substantial landed means. One of them owns an eating-place in Cumbum and runs a tamarind business and the other is the household of the secretary of the co-operative society.

The Future for Labour Service

What are the factors that militate against the continued existence of the system of labour services?

Very briefly, the most important are the growth of commodity economy in agriculture, of intensive farming for profit, and of cultivation that requires large-scale employment of agricultural workers at tasks crucial to the crop as well as the need to ensure higher levels of productivity in agriculture. A modern farming system of the type in Cumbum Valley cannot be sustained by forms of employment that depend on a personal relationship between the worker and the employer or on labour service. It requires a large, settled force of wage labourers. The huge increase in the demand for labour, by landlords *and* the peasantry, has been accompanied by the growing importance of forms of labour hire that work towards impersonalizing the relationship between employers and individual workers. Labour cannot be mobilized in the old way for the new agriculture, and, as the social nature of production asserts itself in the agrarian economy, so also do old bonds decay.[15]

An increase in the sources of demand for agricultural labour, that is, in the number of employers, can work to weaken labour services, particularly in cases where a higher demand for labour leads to higher levels of employment per worker. Multiplicity in the sources of demand can also serve, as has been noted, as an impetus for new forms of labour hire.[16]

[15] On the effects of agricultural commercialization on hired labourers in rural areas, in terms of increased job mobility, the erosion of *jajmani* relations, and labour migration, see Dasgupta (1977, p. 16).

[16] *Per contra*, there are village situations where the presence of a despotic dominant landlord works to preserve labour service and attached labour; see, for examples, Bikramaditya Singh, in

Another important factor promoting new forms of labour hire is the spread of modernity—the penetration, however limited, of civil society in the village. For one thing, literacy and education, including college-level education, have grown. There are degree-holders, and a doctor, a lawyer, and a student in a research department of the Madurai Kamaraj University living in the village today. For another, recent years have witnessed the spread of information through books and newspapers, the spread of cinema, radio, and now television. There have also been major developments in modern methods of communication and in the extension of public transport. The bus, in particular, is a great democratizer: no one can choose the caste of the person who sits next to him or her in a bus and no untouchability can be practised against a harijan conductor, employed through reserved quotas, when he hands a ticket or change to a passenger. And while the judiciary, the revenue, and development administrations, and the institutions of parliamentary democracy and local self-government offer no permanent succour to the poor or radical change in their lives, and stamped though these institutions are with the seal of the dominant class and tainted by corruption as well, they too have played an important role in the advance of modernity.

Another factor that works against unfreedom is mass action, or the intervention of mass organizations such as trade unions of workers and peasants. The latent class tension in the atmosphere of the Cumbum Valley is palpable; the actual unionization of peasants and agricultural labourers, however, is poorly developed. There was a powerful tenants' movement in the Valley, particularly in and around Uthamapalayam in the 1940s, and agricultural workers were briefly unionized in wage struggles in the southern villages of the Valley (not in Gokilapuram) in the early 1970s. Many workers from the Valley have migrated to cardamom estates in Kerala, and the sounds of steel from trade union struggles in the plantations of Kerala do filter through to the Valley. The task of unionization in the Valley, however, still remains to be undertaken.

Migration can play an important role in dismantling the system of labour service,[17] and can instigate and catalyse the development of proletarian ideas and qualities in the labour force in three ways. First, it is a means by which to change the conditions in which a worker sells his or her labour power. A time-tested method by which the poor attempt to escape medieval forms of oppression that derive from their position in the village hierarchy is to abandon the village altogether.[18] Migration confers anonymity on the worker; it enables

Mundle (1979); Nagaraj (1981). See also Guhan and Mencher (1983) on Iruvelpattu village in South Arcot district, Tamil Nadu.

[17] On this, see Lenin (1972*a*, pp. 248–51), and the commentary on Lenin's analysis in Trapeznikov (1981). Trapeznikov writes: 'Usury and labour service were not uniformly prevalent everywhere. While the strongest survivals of feudalism dominated in the agricultural area of Central-Black-Earth-Russia, types of purely capitalist husbandry close to farming were widespread in the border lands, which formed an area of internal peasant colonisation. An immense role in advancing this process was played by the migrant movement . . . ' (p. 49).

[18] Here is Max Weber's assessment of the phenomenon: 'Local employment is historically and mentally associated with traditional power relations—it is the urge for personal freedom that drives the worker to employment away from his home. They sacrifice their accustomed conditions

the worker's participation in a labour market in which, unlike in the village, his or her labour power is not stamped 'bonded' or 'untouchable' (see Chapter 9; Appendix 5). There is also what may be called the purely economic advantage to be had from migration, when wages are higher at the migrants' destinations than in their villages. Secondly, the migrant labourer feeds information about the conditions of life and work and about labour organization outside the village back to the village poor. Workers who migrate from Gokilapuram to cardamom estates in Kerala, for instance, bring back news of higher wages, of unions, and of the rights of workers under the law (see Chapter 5; Appendix 5). Thirdly, labour migration can be a major source of monetization of the household economies of the rural poor, through the medium of migrants' remittances. In dry and backward regions of Tamil Nadu, where there is very little cash-paid employment available for agricultural labourers, cash remittances from migrant family members often form a very high proportion of the total cash income of village-based households.[19]

What are the factors that militate against the *decline* of labour services? In the first place, poverty, unemployment, and land hunger—and a level of wages that is unrelated to productivity, need, or wage levels in urban occupations. There is the fact of continued land monopoly, of the socio-political domination of the landlords in village life. Women in the village are subject to conditions of harsh discrimination. There is the continued existence of the caste system. The criminal practices of untouchability in tea-shops and of having separate barbers and washerpeople for harijans persist unabated in the social life of Gokilapuram. Child labour, usury, and debt bondage, and various forms of extra-economic coercion also persist. There is very low literacy among the rural poor and those who *are* literate have achieved very low levels of education.[20] Commenting on this in a conversation in 1980, P. Sundarayya made the incisive—and deeply disturbing—observation that 'even for the next 50 years under this system, we shall not be able to provide every child in the villages of our country with 10 years of continuous education'. With these also come a lack of penetration of modernity, of scientific thinking on a general scale, and an absence of a general democratization of rural life.

Nevertheless, one simple fact remains. The huge demand for labour in Gokilapuram and the Cumbum Valley—and, in general, the pressure exerted on the market for hired labour by the new technology in agriculture—cannot be met by a system of labour service, by a system that 'presupposes the personal dependence of the one hired upon the one who hires him'. Wage labour sustains the most crucial aspect of economic life in the village—production itself—and it is on forms of wage labour that we focus in the next chapter.

in their aspiration for emancipation: their apathy is shattered. The oft lamented "mobilisation" of the rural labourer at the same time sets in motion the beginnings of class struggle' (Weber 1979, pp. 193–4).

[19] See Swaminathan (1985) on migration from Dindigul taluk, Dindigul Quaid-e-Millath district.

[20] On illiteracy and bondage, see Indian School of Social Sciences, Calcutta (1976, pp. 7–8 and elsewhere).

9

Aspects of the Labour Process in Agriculture

This chapter discusses, at some length, wage systems and aspects of the labour process in a series of agricultural operations in the village. The chapter consists, in the main, of a series of case studies of some of the major job options in agriculture that are available to agricultural labourers. In each case, I have discussed the nature of the task, methods of recruitment of workers to the task, the system of payment of wages for the task, and the significance of the task in relation to the overall labour calendar of agricultural labourers. There is also a discussion of employer–employee relationships as manifested in each specific task.

There is a very wide variety of forms of labour, of institutions of labour hire, and of forms of remuneration in agriculture in South Asia.[1] Tasks that are done by men in one region are done by women in others, tasks that are piece-rated in one region are time-rated in others, and tasks that are paid for in cash in one region are paid for in kind or in cash–kind combinations in others. There are also variations in forms of labour hire within contiguous agricultural zones—and, as we shall see, even within a single village—that are associated, *inter alia*, with different types of land, crops, and operations.[2] The local names for specific forms of labour hire can differ in neighbouring villages: the terms for irrigation workers, to take an example, are different in many villages of the Cumbum Valley. Within the same village, the same task can be done on piece-rates in some fields and on time-rates in others.

Institutions of labour hire—the terms and conditions of employment in different tasks—have been sensitive to technological change in agriculture and to specific aspects of that change.[3] At the same time, while adapting to new conditions of production, new forms of labour hire also retain and sometimes draw on traditional village-level institutions.[4] Changes in the terms and

[1] See the discussions in Darling (1954); Clay (1976); Bardhan and Rudra (1981); Nagaraj (1981); Rudra and Bardhan (1983); K. Bardhan (1983); Binswanger and Rosenzweig (1984); S. V. Rao (1984); Rodgers and Rodgers (1984); C. R. Reddy (1985); Drèze *et al.* (1986); Rudolph and Rudolph (1987, ch. 13).

[2] In his introduction to Max Weber's study of East Elbian rural labourers at the end of the 19th cent., Keith Tribe writes that 'no matter how small an area was selected, wide differences in payment and conditions of employment could be found' (1979, p. 174).

[3] For a discussion of this point, see Clay (1976); K. Bardhan (1983); Binswanger and Rosenzweig (1984).

[4] See Hayami (1981) on what he calls the 'congruence' of new systems with old ones, and on how the 'traditionally sanctioned principle in a community makes a certain arrangement less costly to establish than others' (p. 18).

conditions of hired labourers are of crucial importance to their real earnings[5] as well as to their freedom.

Even at a given level of mechanization and technical development, the labour process over a single production cycle in agriculture is far from uniform: the biological requirements of a crop, the scale and type of agriculture, established farming practices, and the impact of seasonal variations impose complex and varying rhythms on the labour process in individual crops. These different factors give rise to different forms of labour hire.[6]

Some tasks are labour-intensive and last only a short period: the crop allows only a few days in which a task that requires a large number of persons to complete can be done. Examples of such tasks are transplanting paddy, and weeding and harvesting different crops. Some tasks, such as cotton-picking, require intensive employment over a rather longer period. Conversely, there are tasks that are spread out over a whole crop cycle: these are crucial to the crop, but are not uniformly intensive, and periods of intense work alternate with periods of relatively light labour. Examples of this are water management in rice cultivation (and to an extent in groundwater-irrigated crops) and general upkeep tasks on banana fields.

Forms of labour hire can vary sharply with the form of ownership of the resource to which labour power is being applied. An example comes from tasks involving irrigation. (In the case of irrigation, the forms of labour vary with the form of ownership as well as with the nature of irrigation from particular sources.) Irrigation tasks connected with paddy cultivation involve the use of a public resource, Periyar irrigation water, by individual cultivators. These tasks are done by a group of irrigation workers who are responsible for irrigation over wide, contiguous segments of the command area (*ayacut*), and these segments cover the plots of many individual cultivators. By contrast, irrigation tasks on land irrigated by wells, where each well is privately owned and irrigation water is privately utilized, are done on specific fields by workers hired by specific cultivators, or by members of peasant-cultivator households. (There is a detailed discussion of irrigation labour below.)

Different operations require different forms of on-field supervision; these give rise to different types of time- and piece-rates. Some operations are supervised directly by employers or their representatives; others, often operations that require a substantial effort in mobilizing labour and organizing on-field divisions of labour, are handed over to groups of workers under a head labourer, jobber, or labour contractor.

Every operation calls for a certain skill. In some operations, such as transplanting, harvesting, threshing, weeding, and forming irrigation basins

[5] On changes in labour contracts and changes in real incomes, see the review of literature in Binswanger and Rosenzweig (1984, pp. 34–5), and the following sources cited by them: S. Bhalla (1976); Clay (1976); Kikuchi *et al.* (1979); Rudra and Bardhan (1981).

[6] On the interaction between specific crop systems and agrarian regimes and the characteristics of hired labour, see Nagaraj's description by zone of the coastal, ghat, and interior regions of Dakshina Kannada district (Nagaraj 1981) and also C. R. Reddy (1985).

for different crops, these are learned by members of agricultural labour families when they are young. Certain tasks, however, require special skills—the tasks of the irrigation worker on surface-irrigated land, for instance, require a detailed knowledge of the system of canals and field channels and of farming practices and stages of crop growth in paddy. There are other tasks that require special skills: there is a group of workers, for instance, who work only in grape cultivation in vineyards in different parts of the Valley. There are skilled workers whose jobs are associated with the spread of mechanization, such as men who operate power sprayers, tractor drivers, and electricians who work at maintaining and repairing electric motor pumps. A feature of manual tasks in agriculture, however, is that skills become obsolete very slowly; skill obsolescence and consequent redundancy are not problems of agrarian labour markets of the type discussed here, characterized as they are by a very low organic composition of capital.

A major occupational hazard of different sections of workers in the village is exposure to pesticides. There are elaborate international guidelines for protecting agricultural workers from such exposure (see, for example, ILO 1979, ch. 3); but in Gokilapuram, even the most elementary safeguards against hazardous exposure to pesticides are non-existent. The most exposed are those who work with diesel-powered pesticide sprayers and who work mainly or exclusively at spraying pesticides. There are also workers in vineyards who have to spray the plants in the enclosed bowers regularly. There are workers who help others mix pesticide solutions (farm servants regularly help the sprayer operators in this) and workers who are hired to dust pesticide powders on the standing crops. Pesticides are mixed in buckets near the field. Workers immerse their bare arms in the buckets and stir mixtures with their hands. Sprayer operators do not wear any protective clothing (overalls, gloves, boots, apron, goggles, mask, helmet), but wear the same clothes as other workers—they are bare-chested and wear lungis tied up short.[7] When the outer boundaries of a field are being sprayed, employers often insist that the sprayer operators spray pesticides *against* the wind, to ensure that pesticide is not carried away into another's field. Workers who handle pesticides are directly exposed to the chemicals by means of inhalation and dermal absorption; the fact that they do not wear gloves or wash and shampoo as systematically as required also exposes them to absorption through ingestion. Workers who do not handle pesticides directly are also potentially exposed to pesticide residues on foliage.

Farming practices in the village have developed in such a way that the employment of hired labour in most, although not all, operations is gender-specific; thus, for example, only male hired labourers work with hoes and

[7] Problems of a similar kind have occurred in recent years in Sri Lanka. Responding to a television advertisement of the Happy, Progressive (but bare-bodied) Farmer spraying chemical pesticides, the Sri Lanka Association for the Advancement of Science moved successfully to have the advertisement taken off television.

ploughs and only female hired labourers weed and transplant fields. Even in an operation such as threshing paddy, where both men and women are employed, there are subtle divisions of labour between men and women on the threshing-floor.

It is clear that operations that are considered 'men's tasks' in some areas are done by women in others and vice versa (Boserup 1976; B. Harriss 1987, pp. 99 ff.).[8] There is also evidence, from other parts of India and from South-East Asia, of changes in the gender division of labour having accompanied the introduction of new technology in agriculture.[9] In Gokilapuram and in neighbouring villages one change that has taken place in recent years is that women are beginning to work as labour contractors for groups of contract workers. These groups consist of women workers at the time of transplanting paddy and of men and women at harvest time.[10] The roles of these women head labourers or contractors in the contract groups are mainly those of organizers and supervisors (see the discussion below on the tasks of the labour contractors). With respect to wage labour in the village, however, a striking aspect of the current situation is the *resistance* of the existing gender division of hired labour to erosion or change despite recent technical change in agriculture. The relative shares of male and female labour time in total labour absorption have fluctuated in recent years (see Chapter 10), but the pattern of allocation of specific tasks between men and women has remained basically unchanged.

There are very few tasks in agricultural production that are caste specific: these are water management on surface-irrigated land (which, although a task crucial to the crop, employs only 18 to 20 men), guarding surface-irrigated land and dry land, and certain other menial tasks that are generally peripheral to actual crop production.

The foregoing are among the factors that determine specific forms of employment and methods of recruitment of workers for different operations or groups of operations. These different forms of employment are characterized by different wage forms (various types of time-rates and piece-rates), by different forms of payment (cash wages, wages in kind, and wages in cash and kind), by various types of wage bargains, and by differences in the periodicity of employment contracts. Examples from Gokilapuram of the last are daily employment, long-term employment with monthly wages as paid to farm servants and tractor drivers, and yearly employment with biannual payments, at harvest time, for irrigation labour on surface-irrigated paddy land. The different forms of labour hire that prevail are also characterized by different relationships between employers and hired labourers and by different degrees of impersonalization of that relationship.

[8] There are, for example, differences between the gender division of labour in Gokilapuram and the village in Tiruchi described in Komuguchi (1984, table V-2, p. 81).

[9] On historical changes in the gender division of labour in a South Indian district, see M. A. Reddy (1983) on Nellore; see also Fredericks (1977, pp. 183 ff.).

[10] See S. V. Rao (1984, p. 767) on female heads of groups of contract workers in Karnataka.

It is not being suggested here that a given level of technical development and a given set of crop requirements dictate the emergence of certain specific forms of labour hire in a deterministic way—variations in forms of labour hire from region to region, and indeed from village to village, are quite enough to dispose of such a surmise. However, although forms of labour hire are subject to continual change, particularly in response to changes in farming practices and in agricultural technology in general, certain wage forms and systems of employment have tended to crystallize in each operation in specific regions and villages.

Forms of Wages

An assortment of wage forms prevails in major tasks, combinations of different types of time-rates (daily rates and rates paid for longer terms of employment) and piece-rates (rates measured by unit of area over which work is done and rates that are proportions of physical output) on the one hand, and different forms of payment (cash, kind, cash and kind) on the other. (An illustration of these combinations is given in Table 9.1.)

Time-rated wages are the payment received by a worker for work done over a specified period of time. The most important and generalized manifestation of time-rates in village agriculture is daily-rated casual labour: a job is specified

Table 9.1 *Forms of wages, illustrated by certain manual tasks in Gokilapuram*

Type of rate	Type of payment		
	Cash	Kind	Cash + kind
Time-rates			
Daily rates	Hoeing (men); weeding (women); basal manuring (men, women)	Threshing (men, women)	Ploughing[a] (men)
Long-term rates		Irrigation-work on *nanjai*; guarding crops on *nanjai* and *punjai* (men, boys)	Farm servants (men)
Piece-rates			
By acre (or other unit of land measurement)	Forming irrigation basins; hoeing on *thottam* (men); transplanting first crop (women)	Transplanting second crop (women); harvest (men, women)	
By measure of physical produce (e.g. amount harvested)	Tamarind-processing (women); tamarind harvest (men, women)		Groundnut harvest (men, women)

[a] Payment in kind for ploughing is in the form of cooked food.

for a day, the duration of the working day is specified, and a wage is paid for that day's work. The duration of the working day that is most prevalent is, as I have discussed in Chapter 6, eight hours. In most daily-rated tasks, the eight-hour working day is becoming the norm in Gokilapuram. There are, of course, exceptions to this general practice; among the important operations, a single working day for threshing paddy, which is a time-rated operation, far exceeds eight hours.

In agriculture, the timing of operations is determined by the stages of growth of the plant itself, and the duration of employment in particular tasks is circumscribed by these stages. Thus in many tasks (such as weeding), enough workers are employed to complete a task in a day or a little more than a day. In other tasks (such as cotton-picking, or harvesting vegetables, or picking jasmine) work extends over many days.

A feature of time-rated wages is the general practice that the same wage is paid for the same kind of work[11]—with the important qualification that men and women are paid unequally. When men are employed to do hoeing on bunds, each is paid the same wage; women who pick cotton are each paid equally; when men and women work in basal manuring, each man receives an equal wage and each woman an equal wage.

There are other types of time-rates, rates that involve a longer time period and generally a job specification that is less narrow than in the case of the daily-rated task. Examples of workers on long-term contracts who are paid at time-rates are farm servants, who are employed at a monthly wage, and irrigation workers, who are employed for a year at a time and receive their remuneration with every paddy harvest.

In Gokilapuram, farm servants are generally employed by households that are involved in more than a single sphere of agricultural or other economic activity, that is, by households that manage different types of land or households that combine field crop cultivation with other forms of business activity. As a matter of fact, all but two of all farm servants' employers in Gokilapuram were landlords; the two non-landlord households were a rich peasant household and a household whose main income was from business activity other than field crop cultivation.[12]

Farm servants have a wide range of duties. First and foremost, they are multi-purpose manual labourers. They do agricultural and non-agricultural labour: they work in production, agricultural produce processing, transport

[11] See Drèze *et al.* (1986) for a review of the empirical literature on this. A rare example in the literature of evidence of wage discrimination in on-field production tasks in the contemporary period is in Nagaraj (1981). In one taluk in Nagaraj's study, beck-and-call labour received a lower wage than other categories of labourers, and the margin of discrimination was 10% to 15%. Nagaraj also provides some documentation on patterns of wage discrimination by crop and by operation in the region: in areca cultivation, there was wage discrimination in all operations and, in paddy cultivation, discrimination took place in minor crop operations.

[12] On the employment of farm servants by farmers with relatively large holdings, see Mundle (1979, p. 111); Basant (1985, p. 390 and n. 4); Breman (1985, p. 263); Patnaik (1987*a*, p. 171).

and livestock maintenance, and they do various domestic chores for their employers. Farm servants also mobilize workers when hired labour is needed on their employers' fields; they often oversee and supervise hired labourers for their employers. They are responsible for monitoring the stages of crop cultivation and the health of standing crops. They help to take decisions regarding the timing of specific crop operations and of other tasks. The labour of farm servants is arduous and they are required to work longer hours than daily-rated casual labourers.

In general, the majority of labourers who are employed on long-term contracts are males. There are, however, regions where women are employed as farm servants.[13] Among farm servants, there is a high proportion of workers from scheduled castes and tribes and from other traditionally oppressed caste groups (see also Basant 1985).

Most studies of the remuneration of workers on long-term contracts show that while a farm servant may have higher *annual* earnings than the average casual labourer, the remuneration per working day of a farm servant is lower than the regular daily wage earned by a casual worker (Mencher 1978; Mundle 1979, p. 114, table 5.33; Basant 1985; Drèze *et al.* 1986; Kumar 1986). There are exceptions to this: ICRISAT studies describe regions in central India where what Ryan and Ghodake (1984) call regular farm servants' 'wage equivalents' are *higher* than the daily wage that is prevalent locally as well as regions where these wage equivalents are much lower (Binswanger *et al.* 1984, pp. 153, 156, 161).[14] Using Farm Management Survey data from the late 1960s as well as Agricultural Labour Enquiry (1956–57) data, Rakesh Basant also attempts to show that the average daily earnings of a farm servant are generally lower than for daily-rated tasks. More detailed analysis by him of Farm Management Survey data for Thanjavur district indicates that this conclusion holds for most crop operations. There were some daily-rated operations for which wages were lower than the daily earnings of farm servants: these were either operations done by women or operations that made up only a small proportion of the total casual employment that a worker normally obtained (Basant 1985, pp. 395–6). From his data gathered from a village in Tiruchi district, Nakamura concludes that a daily labourer would have to gain wage employment for 320 days a year to earn the same annual income as a farm servant (Nakamura 1976). Some empirical observations in the literature that have emerged from studies of specific regions are of particular interest here. ICRISAT studies suggest that employers collude to fix regular farm servants' wages; the Bardhan and Rudra studies of agrarian relations in West Bengal indicate a positive correlation between the wage levels of long-term labourers and the length of their service with a particular

[13] See M. A. Reddy (1983) on women farm servants (the local term is *maleta*) in Nellore district. The institution is described by him as 'unique' to the district.

[14] Such studies, of course, make specific assumptions about the hours of work of regular farm servants in order to arrive at wage equivalents.

employer; and Nakamura's study in Tiruchi district indicates that the remuneration of the long-term labourer varies with the worker's marital status (Nakamura 1976; Rudra and Bardhan 1983, p. 17; Binswanger *et al.* 1984). In Gokilapuram, a farm servant's remuneration is a monthly sum, fixed in cash and paid in cash and kind. Most farm servants received occasional doles of cooked food, although they did not receive these regularly or as a matter of right. Farm servants also received a *dhoti* and towel annually. The daily wage equivalent of the remuneration of farm servants and irrigation workers on year-long contracts in Gokilapuram was below the wage rate for daily-rated, cash-paid tasks for men.

In terms of the freedom to sell their labour power, farm servants range from bonded labourers to workers who are relatively free to change their employers. As has been discussed in Chapter 8, farm servants are not necessarily workers who are completely unfree; the relationships between employers and farm servants are not necessarily complex, traditional relationships.[15] Even in areas where the current word for farm servant derives from more archaic systems of long-term attached labour, contemporary systems of employment of farm servants can be qualitatively different from earlier systems.[16] There are farm servants who have not taken loans from their employers that have to be repaid with interest and who are, in this and other respects, relatively free. Even such workers, however, are likely, at any point in time, to have taken cash or grain advances on their wages, and are not free to leave their employers until these are repaid. Other members of a farm servant's family are likely to work for wages for the same employer; in Gokilapuram, at least one other member of every farm servant's family in the village also performed labour services for the employer.

Piece-rates are of varying types: they vary in the method by which they are calculated (x rupees for every unit of area transplanted; y rupees plus z measures of groundnut for every bag of groundnut harvested; and so on); they also vary in that some piece-rates are shared by (large or small) groups of workers, some are earned by households, and some by individual workers.

In most piece-rated work, the duration of work is implicit in the job specification; in this sense piece-rates are a transmuted form of time-rates.[17] Thus, while groups of transplanting workers are paid according to the extent of land they transplant, it is implicit in the contract that, weather conditions being favourable, a particular field will be transplanted in one, or at the most two, days. Just as the intensity of work is implicit in the wage form itself, so is the quality of the work. In general, the tasks where piece-rates prevail are tasks that require intensive work in a (short) specified period of time and for which

[15] As pointed out, for instance, in Binswanger *et al.* (1984).

[16] The descriptive literature on this is extensive: see e.g. Breman (1974, 1985); Djurfeldt and Lindberg (1976); Gough (1981); Binswanger *et al.* (1984); S. V. Rao (1984); Hatti and Raagaard (1985); C. R. Reddy (1985); Saldanha (1989).

[17] The discussion here draws on Marx (1977, pt. 6, chs. 19–21), particularly chs. 20–1, where he deals with time-rate and piece-rate wages.

the *quality* of work can be determined by the eye. Thus harvesting and transplanting are tasks that are piece-rated, where quality can be determined by looking at a field (are the seedlings planted upright and neatly? is every plant cut and bundled into sheaves?). The same applies to another piece-rated operation, tamarind-processing, where the employer can quickly judge whether each pod has been shelled, decorticated, and defibred.

In his discussion of wages paid at piece-rates, Marx points out that 'since the quality and intensity of work are here controlled by the form of wage itself, superintendence of labour in great part becomes superfluous' (1977, p. 518). The task of superintendence is handed down—to a middleman, as in the example of the sweating system in England given by Marx, or, for example, in sectors of the *beedi* and quarrying industry in Tamil Nadu. The task can also be handed down to a person who has emerged from the ranks of the workers themselves, a head labourer who is also labour contractor and jobber. In the *avatara* in which the jobber appears in certain operations in the agriculture of Gokilapuram, he or she is called the *kothukkarar*.

Under a piece-work regime, it becomes the personal interest of the worker to work as intensively as possible and to lengthen the working day, since this means an increase in personal remuneration. This also applies to group remuneration where there is competition between groups and there is more than one field to be worked on. (In agriculture, however, indefinite intensification is circumscribed by the periodicity of specific crop operations.)[18] While people working in similar tasks for time wages are paid equally, it is in the very nature of piece wages that individuals and groups earn different amounts. With regard to the actual earnings of workers, 'there is, therefore, great variety according to the different skill, strength, energy, staying power, etc.' (Marx 1977, p. 520) of individuals and groups of workers. Workers' earnings also vary crucially with the amount of work that the labour contractor, head labourer, or individual is actually able to obtain in a season.

Working for piece-rates does not, of course, make the hired labourer any less a hired labourer (or the proletarian any less a proletarian). Piece wages in the abstract are a specific form of the *general* wage labour relationship, although they emphasize the aptitudes and capacity for work of the *individual* labourer. Marx writes that this form 'does not alter the general relations between capital and wage labour' and goes on to say that 'the wider scope that piece-wages give to individuality, tends to develop on the one hand that individuality, and with it the sense of liberty, independence and self-control of the labourers, on the other their competition one with another' (1977, p. 520).[19]

Wages are paid in cash, in kind, or partly in cash and partly in kind. A detailed and useful discussion of monetized and non-monetized agricultural wages and of their relative prevalence appears in the work of the Soviet

[18] There is a discussion of the limits imposed by biological processes in agriculture to extending the working day and increasing absolute surplus value in Georgescu-Roegen (1976).

[19] See, in this regard, the discussion below of piece-rates in groundnut harvesting.

economist V. G. Rastyannikov (1981, pp. 63 ff.). The major features of his discussion are of relevance here. Wages in cash are likely to prevail in areas where commercial, non-food crop production is predominant and in areas that 'are under the cash nexus influence of urban areas, industrial establishments and plantations'. Rastyannikov also makes the point that 'the specialisation of some regions in the production of non-grain food crops and industrial crops . . . reduces the range of payments in kind' (1981, p. 63).[20] The transition from forms of natural economy based, for instance, on cereal production to agricultural systems where developed commodity–money relations are prevalent implies a transition from *traditional* forms of payment for hired labour to *wage* payments for hired labour.[21] This transition is not, however, necessarily accompanied by a transition from wages in kind to wages that are fully monetized. At the same time, where there is a transition towards developed commodity–money relations and towards wage payments, the share of monetized wages in the total income of a hired labourer is likely to become bigger. The literature documents wide variations in the extent of monetization of wages between regions and across villages within regions.[22]

Wages in kind take different forms. They are generally paid in the form of a commodity that can directly enter the worker's consumption fund; thus, they are more likely to be paid in the form of paddy or coarse grain or groundnut than in the form, say, of cotton.[23] Although wages are generally paid in the form of a commodity that can be directly consumed, a worker is likely, in practice, to sell a portion of the commodity on the market in order to gain cash.

Wages in kind can be paid in the commodity that is being produced: examples are paddy or millet as payment for harvesting or threshing paddy or millet, and part payment in groundnut for harvesting it. Another common form of payment in kind is payment in cooked food. In Gokilapuram, as in other places, no operation is paid for solely in cooked food; where there is a cooked-food component in wages, it invariably supplements wage payments in cash. The practice of paying wages in kind in the form of cooked food as a supplement to cash wages still continues, in most cases, for ploughing; it is no longer common practice to provide cooked food with cash payments for work in other operations.[24]

[20] He also cites the study of farms in Punjab by Ashok Rudra that suggests that the proportion of wages paid in cash varied directly with the size of farm (Rastyannikov 1981, pp. 66–7).

[21] On this point, see Rastyannikov 1981, p. 69.

[22] See e.g. Dasgupta (1977, pp. 13–14); Rastyannikov (1981, pp. 63 ff.); Rudra and Bardhan (1981, pp. 28 ff.); Binswanger *et al.* (1984, pp. 153 ff.); Rodgers and Rodgers (1984, pp. A-21 ff.). Parthasarathy (1977) cites evidence from the 1970s on the virtual disappearance of wages in kind in parts of West Godavari district in Andhra Pradesh, where even the harvesting and threshing operations were paid in cash.

[23] Although I have been shown village-level data from a dry village near Dindigul in neighbouring Quaid-e-Millath district where low yields of cotton were raised and where, on occasion, cultivators paid workers partly in cotton, much to the workers' dissatisfaction.

[24] On cooked-food components in wages, see Djurfeldt and Lindberg (1976, p. 129); Breman (1978, 1985); Rastyannikov (1981, p. 67); Rudra and Bardhan (1981, p. 28); J. Harriss (1982);

Within a village, the crops and operations for which wages in cash and wages in kind are paid are likely to be distinct. In Gokilapuram, wages in kind are paid for certain operations involved in cereal production (discussed later in the chapter) and, to a certain extent, in groundnut production. Workers in Gokilapuram, as elsewhere, hold on to wages in kind, particularly wages in kind for harvesting and threshing paddy and for transplanting paddy in the first paddy season. As we have seen, real wage rates for cash-paid, daily-rated operations remained stagnant or declined between the late 1940s and the late 1970s. Over the same period, the total wage earnings of an individual worker at harvest or transplanting time are also likely to have declined. In these circumstances, wages in kind are perceived by workers to provide some protection, albeit inadequate, against inflation. This perception has some justification in the case of wages that are time-related and paid in kind, as are wages for threshing and for irrigation work. Wages for these operations have risen over the years. In the case of harvesting and transplanting, however, the amount of grain paid as wages per acre has not changed and the amount of grain earned per worker has declined. In the case of such operations, wages in kind have not in fact been any protection against an erosion of real wages.

There are other non-monetary components of the aggregate incomes of hired labourers, components in kind that do not derive from wages in kind. These can be of different types. There are traditional payments that are added to wages by individual employers, such as the additions to harvest wages that are discussed below. There are rights that are granted to agricultural labourers by the community of proprietors; an example is the right to fish in the canals that was granted to irrigation workers in the past. There are other types of rights to common property resources, such as the right of potters to take clay from near the river bed. In Gokilapuram, there are also rights allocated to specific families to sweep the roads for fallen paddy at harvest time. Farm servants, as we have seen, receive towels and *dhotis* at festivals, particularly at Deepavali and Pongal. For farm servants, such articles are generally a part of regular remuneration. There are other workers too, who have special relations with their employers, and receive similar articles as gifts at festivals. In Gokilapuram, such non-wage supplements in kind to the total earnings of hired labourers are certainly on the decline.

Methods of Recruitment

Given the high levels of unemployment in village India, it is reasonable to expect that in the labour market, it is workers who seek employment and not employers who have to search for workers to employ. A recent contribution to

Rodgers and Rodgers (1984, pp. A-21 ff.). Rahman and Das make the interesting observation that in their study area, employers break the working day into two shifts in order to avoid including a cooked-food component in the wage (1982, pp. 14–15).

the literature on labour markets in Indian agriculture, Drèze, Leruth and Mukherjee (1986), which provides a useful review of the literature on job search in Indian villages, asserts a contrary view: 'in India the search for casual labour is performed by employers rather than employees' (pp. 28–9). Again, the paper states the following: 'Search in the village labour market is invariably by employers. We are yet to find a study with evidence to the contrary in India' (p. 28). This is *not* true of Gokilapuram. When a worker expects that there will be work on an employer's field on a particular day, he or she goes to the employer's house the day before (or early in the morning on the day of work) to ask for employment. Agricultural labourers over whom an employer has a right of first call as well as workers who are relatively unattached follow this practice; the former are generally likely to get work if there is work to be had, while the latter may have to go to more houses than one before securing a job for the next day. A very important form of job search is when a head labourer or labour contractor (*kothukkarar*) seeks employment on behalf of a group of workers from different cultivators. This form of job search involves crop operations that account for a high proportion of all hired labour, namely, transplanting and harvesting paddy and harvesting groundnut. In addition to the foregoing, there are agricultural labourers who, when they do not have employment on a particular day, go to a field where they know work is about to begin and ask the employer or his agent if they can fit one more person into the day's roster.

In Gokilapuram, employers who draw regularly on large numbers of wage labourers have a person to mobilize workers for them whenever they are needed. This can be a farm servant (*pannaiyal*), or a person who manages cultivation for the employer without actually doing manual labour himself, or an agricultural labourer who is in some way attached to the employer. These people have a regular roll of agricultural labourers on whom they call; many agricultural labourers also seek such persons out, in order to get employment when it is available. Some smaller employers, such as persons from among the middle and hard-labouring sections of the peasantry, go themselves to the houses of agricultural labourers when they need them and recruit them for daily labour.

The fact is that when large numbers of workers have to be assembled for work, there is a need to mobilize workers efficiently even when there is a surplus labour force. Employers also seek to employ the best workers and make efforts to ensure this. For these reasons, employers who employ large numbers of workers may have a person to mobilize workers for them; it would be misleading, however, to interpret this as evidence that the task of 'search' on the labour market is one undertaken by employers rather than by hired labourers. In Gokilapuram, forms of labour hire have developed in such a way as to provide an abundant supply of labour throughout the year, including peak agricultural seasons. The village labour market is also characterized by very high levels of unemployment. In such circumstances, the burden of finding a job falls squarely on the worker. The employer does not have to search for a labour force as such; his task involves employing the best possible workers, employing them on time, and deploying their labour efficiently.

There are, of course, regions where the labour market tightens during peak seasons. An example is in Gudalur, in the south of the Cumbum Valley. Gudalur has a more crowded agricultural calendar than Gokilapuram: there are two crops of paddy a year on surface-irrigated land, two crops a year on dry land, and groundwater-irrigated land is in use throughout the year. Plantation work also draws off a substantial number of workers. As a consequence, the market for hired labour is active and marked by diversity. Another feature of cultivation in Gudalur is that the cultivated fields (particularly dry land and groundwater-irrigated land) are far-flung: some are 6 kilometres and more from the village settlement. In the late 1970s and early 1980s big employers in Gudalur generally employed a person (often a woman), paid separately for the job, to mobilize day labourers. Fluctuations in wage levels were more pronounced in Gudalur than in Gokilapuram, fluctuating within certain village-level limits. During seasons of peak demand in 1976–7, workers could bargain for higher wages for work in distant fields, to which workers are generally transported by cart or tractor trailers, but sometimes travel on foot. Against a usual rate of Rs 6 per day per man in 1976–7, there were days (generally days on which employment was available concurrently on surface-irrigated land, dry land, and groundwater-irrigated land) on which men could earn Rs 7.50. It was not the practice in Gudalur for employers to offer workers employment in the lean season in return for workers assuring employers they would make their labour power available in the peak season. In recent years, there has been a shift in cropping pattern in Gudalur. More coconut, which is a crop that cuts back sharply on current levels of employment, is being planted on lift-irrigated land, and employers look forward to a consequent shift in their favour in the labour market.

Wage Systems in Gokilapuram

Cash-Paid, Daily-Rated Casual Labour

This is a widely prevalent form of wage labour, covering hoeing work for men and weeding for women in the cultivation of every crop, and a number of other operations as well. These are some of the major operations paid at a daily rate in cash.

In rice cultivation, the major cash-paid, daily-rated operations are basal manuring, done by men and women (men transport and dump the manure, women scatter it on the fields); the construction and maintenance of field bunds, which is hoeing work done by men; and weeding, which is done by women.

On unirrigated land, the major tasks are basal manuring; the construction and maintenance of field bunds; and weeding.

Where wells are the source of irrigation, the system of cash-paid daily labour comes into full play. In all non-cereal crops (and on many fields, in cereal crop cultivation as well), almost every major operation (exceptions are noted elsewhere) is performed by daily-rated, cash-paid casual labourers. Some operations are common to all groundwater-irrigated crops: basal manuring, the construction and maintenance of bunds, and weeding. The other important operation done on all groundwater-irrigated fields is a hoeing job done by men throughout the period of plant growth, the task of forming irrigation basins for well-water to flow through to the plant. This is a tough task, and requires experience with particular crops; since the spacing between plants is patterned differently for different crops, each crop requires a particular type of irrigation basin to be dug. Examples of other daily-rated, cash-paid operations in the major crops on groundwater-irrigated land are:

- on banana fields: digging pits and planting trees (men); heaping and putting manure at the base of the tree (women); cutting dead leaves and side shoots (men); clearing the field of the trees after harvest (men).
- on cotton and vegetable fields: transplanting where this is required (women); harvesting (the task is done by women; men help, in smaller numbers, in packing crates or sacks with the produce); clearing the fields of plants (men).

As mentioned earlier, weeding in every crop is done by daily-rated, cash-paid women. Few things create as much fear for their future among agricultural labourers as the possibility of weedicides pushing women out of employment in this task. It is interesting to compare Gokilapuram in this respect with Gudalur, a village that has an even busier market for hired labour. Gudalur is the only village where I have seen paddy weeding being given out on piece-work. Weeding is a task where the quality of the day's work is important and which often involves close supervision, and employers do not generally want to employ workers under a contractor for the task. Unlike transplanting or harvesting, the overall quality of a day's work cannot be judged by just looking at a field after it is weeded. But the weeding season in paddy coincides with operations elsewhere on dry land and groundwater-irrigated land and piece wages are offered to women to attract them to the job.[25]

On tamarind trees, the two major operations are both at harvest time and are both daily-rated and cash-paid. Men climb the trees and beat the fruit down with sticks and women gather the fruit in baskets and carry it away to be transported. Other tree crop operations (harvesting mango and jackfruit) are

[25] 'It is easy to tell whether a field is being weeded on daily wages or on contract,' one landlord said. 'When it is on daily wages, only one arm has cloth tied on it [women wear cloth bandages to prevent their arms from being cut by the edges of paddy leaves]; when it is on contract, both arms have bandages, because they weed with both hands to earn higher wages.'

relatively minor in Gokilapuram; they are done by male day labourers. For all these operations, the duration of work is eight hours.

The wage rates for all these operations was Rs 6 for men and Rs 2.50 for women in 1976–7. (By 1981–2, they had risen to Rs 7.50 to Rs 8 for men and Rs 3.50 to Rs 4 for women.) The monetization of wages for these tasks is complete: wages are paid entirely in cash, with neither a grain component nor a cooked-food component.[26]

Payment for such tasks was generally made on the same day or, if the work took more than a day, on completion of the task; prolonged non-payment of wages was not common. The employers who paid on time were generally the richer employers. Given their tight economic circumstances, the poorer members of the peasantry were often poor paymasters as well. Peasants were also likely to be hard taskmasters, more likely to be on the field labouring with hired labourers, working along with them, supervising them, and attempting to get as much work as possible for the wages that they paid.

What role do such operations play in the overall labour schedule of landless agricultural labourers? Table 9.2 presents the survey data on the number of eight-hour days spent in paddy weeding and 'other' operations in non-cereal crops irrigated by groundwater, in tamarind and in 'other' crops as a percentage of total agricultural time in all agricultural operations (other than days of

Table 9.2 *Eight-hour days at major cash-paid, daily-rated tasks and ploughing as a percentage of all labour days[a] in agriculture, by sex and caste, landless agricultural labour households, Gokilapuram, 1976–1977*

Sex	Scheduled caste households	Non-scheduled caste households	All landless agricultural labour households
Men	63.3	56.4	60.6
Women	49.8	40.0	45.4
All workers	54.8	45.4	51.2

Note: The crop operations considered here are weeding and field preparation in paddy cultivation, and all crop operations in the cultivation of irrigated non-cereals, tamarind, and 'other crops'.

[a] 'All labour days' excludes days at farm servant labour and guarding crops.

Source: Survey data.

[26] In contrast, for instance, to conditions in another relatively commercialized village, Nesal in North Arcot, surveyed by John Harriss (he calls the village 'Randam'), where, even for weeding and hoeing work, gruel was given to the workers in addition to a cash payment (Harriss 1982, p. 120). In some tasks, such as tomato- or chilli-harvesting, women are permitted to take home some damaged tomatoes or overripe chilli—these are bonuses that go to the evening's cooking-pot, not part of the wage. See also nn. 20, 22, 24.

Table 9.3 *Earnings from major cash-paid, daily-rated operations and ploughing as a percentage of total agricultural earnings,*[a] *men, women and all workers, scheduled caste, non-scheduled caste, and all landless agricultural labour households, Gokilapuram, 1976–1977*

Sex	Scheduled caste households	Non-scheduled caste households	All landless agricultural labour households
Men	66.0	65.0	66.7
Women	50.6	40.2	46.0
All workers	58.3	51.4	55.4

Note: The crop operations considered here are weeding and field preparation in paddy cultivation, and all crop operations in the cultivation of irrigated non-cereals, tamarind, and 'other crops'.

[a] 'Total agricultural earnings' excludes earnings from farm servant labour and guarding crops.

Source: Survey data.

pannaiyal labour and crop guarding). Table 9.3 shows the earnings from the same operations as a percentage of total earnings from agricultural tasks (again excluding *pannaiyal* earnings and earnings from crop-guarding). This is not an exact measure of the significance of daily-rated tasks: on the one hand it includes the ploughing operation (which is paid in cash and cooked food) in the crops mentioned, and on the other it excludes daily-rated, cash-paid operations in coarse grain and groundnut. Nevertheless, it provides a broad idea of the order of magnitude of the role of such operations in the overall labour calendar.

The tables show the relatively high proportion of days spent in and earnings from daily-rated, cash-paid casual labour in the labour schedule of agricultural labourers: it was more than 50 per cent in both cases. Some of the other conclusions from the data are the following:

1. The proportion of days spent in and earnings from daily-rated, cash-paid casual labour was higher for men than for women. This reflected, among other things, the importance of hoeing work and tamarind labour in a man's working year and the fact that a very important operation in a woman's working year, transplanting, was a piece-rated operation and one in which men hardly participated.

2. The proportion of days spent in cash-paid daily employment was higher for scheduled castes (men and women) than for non-scheduled castes. This derived from the fact that agricultural labourers of the scheduled castes worked more in tasks that involved going out of the village—in the present case, in crops grown in groundwater-irrigated fields and in tamarind cultivation (see Chapter 6). Within paddy alone, however, scheduled caste women worked

more in transplanting in contract gangs than non-scheduled caste women, and the latter spent a higher proportion of days of labour as day labourers in that crop.

3. Among women, there was a fair degree of coincidence between the proportion of days of labour spent in daily-rated, cash-paid tasks and the earnings from such tasks. Among men, however, the divergence was clear—the proportion of earnings from such tasks was higher than the proportion of total labour time spent in them. The reason is probably that when men worked in piece-rated tasks (of which the major one was harvesting paddy) and at threshing paddy, the wage per eight-hour day worked out to less than they would have earned as cash-paid daily labourers. This was particularly true for non-scheduled caste men (for whom the divergence was the greatest), since at harvest time they were likely to find employment in contract groups that could not contract for as much harvest work as a scheduled caste, particularly Pallar, labour gang.

By 1986, there was a shift in employment patterns, particularly among women, with a decline in the share of cash-paid, daily-rated tasks (see Chapter 10).

Ploughing

Ploughing (ploughs are pulled by a bullock pair driven by a man) is an example of an operation that, when done by a hired casual labourer, is daily-rated with wages paid in cash *and* kind.

There are different types of ploughing. On surface-irrigated land, before the field is flooded, the soil is turned with a wooden plough with an iron tip; wet-land ploughing is done with an improved ploughshare, with a metal scoop; on some fields puddling is done with a bullock-pulled cage-wheel which has a seat perched on it; levelling is done with a wooden plank pulled by bullocks. Dry land and groundwater-irrigated land are ploughed with animal-driven ploughs; the crude metal-tipped, wooden plough is used.

Ploughing is tough work, requiring skill at controlling and guiding the animals. Although the duration of work varies with the size of the field to be ploughed and the number of ploughs that are driven, the average working day at ploughing is about five to six hours. Ploughing is a daily-rated operation where a late-morning or early-afternoon meal—generally rice with some vegetable or pickle—is given along with a cash wage. Men were paid Rs 6 in cash as well as food when they drove the employer's bullocks and plough and Rs 15 with food when they were hired and brought their own animals and ploughs.

Bullock-powered ploughing is an operation that is in decline, ousted by tractor ploughing, and much less work is available now for men at ploughing than was the case before tractorization. On groundwater-irrigated land, where ploughing is done before irrigation basins are dug, tractorization (ploughing with disc ploughs and nine-pronged tillers) is widespread. On rice-growing surface-irrigated land, even less-well-off peasants sometimes hire tractor time, and in general ploughing is done exclusively by animals and ploughs only on those fields that are inaccessible by tractor or which are really too small to be ploughed by a tractor. On surface-irrigated land, however, some ploughing operations continue to be done by animal-driven ploughs: a single wet-land ploughing, nursery ploughing, and levelling are done by men with animals even where the employer actually owns a tractor. On dry land, ploughing is generally done by plough animals and men; in Gokilapuram only those who own tractors plough their unirrigated fields with disc ploughs; the others do not bother to hire tractor-time on dry land.[27]

With the decline of bullock-powered ploughing, there has been a decline in the number of agricultural labour households that maintain their own draught animals. The cost of maintaining such animals has become prohibitive when measured against the number of days of employment available at the operation (and this is combined with the fact, see Chapter 5, that tractors have displaced animal power substantially in carting and the second threshing of paddy as well). The loss of draught animals by a peasant household is often a clear indication of impending depeasantization (see the case study of Selvam's household in Appendix 6), and in an agricultural labour household, of its further proletarianization.

The decline of ploughing with draught animals has also contributed to creating relatively high daily rates for persons who do in fact maintain their own animals on the days that they actually receive employment: the rate went up to Rs 20 in 1981–2,[28] and more than Rs 40 in 1986.

Contract Labour: Transplanting and Harvesting Paddy[29]

The *kothu* system, which is a specific form of contract labour, is prevalent in two operations in paddy cultivation, transplanting and harvesting. These

[27] In Gudalur, when groundnut is sown on unirrigated land, the ploughing that accompanies sowing is generally done manually with plough animals, even by those who own tractors and have done the initial ploughing with them.

[28] Another example from Gudalur: on days when ploughs for hire were required at distant unirrigated fields in 1981–2, demand conditions could, on occasion, pull the daily cash rate up to Rs 40 per day.

[29] There is a rich literature on changes in systems of labour hire in rice-harvesting and -transplanting in South-East Asia, and on changes in that region from wage systems that are 'more open and generous systems to less open and generous ones' (Kikuchi *et al.* 1984, p. 125). See e.g. Bautista (1977, pp. 117–22); Fredericks (1977); Horii (1977); Takahashi (1977, pp. 98 ff.); Kikuchi *et al.* (1979, 1984); Kikuchi (1981). See Hart (1986) for a re-evaluation of the labour process and employer–worker relations in rice cultivation in Java. See also Wickramasekara (1977) on Sri Lanka; Clay (1976) and Rahman and Das (1982) on Bangladesh. The very useful paper by Clay shows that transition in the labour market and in payment systems 'is associated with a change in modes of payment which, in a cross-sectional analysis, is reducing payments to labour below the level prevailing under the traditional payment system', (1976, p. 438). On migrant labour for harvesting, see Dasgupta (1977); on contract labour in paddy operations in Thanjavur district in

operations accounted for about a quarter of the time spent at agricultural tasks by all landless labourers. A *kothu* is a group (or 'gang') of contract workers that works at piece-rates under a labour contractor or jobber—the *kothukkarar*—who is himself an agricultural labourer by origin. Piece wages are fixed per *kuzhi* (0.6 acres) or *sey* (2.4 acres).

Across the village as a whole, from the first field to be transplanted or harvested to the last, the season for these operations lasts for about 20 to 30 days. For a single field, however, there is a two- to three-day period in which the crop must be transplanted (or reaped). Transplanting and harvesting, then, are short-term operations, requiring intense labour and, consequently, a major effort in mobilizing and deploying it. The need to mobilize large numbers of workers for intensive labour is particularly pressing during the first crop. During this period, rain can divide the season into short stretches of days on which work can be done and gaps in the rain must be sought to get work done quickly.

There were five contract labour groups in the village; a brief description of four of them follows.[30] Groups were organized, broadly speaking, from particular neighbourhoods or localities.[31] An important feature of the groups was that four out of five in the village were multi-caste groups which, though led by caste Hindus, included members of the Chakkiliyar and Parayar castes, who were traditionally barred from working in transplanting and harvesting.

The Pallar group, led by labour contractor K. Ramasami, was, and remains, a single-caste group; men and women of the scheduled Pallar caste are considered paddy labourers *par excellence* in the village and, although the identification of this caste with labour in the paddy-fields is not as close or exclusive as it used to be, this is still one of the busiest groups. Twice a year, at harvest time, workers of the Maravar caste from Kamayakavundanpatti village, which is about 10 kilometres to the south, join the group and work under its scheduled caste labour contractor.

the 1950s, see Sivertsen (1963); and on contract labour groups in peak seasons in Coimbatore, see Kandaswamy (1964). See also Nakamura (1982) and Komoguchi (1984) on harvest labour in Tiruchi district.

30 The material for this section comes from the survey of 1977 and from case studies conducted in 1981–2, in the first paddy season. Of the five labour contractors who were active in the village, I know four. In both seasons, I followed one group (whose members were all from the scheduled Pallar caste in both seasons) across every field that the group worked on in the season, collecting material on the number of persons in the group every day and on each field; the socio-economic status of the employer; the size of the plot and its location; the number of hours of work each day and on each plot; workers' remuneration, including earnings per field and per day; and the tasks and remuneration of the labour contractors. During the harvest season, I have matching data for one more group, which was a multi-caste group, and in transplanting I have similar, though less comprehensive notes on the second group. The quantitative data on the systems of group labour refer to the first paddy crop in 1981–2, except where I have specified otherwise.

31 Rahman and Das (1982) note that in their study area in Noakhali district, 'the broad principle of neighbourhood applies to membership' in a *sardari* group, which is a group of contract labourers (p. 11).

A second group was led by Ponnayya Thevar, of the Kallar caste. This was a multi-caste group, with some domination in its membership of Kallars and persons of the Potmaker caste, who live geographically close to the Kallars and who have traditionally 'close' relations with the Kallars of Gokilapuram. There were also workers of the scheduled Parayar and Chakkiliyar caste and of the Achari and Nadar castes in the group. It was uncommon for Maravar workers to join the Kallar-led group; in our reference season, a few Maravar workers joined the group for two days in the middle of the season and left after that.

A third group, led by Paraman Chettiar, a Telugu Chettiar, was the one that most Maravar workers joined; this group was also a multi-caste group, drawing on other castes, including a few scheduled caste workers as well. A fourth group was led by a woman, Azhagammal. This was also multi-caste; a feature of this group was that the labour contractor drew on migrant workers from the dry villages in the east where no paddy is grown and who came to Gokilapuram for employment in the transplanting season.

The first three groups were the most active in the village, the ones that could always mobilize workers, that had some standing in the village, and which most major and medium landlords called to their fields.

The labour contractor (*kothukkarar*) who leads the group of workers (the *kothu*) who work at piece-rates, is to be distinguished from a head labourer and from a rural works contractor who contracts for different kinds of rural civil construction work. While a labour contractor here is a person who has come from the ranks of agricultural labourers themselves, he supervises the group at transplanting and harvest time and does not work as a labouring member of the group. (He often works during the groundnut harvest season, however, when individual households in the group compete for higher earnings.) This labour contractor is also to be distinguished from contractors who, for example, bid for contracts to build village roads and culverts. Although the labour contractor has a regular group of workers ready to work under him during specific seasons, unlike the rural works contractor he provides no working capital at all.

The most important functions of the labour contractor are, of course, to mobilize workers in two seasons and to have the qualities of an organizer, to be able to keep the group together, working together.[32] The labour contractor must also keep an eye on the crop, judge when it will come to transplanting and to harvest, and contract with owners for employment. He must maintain 'good relations' with some big employers to ensure regular work each season and, when he contracts for employment, must be able to stagger the work for the

[32] There are points of similarity between the labour contractor and the gang master of the labour gangs that Marx has vividly described: 'He is the recruiting-sergeant for the gang, which works under him, not under the farmer. He generally arranges with the latter for piece-work, and his income, which on the average is not very much above that of an ordinary agricultural labourer, depends almost entirely upon the dexterity with which he manages to extract within the shortest time the greatest possible amount of labour from his gang' (1977, pp. 649–50).

groups, setting the volume of work against the members in the group and a notion of how many acres can be done a day. A labour contractor must be able to extract work from each person in the group, assessing the intensity of work that each person is capable of and, at harvest time, must be able to divide labour between persons, monitoring and shifting the division of labour as work goes on. He must be able to calculate individual wages—and the calculations of the labour contractors whom I followed were invariably accurate—and ensure that wages are paid without delay. In the seasons when there is work for a group, a labour contractor is a busy person indeed.

Transplanting

What follows is a brief description of the work of women in a group at the time of transplanting. First, they uproot the seedlings in the seedling nursery, bundling them into tight seedling bundles of 200 to 400 seedlings each. They leave the bundles in the field, and these are carried to the main field by the male workers. Carrying the bundles from the nursery to the main field is the only task done by men, and there are only one or two men working with the transplanting group (the labour contractor himself does not work in the field). After the seedlings have been uprooted and shifted to the main field, the women in the group move there to transplant the seedlings. While they transplant, one worker (or more, depending on the size of the field) distributes the seedlings to women as they transplant. The women who distribute the seedlings are not members of the group of contract workers; they are generally women of the cultivator's household (when the plot belongs to a peasant or small cultivator) or women directly employed for regular daily rates by employers.

A labour contractor surveys the fields to judge when fields will come to transplantation and, seeking out the landlords first, contracts for work with individual employers (he has some regulars among the employers), sometimes contracting a week in advance of actual transplanting. Each labour contractor generally has in his head a list of households from which women will join his group (the labour contractors whose groups I followed were both illiterate) and he announces in different areas the day on which the season will begin and the field to which the workers must come. After that, he does not really have to seek workers since women come to seek work with the group. This is particularly true of the first crop, which is the first continuous period of employment for women after a long lean season. He also employs some men for transporting the seedling bundles.

In Ramasami's group (the exclusively Pallar group), after the first two days (on which 11 women and 30 women were in the group), the size of the group began to swell, with an average of 69 women, varying from 51 at its smallest to 94 women at its biggest. The size of Ponnayya Thevar's group ranged from 36 to 83 women, with a modal size of 52 women.

The Pallar labour contractor, Ramasami, also attempts to accommodate elderly women in the group during transplanting time, women who would not ordinarily get work in daily tasks because of their age. Working in the group of contract workers is often the only agricultural labour that they get in the year; their relative weakness is compensated by the younger women in the group. Of the approximately 115 women who worked in Ramasami's group at some time during the reference season, I recorded the ages of 109 (Table 9.4). The results showed that there was a remarkable incidence of women over 45 in the group (they constituted about one-third of the group), and that 15 women who were more than 60 years old also worked in the group. (These older women are likely to have worked for fewer days in the season.)

Table 9.4 *Ages of women in Ramasami's (scheduled caste) transplanting group, Gokilapuram, first paddy crop, 1981–1982*

Age group (in years)	No. of women
16–19	14
20–24	7
25–29	9
30–34	13
35–39	14
40–44	16
45–49	5
50–54	9
55–59	7
60–64	9
65–69	0
70–74	1
75–79	5
TOTAL	109

Source: Case-study data.

The duration of group contract work depends on two factors—the number of women in the group every day and the number of fields on which the labour contractor can obtain work. In a transplanting season of about 30 days, the Pallar women's group received employment for only 11 days. The hours of work per day varied between 5 hours 30 minutes and 9 hours 15 minutes. The average working day was 7 hours 26 minutes. It is an indication of the pressure of population on the system of group contract labour that the average working day was shorter than eight hours.

Turning to the rates of remuneration of the workers in the group, the practice is for piece wages to be fixed in rupees per unit of land in the first season and in measures of paddy per unit of land in the second season. The employers say: 'The second transplanting comes after we have grain from the first harvest, so we pay in grain; the first comes after the lean season and so we pay in cash.'

In the reference season, the rate per *sey* or 2.4 acres (which is just 0.07 acres short of one hectare) was Rs 220. In the preceding season, that is, the second paddy season in 1980–1, the wage rate in kind was 180 measures of paddy per *sey*, or 217.5 kilograms of paddy per hectare. The money is shared as follows: the labour contractor keeps a certain amount for the men who transport the seedling bundles (generally at the rate of Rs 20 per *sey*); the rest is divided equally among the women, with the labour contractor getting one share as well. The actual earnings per day vary with the amount of work available and the number of women in the group. During the 11 days that Ramasami's group received work, the earnings varied from Rs 1.90 to Rs 6.80 per woman, and the average wage was Rs 4.35. This average wage was 35 paise above the wage rate for women at cash-paid, daily-rated tasks during that season.

The labour contractor also receives additional money—tips—from the employer. The amount generally varies positively with area harvested, although there are no fixed rates. For the reference season, Ramasami got a total of Rs 133.75.

Harvesting

Let us assume that the day is a dry one, and harvesting is uninterrupted by rain. Men and women move on to the field at about 8 a.m. and begin to reap the crop, laying the stalks behind them in straight swathes. As the sun gets higher and dries the swathes, some of the group are diverted into the task of making straw ropes. Others are then given the task of bundling the stalks into sheaves and tying them down with straw ropes. By the time the last of the reapers finishes cutting the crop, the tying of sheaves is well under way and some persons are still making straw ropes. The last task is to carry the sheaves to the employer's threshing-floor or to a waiting cart or tractor trailer. If wages are paid the same day, some workers of the group thresh the grain for the day's wage. All tasks are done by both men and women.

Harvesting calls for intense labour; on the field, it also requires a well-tuned division of labour that shifts through the day, from the start of the day when everybody starts by reaping the crop, to the end of the day when all the workers lift the sheaves off the field. The person who organizes work on the field, continuously reassigning workers to jobs, is the labour contractor. During transplanting, once work is begun he does not have to remain at the field through the day; during harvest time he has no time to be off the field.

As in transplanting, the overall duration of the work depends on the number of persons in the group and the number of fields that a labour contractor can contract for. In addition, in the first season, an intensive working day can quickly be called to a halt by rain. In our reference season, the working day varied from 3 hours (on a day cut short by rain) to 12 hours; the average working day was 7 hours 45 minutes, still short of eight hours.

The five labour contractors who organized groups for transplanting paddy organized groups for harvesting as well, although, at harvest time, they recruited men and women. The ratio of men to women in the group varied from day to day.

Piece wages for harvesting paddy are calculated, for both crops, in measures of paddy per *kuzhi* or per *sey*. The rate varies according to the distance that workers must carry the sheaves, the distance from the field to the road or threshing-floor. There is another, less certain factor that intervenes in the immediate determination of wage rates: the weather on the day of harvesting. When the day is damp and the crop has to be harvested, the labour contractor bargains with the owner—who stands at the side of the field anxious to get work done, while the members of the group (who normally rush to the field to start work) sit by the field bund, waiting for the bargain to be over—for a higher piece-rate. This bargain is informed by a knowledge on both sides of conditions in the village and the village labour market, and can raise the rate by only 5 to 6 per cent of the regular rate, but even that is a welcome raise for the members of the group. In general, the range of wage rates paid to harvest groups is very wide; in the reference season it varied between 42 and 96 measures per *kuzhi* (or between 203 and 464 kilograms per hectare).[33]

There is also a series of other—extra, 'traditional'—payments to the group which, when taken together, represent an important addition to their wages.[34] These are:

- *Suthanthiram*. After the harvest, each employer gives the group nine measures (10.9 kilograms) of paddy to be distributed to the group. Nine measures are given regardless of the extent of land harvested, and a person who has employed the group on two separate occasions on different fields pays *suthanthiram* each time.
- *Inam*. The employer gives *suthanthiram* 'of his own accord'; the group asks for and receives the *inam*. *Inam* is also nine measures.
- *Vazhttu*. After the *inam* is given, the labour contractor walks up to the *marakkal* (volumetric measure), folds his hands and prays loudly that paddy always grow well on the employer's field. The group receives another nine measures from the employer as *vazhttu*.
- *Kothukkuli*. The labour contractor is paid three measures (3.6 kilograms) of paddy by the employer. This is kept by him.

[33] E. J. Clay's data from Bangladesh show very wide differences in rates of payment to harvest workers who were paid in terms of a share of the gross crop. He records a 25% difference in the share paid to groups harvesting within 100 yards of each other (1976, p. 437). Clay also suggests that factors such as the location of a plot, the balance of supply and demand for labour on a particular day, the size and quality of the plot to be harvested, and the levels of yield can all play a part in determining rates of payment for hired labour in harvesting (pp. 437–8).

[34] On *suthanthiram* and other traditional payments, see S. Menon (1983, p. 158).

- *Aruvappiri*. A sheaf is tied together that is big enough to yield, on the average, eight measures (9.7 kilograms) of paddy when threshed. This paddy is sold by the labour contractor and the money distributed to the *men* in the group. This is the only time during harvest when men and women are not paid equally: women receive nothing of the *aruvappiri*.
- *Kothu kathirkattu*. The labour contractor is given a sheaf of grain that yields about eight measures of paddy when threshed. This is kept by him.

In paying the workers, the labour contractor adds up the grain wage received at the regular piece-rate and the amount received as *suthanthiram*, *inam*, and *vazhttu*, and divides the amount equally among all the members of the group, counting himself as one member, with equal shares paid to men and women. The *aruvappiri* is sold and the money distributed to the men in the group.

The average wage per day (inclusive of extra payments) during the reference season in Ramasami's group and Ponnayya Thevar's group is given in Table 9.5. The table shows, first, that harvest wages represented a grain payment that, when converted into cash, was higher than the wage rate at cash-paid, daily-rated operations. The cash equivalent of the average earnings per woman worker was Rs 6.58 in Ramasami's group and Rs 7.40 in Ponnayya Thevar's group. The cash equivalent of the average earnings per male worker was Rs 7.70 in Ramasami's group and Rs 8.87 in Ponnayya Thevar's group. By contrast, the wage rate for daily-rated, cash-paid operations during that season was Rs 4 for women and Rs 7.50 for men. Secondly, the data show that the difference between the wage paid for daily-rated, cash-paid operations and the average daily earnings from harvesting was greater for women workers than for

Table 9.5 *Average daily earnings, harvest groups, Gokilapuram, first paddy crop, 1981–1982*

Sex and wage equivalents	Average earnings per day	
	Ramasami's group (scheduled caste)	Ponnayya Thevar's group (multi-caste)
Women		
Measures	3.72	4.18
Kilograms	4.50	5.06
Rupees	6.58	7.40
Men		
Measures	4.35	5.01
Kilograms	5.26	6.06
Rupees	7.70	8.87
Women's wage as % of men's wage	85.6	83.4

Source: Case-study data.

men. Thirdly, the average daily earnings of workers in the scheduled Pallar caste group were lower than in Ponnayya Thevar's group. A reason for this is that with the exodus of Pallar workers from the group for threshing, Ramasami's group had to depend more on the very young, while Ponnayya Thevar's group did not lose as many people to the threshing operation, and employers tried to find the group with the strongest workers for their fields. Fourthly, although the basic wages and *suthanthiram*, *inam*, and *vazhttu* are distributed equally, the amount that men received as *aruvappiri* put their earnings up by more than a rupee over a woman's daily earnings, although (and this is the fifth point) the divergence between male and female earnings was much lower in harvesting than in other tasks.

What proportion of gross produce goes as harvest wages? This will clearly vary with the yield on a field and the piece-rate that is fixed for it. In Table 9.6, we have taken a range of piece-rates (the highest, the lowest, and the median rate) and three yield rates (a very high yield, an average 'good' yield, and an average 'poor' yield) and have calculated wages and other payments at harvest time as a proportion of total yield, using each of the nine combinations of yields and piece-rates.

Table 9.6 *Harvest wages as a proportion of gross output, Ramasami's scheduled caste group, Gokilapuram, first paddy crop, 1981–1982*

Wage-rate (*marakkal* per *kuzhi*)	Yield (*kalam* per *kuzhi*)	Wages as % of gross output	Wages + extra payments as % of gross output	
			excluding payment to *kothukkarar*	including payment to *kothukkarar*
32	24	8.3	11.6	13.0
32	30	6.7	9.3	10.4
32	32	6.3	8.7	9.8
20	24	5.2	8.5	10.0
20	30	4.2	6.8	7.9
20	32	3.9	6.3	7.4
14	24	3.7	6.9	8.3
14	30	2.9	5.5	6.7
14	32	2.7	5.2	6.3

Note: For weight and area measures, see Glossary.

Source: Case-study data.

First, there was a fairly wide range of values between which the proportion of all harvest payments to gross paddy yield varied, between 13 per cent for the extreme high wage rate and low yield combination to 6 per cent for the extreme low wage rate and high yield combination. What is important to note is that unlike in regions where harvest wages are a fixed proportion of gross yield, there is a more complicated and finely calibrated system prevalent here. Secondly, the proportion that corresponded to an average combination (medium yield, middle-level wage rate), 8 per cent, was not a high proportion compared to other regions. Thirdly, extra or traditional payments to the group and to the labour contractor represented a substantial amount, ranging between 4 per cent and 5 per cent of gross produce.

The Significance of the Contract Groups: Workers and Employers

In Chapter 6, we saw the significance of the paddy transplanting and harvesting operations in the labour schedule of agricultural labourers. To recapitulate: in the survey year, transplanting paddy occupied about 19 per cent of the year's agricultural employment (in eight-hour days) for women. Harvesting paddy occupied about 13 per cent of the agricultural labour time of men, 12 per cent of the agricultural labour time of women, and 13 per cent of the labour time of all landless agricultural labourers. The two operations together took up 25 per cent of the agricultural labour time of all landless agricultural labour households.

Almost all of this work was done by groups of contract workers. In the 1981–2 reference season, on a rough estimate at least 85 per cent of the area was transplanted by the groups. Some peasants and a few small landlords employ daily labourers for this task; their fields are small and they have the time for close supervision. This does not mean that all small peasants do the work with family and daily-rated hired labour: I have in my records small peasants who employed worker groups, including even a Pallar tenant-cultivator with an operational holding of 0.6 acres. For harvesting, group contract labour is more universal, and I did not see a single field worked by means other than that of group labour in the reference season.

Who employed the groups? Table 9.7 gives the share of different employers, in terms of the area transplanted and harvested by the group, as a percentage of the total area transplanted and harvested over the whole transplanting or harvesting season. Landlords heavily predominated as employers, with the peasantry following as a distant (though still significant) second for Ramasami's Pallar group. In the case of Ponnayya Thevar's multi-caste group at harvest time, peasants formed a second and very important component of employers. This is a reflection of the fact that most big landlords still prefer to employ the group that has traditionally dominated the transplanting and harvesting operations in the village, namely, the Pallar group.

Features of Contract Labour

Contract labour represents, in an important sense, a 'free' wage labour form in the village, a form of labour hire in which workers work in a group under a labour contractor without being obliged to negotiate directly with an employer. I sought to gain information on workers' attitudes to contract labour directly

Table 9.7 *Who employed the contract group? Data for Gokilapuram, first paddy crop, 1981–1982* (%)

Class of employer	Transplanting season	Harvesting season	
	Ramasami's group (scheduled caste)	Ramasami's group	Ponnayya Thevar's group (multi-caste)
Landlord	72.5	72.3	56.7
Peasant	17.0	14.5	39.2
Business activity other than crop production	9.3	7.5	4.1
Salaried persons, professional	1.3	5.7	0.0
All employers	100.0	100.0	100.0

Source: Case-study data.

from labour contractor Ramasami. When asked whether, despite the labour in the group being laborious and demanding, workers welcomed contract labour, Ramasami said firmly that there was no doubt that workers preferred it to searching for individual daily employment. He gave a clear-cut, point-by-point list of reasons for this preference.

First, 'they are sure of getting work without wandering about from house to house looking for it.'

Secondly, 'although the work may last for longer hours, we can arrange the work time to our preference. For instance, during harvest time, from 10 a.m. to 12 noon, I may send them home; they go home, do any housework that they may have on hand and look to the children, and work again when the sun is not as hot.'

Thirdly, 'apart from all that, they can expect one measure more as wages than they can from daily wage labour.'

Fourthly, 'as far as women are concerned, the harvest is the only time in the year when they are paid equally with men.'

Fifthly, 'when they come for contract work, they do not have to beg a *mudalali* (landlord, employer) for a job. When they do not go on contract work, they have to go a day before to the *mudalali* and ask "*mudalali, mudalali*, is there work in your fields for us tomorrow?" Not only must they beg and grovel before an employer, they must do o.c. work [work without pay] on top of it.' (Ramasami also said, however, that members of the group may have to perform minor labour services for landlord employers: 'A big landlord may ask us to carry fertilizer bags to the fields. On the threshing-floor, he may ask some of us to collect the sheaves or to stack the hay. Although we will not do work in the fields without payment, these tasks are done without payment.')

There is another aspect of contract labour that deserves note. Although there is competition between groups for employment, it is perhaps not reading too much into the situation to see, within the group itself, the germs of collective labour. Work is shared in the group, the younger and the tougher workers have to work harder than those who are old and weak. To turn to labour contractor Ramasami once again:

Question. Do those who are given tougher work resent it?

Answer. Sometimes a youth will complain that an old man gets the same wage for an easier job. In the beginning, I explain to him that he must not complain—that old people also need money, but cannot do the same work. If he continues to complain, I curse him—and I am capable of fine cursing—and tell him to get on with his work. If he still complains, I send him home.

Question. Will he be taken again the next day if he asks you to, or will you refuse him employment in your group?

Answer. No. I will not do that, I could not do that. They also must live. I cannot ruin them and their families by not allowing them to work. They will also quickly learn how to work together.

Ramasami responded immediately and enthusiastically to a question about whether there was joy in the collective labour of the group. When working in the contract labour group, he said, no person need cheat the other to be assured employment and wages. They work joyfully and together. Of the harvest group, he said, 'It is a task where men and women work together, labouring jointly on the fields. Young boys and girls work together, they sing songs of the harvest. At the end of the day, they receive the same wages—the harvest is a time of joy.'

Changes in the Contract Labour System

On changes in the contract labour system, my information is from interviews with the labour contractors and with old agricultural labourers; the information that they provided was checked with other agricultural labourers and members of landlord households. The best informant on the contract labour system was, inevitably, an old Pallar agricultural labourer, V. Subramaniam, who is now dead. He was known in the village as 'Member' Subramaniam, because he was the harijan member of the panchayat.

Contract labour in transplanting and harvesting is not an entirely new form: no one can remember a time when the system of group contracts did not exist in the village, and Member Subramaniam remembers his elders speaking of the system as a regular feature of village agriculture when they were young. The *coverage* of the system, however, is greater today than it was in Member Subramaniam's youth, particularly in transplanting. Today almost the whole village is transplanted by the groups, and this has been the case with harvesting for much longer. Among the workers, an important change is that the operations that are done by means of group contracts are no longer dominated by Pallars. People of all castes have groups that they can join and a significant feature of this drawing-in of all castes is that workers of the Parayar and Chakkiliyar castes, who were earlier excluded, now actively participate in different groups. They generally work in Ponnayya Thevar and Azhagammal's groups, which are both multi-caste, and they work along with caste Hindus at transplanting and harvesting time.

There are changes in the organization of contract labour in transplanting that reflect the fact that in the contemporary period, the tasks involved in the whole operation have been integrated and the task itself speeded up. When Member Subramaniam was young (he was about 75 years old in 1981) separate wages were paid for uprooting and bundling the seedlings, and for transplanting the seedlings. The first task was done either by the women of the group or by women employed specifically for the purpose; in any case it was paid for separately. The task of transplanting the seedling bundles and distributing the seedlings on the main field was done by persons employed separately on daily wages (distributing seedlings still is). Another change relates to the monetization of wages in the transplanting operation; in Member's youth, transplanting in both seasons was paid for in grain: 'today the employers say that they have no grain after the lean season and pay the *kodai* season wages only in cash'.

The major change in the system has been the increase in the pressure of the labouring population on the groups. This is evidenced by the existence of more groups, more persons per group, fewer days of work for the individual worker, and lower earnings per day. A related change has been that as the supply of labour increased, the gap between wages paid at daily rates and earnings through piece-rates has narrowed. In Member Subramaniam's youth, there were not more than 40 women in the Pallar—then the largest—transplanting group; in 1981–2, on the peak day, there were 94 women. In the transplanting season, this group could expect work for about 30 days, while in the first paddy season of 1981–2, Ramasami's group could get work for only 11 days. In earlier days a woman got 8 to 12 measures of paddy a day for transplanting, or the equivalent, in real cash terms, of Rs 14.16 to Rs 21.24 during the first paddy season in 1981–2. As we have seen, the average earnings per woman per day in the reference season were only about Rs 4.50. For harvesting, while individual members of the group got about 12 measures a day at the beginning of the 1960s, in the reference season, they received from 3.72 measures (the lowest average women's wage in my data) to 5.01 measures (the highest average men's wage in my data).[35]

Seasonal Migration: Harvesting Groundnut in Gudalur

From the last week of July to the last week of August, dry-land groundnut comes to harvest in southern villages of the Cumbum Valley. The centre of this

[35] One landlord explained to me how inflation in fact compensated for falling real wages, thus: 'Although the coolie-workers get less grain per person today than earlier, it works out to the same, since today they can get more rupees for a measure of grain than they could have got in those days.'

cultivation is in Gudalur (concentrated in the revenue village of Melagudalur); the villages of Cumbum and Pudupatti are secondary areas of unirrigated groundnut cultivation, and the task in all these villages put together involves harvesting about 8,905 acres of groundnut.

This task of harvesting groundnut is undertaken mainly by seasonal migrants, groups of workers who come under labour contractors from different villages of the Valley and from other taluks; there are groups from as far away as Usilampatti, Thirumangalam, and Dindigul taluks. In July–August 1981, which was a season of relatively low migration because of a damaged crop, I counted about 4,750 workers in and around Gudalur in addition to about 1,000 to 1,200 non-working dependants who had been brought along with the groups.

This is a case study of labour contractor Ramasami's group of Pallar workers from Gokilapuram when they migrated to Lower Camp in the harvest season of July–August 1981 (Ramasami takes a group from Gokilapuram to Gudalur every year). The group was joined later in the season by a group of workers, also of the Pallar caste, from a village near Chinnamanur. Lower Camp is a settlement south of Gudalur town, located at the point where the hill road to Thekkady begins, in the revenue jurisdiction of Melagudalur village. The group consisted of some 39 households, with 42 men workers and 44 women workers and 19 non-working dependants.

Ramasami's group is formed in the village. He goes to Gudalur a few weeks before harvesting is due, checks the state of the crop across the fields of Melagudalur, and begins to meet employers with whom he has been in touch, contracting for a substantial part of the work that the group will do even before its members actually leave Gokilapuram. On 27 July, the group left Gokilapuram for Lower Camp, where there is a place away from a road, near a disused government dispensary, where they pitched camp for three weeks. Every year, before pitching camp, the group goes to the Bhagavathiamman temple that is nearby, where the labour contractor, as head of the group, breaks a ritual coconut and offers flowers and prays for a good season.

The Gokilapuram group is a well-established one. Members of the group open accounts at a provision shop and at a local eating-place and the provision-shop owner, eager for their three-week custom, loans them asbestos sheets for making lean-tos for some in the group. Others spread themselves out on the verandah of the dispensary or make thatch shacks with wooden supports, and at the centre of the camp a row of fireplaces appears, dug into the ground and demarcated with stones, one for each household. The camp is also close to the upper reaches of the Periyar, at the picturesque stretch near the hydro-electric station where the river first enters the Valley from the hills.

There was one special feature of the crop in our reference season: the groundnut crop in Gudalur had been affected by a crop disease.[36] The disease

[36] A recent study has shown sizeable fluctuations in production to be characteristic of groundnut cultivation in Madurai district; see Krishnakumar (1989).

affected the area unevenly: most fields were affected (although some fields had normal yields), and there were individual fields where certain portions were badly damaged and other parts normal. This disease, as we shall see, led to significant changes in the wage system that year, and to certain new types of wage formation.

In the groundnut field, workers first uproot the groundnut plants and then stack them. A small trench is dug in the ground and a wooden hoe-handle laid crosswise over the trench. The plants are then grasped and the roots beaten over the stick to separate the pods. The pods are cleaned, winnowed, and piled, and the groundnut plants are stacked separately.

The amount and intensity of labour of the workers over a working day is dependent on the number of persons working in the group, the extent of land that the labour contractor has contracted for them to harvest, and the state of the crop. The state of the crop refers, among other things, to whether the crop is a good one or affected by disease, to the relative spread and the interspersal of disease-affected and unaffected plants, to the number of pods per plant, their size and ripeness, and to the dampness of the soil and plant at the time of harvest. (When soil and plants are damp, it is difficult and messy to separate the pods from the roots.)

Groundnut harvesting is marked by long hours of work. The hours on the field varied, in the case of Ramasami's group, from 4 to 11.5, with an average working day of about 8.5 hours and a modal working day of 9 to 10 hours. Another feature of the working day here is the long distances that workers must travel to get to the fields. The unirrigated fields of Melagudalur are widely dispersed, and workers walk from 5 to 15 kilometres a day to and from the fields. They leave the camp a little after 6 a.m. and, on a full working day, all of them are back at camp only after 7 p.m. The group as a whole stayed at Gudalur for 21 days, although each worker did not work every day.

Remuneration

Groundnut harvesting is paid by the piece. In a normal year, a cash rate is fixed for every *marakkal* (volumetric measure) of groundnut harvested, and to this is added a payment in groundnut, whose value is about a quarter of the cash wage. The actual rate, which varies within village-level limits, is fixed by the labour contractor and employer after the labour contractor has seen the field. But our reference year was *not* a normal year, since the crop was affected by disease and this affected the wage system significantly. In the reference year, there were two types of piece-rates:

- a cash-paid piece-rate per unit area harvested (referred to as the 'surface piece-rate'); and
- a cash-and-groundnut-paid piece-rate, fixed per bag, or volumetric measure, of groundnut harvested (the 'yield piece-rate'). This yield piece-rate is the method of wage payment in a normal year, mentioned above.

Surface piece-rates varied from Rs 50 to Rs 90 per *kuzhi* (average rate: Rs 66) and yield piece-rates varied from Rs 8.50 to Rs 9.70 per bag (modal rate: Rs 9 per bag) plus about a quarter of the cash rate in groundnut.

Surface piece-rates were shared by the group as a whole, while yield piece-rates were earned by the household or individual. When the work was paid at surface-rates, Ramasami divided the workers between different tasks and was on the spot to supervise them. When the rate was a yield-rate, the very nature of the work generated an individual or household incentive to work, and speeded up work. Under the yield-rate regime, the household divides the labour time of its members between different tasks (uprooting, separating the pods, cleaning and winnowing them, and stacking the plants); in single-worker households, the one worker does each task himself or herself.

Where the crop was poor, the labour contractor bargained for a surface piece-rate and, where the crop was normal or better, a labour contractor bargained for a yield piece-rate. Within each of these rates, there was a further small margin for bargaining, depending on the state of the crop. Wages, in other words, were set after a two-stage bargain between the labour contractor and the employer. First, the employer and labour contractor decided whether the group would work for a yield-rate or a surface-rate. Secondly, the labour contractor and the employer bargained to arrive at the precise wage-rate.

To judge the state of the groundnut pods that are underground by looking at the state of the leaves of the groundnut plant is difficult and requires experience and skill. When a labour contractor was uncertain about the extent of damage to a crop and thought that the employer was better informed than he and was likely to understate the damage to the crop, the aggregate amount that would be paid to the group as wages was a consideration with the labour contractor when he chose a surface-rate in preference to a yield-rate.

But even where the employer and labour contractor were equally well informed about the state of the crop, a labour contractor bargained for a surface-rate when the crop was damaged. Why did the labour contractor do so, when it was possible for him to bargain for a yield-rate that would give a total group earning that was equal to the total group earning in a surface-rate regime? There were, after all, reasons to opt for a yield-rate. The employer always preferred a yield-rate, since it provided the workers with an incentive to harvest every single pod with relatively no supervision. There were reasons for the labour contractor to prefer a yield-rate as well. There is less supervision work for him in a yield-rate regime. In addition, the labour contractor is committed to promoting the reputation of his group as one that does a thorough job of harvesting and yield-rates ensure that all pods that can be harvested by the workers are harvested. The motivation for the labour contractor bargaining for a surface-rate on a field hit by disease that year was not the enhancement of *total* group earnings (since the same aggregate earnings could have been obtained by means of a yield-rate), but the interests of group cohesion and intra-group distribution. If workers had worked for a yield-rate on a field that had been badly affected, the able-bodied members of the group might have fought with each other at the patches where healthy plants clustered, and the share accruing to the less able-bodied would have been too small to have been acceptable.

The following exercise attempts to assess relative bargaining positions under the two wage regimes:

Let x_1 and x_2 be the yield in bags per acre on land where wages are paid at surface piece-rates and yield piece-rates respectively. In general, $x_1 < x_2$. Let the wages per harvested bag under the two systems be w_1 and w_2 respectively. I have taken, at random, five combinations of fields at which the group worked. Each combination represents two fields, one at which a surface piece-rate was paid and the other at which a yield piece-rate was paid. Table 9.8 gives the details on yield and wage rates for each of the five combinations.

Table 9.8 *Yields and wage rates on different types of fields, groundnut harvesting season, Gudalur, 1981*

Field combination	x_1	x_2	w_1	w_2	$w_1\frac{x_1}{x_2}$
1	6.15	14.17	19.50	8.50	8.46
2	8.33	15.76	10.00	8.50	5.29
3	6.15	13.11	19.50	9.00	9.14
4	8.00	15.44	13.54	9.70	7.02
5	9.17	13.40	10.00	8.50	6.84

Source: Case-study data

If the employer were to pay that wage rate under the surface piece-rate regime that would render the worker indifferent between the two systems, then it would be the case that

$$w_1 \; x_1/x_2 = w_2.$$

But, according to the table, in every case but one,

$$w_1 \; x_1/x_2 < w_2$$

(and if we were to take a larger number of combinations, this would be even more emphatically demonstrated). This indicates that an employer was able to pass on to the worker a part of the loss that he incurred as a result of plant disease.

Earnings of Selected Households

Table 9.9 shows the average daily earnings, under the surface piece-rate system, the yield piece-rate system as well as the combined average daily

Table 9.9 *Earnings of 10 selected households, groundnut harvesting, Gudalur, 1981* (Rs)

Serial No.	Composition of household workforce	Average earnings per capita		
		Surface piece-rate	Yield piece-rate	Combined surface- and yield-rates
1	1F	8.35	9.51	9.00
2	1M + 1F	5.90	9.38	8.20
3	1M + 1F	4.72	9.22	7.47
4	1M + 1F	7.32	8.74	8.21
5	1M + 1F	6.82	10.52	9.04
6	1M + 1F	6.98	13.51	11.70
7	2M	6.24	7.44	6.80
8	2F	7.55	9.25	8.72
9	3M + 1F	6.85	6.03	6.33
10	4M + 1F	4.64	5.10	4.86
	Mean	6.54	8.87	8.03
	Coefficient of variation	0.17	0.25	0.22

Note: M = male; F = female. For definitions of 'surface piece-rate' and 'yield piece-rate', see p. 217.

Source: Case-study data.

earnings per capita for 10 selected households. These households are of different types: five are households with one male and one female worker each; two households have four or more workers each; two have two workers each, of which one has two females and the other two males; and the last is a single-worker household.

The average per capita earnings per calendar day were Rs 8.03, against a daily wage rate that year of Rs 4 for women and Rs 8 for men. It must be noted that the hours of work, distance from the fields, and the fact of having to set up camp make for greater pressure on the migrant worker here than in the course of regular daily labour in the home village.

The average earnings from yield piece-rates was significantly higher than from surface piece-rates. This reflects the fact that the yield on fields that are surface piece-rated are lower, as well as the fact that workers work harder when there is a yield piece-rate.

When work is paid for in yield piece-rates, the system of working by the piece comes into full play. It is competitive and it brings out sharply the differences between individuals and individual households—in their capacity for work, work discipline, skill, and strength. That yield piece-rates highlight the differences between households is also reflected in the fact that the coefficient of variation in per capita earnings under this system (0.25) was higher than under the surface piece-rate system (where the coefficient of variation was 0.17).

Significance of the Task

From our 1976–7 records, as we have seen in Chapter 6, groundnut operations as a whole occupied about 3 per cent of total agricultural labour time in terms of calendar days, and a little under 4 per cent in terms of eight-hour days. It is a more important component of a scheduled caste landless agricultural labourer's labour time than that of a non-scheduled caste landless labourer, and of male labour time than female.

Group work at Gudalur represents a relatively free wage labour form. Migrant workers in Gudalur are relatively free of the hierarchies and socio-economic environment of their home village; they are also not members of the social arrangement in Gudalur. For the time they are in Gudalur, harijan workers against whom untouchability is practised in their home village eat and drink in the local tea-shops and eating-places along with other workers; their distance from the home village and the custom they provide the local shopkeepers locate them in a sort of caste limbo.[37] The migrant workers do not perform labour services for their employers. Individual workers have even less contact with their employers than in contract labour in the village—often members of the group do not know the owner of a field. The pressure of work during the groundnut season is such that even the labour contractor, while maintaining contacts with individual employers, does not have to do odd jobs and labour services for these employers to maintain 'good relations' with them.

The weeks in Lower Camp are also a real break from the monotony of life in the village. The first week in camp—before exhaustion really begins to set in and the damp gets into everything and parents begin to miss their children and the meagre creature comforts of their homes in the village—is festive, joyful. There is a thrill in living roughly sheltered and out-of-doors with other worker households, in a camp set near the foothills in a place of arresting scenic beauty; walking long distances, fording the river, and crossing bridges to get to fields at the very foot of the hills and along their lower slopes; working together and in large numbers, away from the caste-bound, too-familiar village; working harder than at Gokilapuram, earning more, and being able to spend a little more and eat a little better than at home; besieging shops and vendors that have set up season-long trade when work is over in the evening—the atmosphere is referred to by the workers themselves as that of a festival.

When the season is over and the members of the group finally leave, the women of the group will be in time for weeding the paddy crop at Gokilapuram.

[37] See Weber (1979) and Breman (1978) on the separation of the migrant worker from the social milieu familiar to him or her.

Water Management and Irrigation Labour

The system of management of irrigation water in the village—the system by which this vital resource is taken to the field—has distinct features; it is also a specific sub-system of the social structure of the village as a whole. Even though plots of land in the command area of sources of surface irrigation (canals, tanks) are individually owned, the tasks of water management and of maintaining irrigation works have traditionally been undertaken not by individual cultivators, but by the community of landowners.[38] Traditional institutions of irrigation control emerged in village contexts in which democratic institutions were ill-developed or non-existent and contemporary village-level water control institutions often bear the stamp of the traditional social order. This section deals with irrigation workers (*neerpaichi*) and guards (*kavalkarar*) in canal-irrigated tracts of Gokilapuram; it includes descriptions of the method of appointing *neerpaichi* and *kavalkarar*, the extent of land irrigated by them, the tasks of these workers and certain features of the conditions of work, and changes that have taken place in the last few years in the irrigation work system. There is also a brief discussion, by way of contrast, of irrigation labour on land irrigated by wells.[39]

The reference period for this section is 1976–7 to 1981–2. As it happened the agricultural year 1977–8 was the last year in which the system existed in precisely the manner that is described here; later parts of this section outline the changes that took place between 1978 and 1981. During this period, there were two irrigated crops grown annually on surface-irrigated land, and the whole area during these seasons was given to paddy.

Gokilapuram is irrigated by the Palayamparavu channel, a right-bank channel that branches off from the river near Surulipatti village and irrigates land in the revenue jurisdictions of five villages—Narayanathevanpatti, Kamayakavundanpatti, Gokilapuram, Karunkattankulam, and the eastern fields of Uthamapalayam village. Within the revenue precincts of Gokilapuram, the canal irrigates a net sown area of 685 acres; of this, 105 acres (15 per cent of the command area) constitute an indirect command area, irrigated through the sluices of the Kalaneerkulam tank. The village is irrigated through a system of field channels. There is a turn system between villages for drawing water, and an intra-village turn system (*murai*) between different segments of the command

[38] Saraswathi Menon writes of 19th cent. Thanjavur district that 'water management and the preservation of the irrigation works was crucial to the cultivation of the district; and it required the coordinated activity of all the landowners. Despite the rise of individual property, throughout the nineteenth century, the management of irrigation water appears to have been largely controlled by the cooperation of the community of landowners, who would decide the hours during which water could be drawn off for particular fields' (1983, p. 501).

[39] The material in this section consists of information collected through interviews conducted in Gokilapuram and other villages in the Cumbum Valley in 1981–2, data collected in the household survey in Gokilapuram in 1977, and case-study material collected between 1977 and 1981.

area of the system within the village. (The Periyar irrigation system in the village is discussed in greater detail in Appendix 7.)

Until 1978, a meeting of all owners of surface-irrigated land (*nanjai*) was called each May and held in the courtyard of the co-operative society in the village. The meeting included all owners and mortgagees of surface-irrigated land; tenants were not represented. Persons from other villages who owned land in Gokilapuram did not generally attend the meeting and they were informed of its proceedings later. The number of households represented at the meeting was over 100, and, with the persons who came to watch what was happening, the total attendance was about 150. At the meeting a committee to represent landowners was proposed; village landlords would have met earlier and decided the composition of the committee. The honorary head of the committee was a representative of the P. T. Rajan family[40] and the other persons on the committee were—and represented the major groups among—prominent Maravar and Chettiar landlords from Gokilapuram. This committee was in charge of the surface-irrigation system in the village, of making representations to the Public Works Department, and, when necessary, of collecting money for the maintenance of the channels. The meeting also appointed two agents, who had to be landlords and of the Maravar caste, to be in charge of organizing irrigation work.

The 'minutes' of this meeting, of the proceedings of the common agreement, were then recorded in a notebook, and the participants signed their names in ratification—an interesting example of how an outward sign of routine democratic procedure can be pressed into service to ratify, after the fact, a decision of the landlords. The custom was for the first and second places in the minutes book to be kept for the signatures of the representative of the P. T. Rajan household and for Savarimuthu Udayar, a rich and prominent landlord with substantial surface-irrigated land holdings, who is resident in the neighbouring village of Rayappanpatti.

The agents, on a day following the meeting, announced among Maravar agricultural labour families that irrigation workers would be chosen. In general, not more than 20 or 22 came, since most of those who were chosen would have been irrigation workers in the previous year. Eighteen agricultural labourers would be chosen to be irrigation workers; all of them belonged, as they do to the present day, to the Maravar caste. Almost all of them were landless, although some had small plots of land (the time of the irrigation worker himself cannot be occupied by the tasks of a cultivator). The modal age of these workers in 1977 was 35 to 40 years; four of them were more than 50 years old.

The working year of the *neerpaichi* began soon after. The picture of the irrigation worker, metal-tipped staff in hand, emerging from his hut and

[40] The family of P. T. Rajan (1892–1974), major landlord, barrister, public figure, and Minister in the pre-Independence Madras Presidency, has vast land holdings in the Periyar command area, including Gokilapuram. The family seat is at Uthamapalayam.

moving out to the sluices on the canal early in the morning of the first water-turn in June, is one of the most representative images of the beginning of the new agricultural year in the Periyar-irrigated tracts.[41]

Workers' Tasks

Under the terms of the common agreement there was little respite for an irrigation worker: he was on call throughout the year on surface-irrigated land, and was not allowed to take on any other waged work in the village even on the few occasions that he might have been able to do so. The tasks of the irrigation workers and crop guards can be broadly divided into those concerned with irrigation and cultivation and those concerned with guarding the risen crop. In Gokilapuram until 1978, both these sets of tasks were done by the same persons.

The main shutters and sluice gates that are on the Palayamparavu channel are opened and closed by workers of the Public Works Department (who are called *laskars*)[42] in accordance with the turns between villages. The internal control of irrigation water in the field channels of the village is the responsibility of the irrigation worker. It is his duty to

- open the sluices to field channels in accordance with the intra-village irrigation turn system (*murai*);
- clear the channel so that water may flow to the field that is to be irrigated;
- open the water inlet, pipe, or channel that leads to each field and regulate water in the field as required by the stage of plant growth;
- dam the water in a channel when the level of a field is higher than the channel, in order that water is lifted to the field;
- block breaches in the field bunds and holes in the bunds made by rats, through which water flows out of the field (when peasants and agricultural labourers construct or strengthen bunds at the beginning of each crop season, the irrigation worker is generally at the field to assess and criticize the work—a shoddily constructed field bund will add substantially to his work-load over the season);
- prevent tampering with field walls and block breaches made by persons who try to drain water from a freshly fertilized field to an adjacent one;
- construct and trim the border wall for the seed-bed of the second crop (the nursery for the second crop is generally raised in a corner of a field before the first crop is harvested in the rest of the field);

[41] Kathleen Gough's fine descriptions of the annual round in her study area in Thanjavur district (1981, ch. 11) capture the significance of the beginning of the irrigation year; she writes that the 'arrival of water was a momentous, ceremonial event' (p. 220).

[42] Before the main channel came into the charge of the Public Works Department in 1922 and was under the control of 'irrigation panchayats', the task of controlling the outlets on the main channel was done by workers who were part of the group referred to in revenue records as village servants. The irrigation worker was called a *neeranikkam*. The *neeranikkam* was from a Pallar family and there was a service *maniyam* (revenue-free land allotment)—of which the family has long been dispossessed—in the name of the *neeranikkam*. (On the decline of *maniyam* land, see S. Menon 1983, p. 486.)

- prevent cattle from grazing on paddy-fields and see that grass from field bunds is not cut by persons other than the owner of the field; and
- guard the crop after it has risen and until the last field has been harvested (this is the *kaval*, or guarding, task).

In addition to this wide range of what may be called the regular tasks of irrigation and guarding, the irrigation workers also do the top-dressing of the paddy crop for all fields, sow the seeds in the nursery for some of the landowners, and do all minor hoeing work repair jobs (*kudimaramath*) on field channels. When major work is to be done on field channels, each owner is asked to contribute labour for the repair of the channel in numbers proportionate to the size of their holding in the command area of the channel. When major repairs are to be undertaken on the Palayamparavu channel, the three villages—Uthamapalayam, Karunkattankulam, and Gokilapuram—contribute labour.[43]

When the common agreement was in existence, in times when there was a shortage of water—this could be at different times of the year, for instance, in June at the beginning of the first crop, or in December–January, during the growth period of the second crop—workers went to the head sluice with the agent, packed empty fertilizer sacks with sand, and made bunds in the river to divert water into the channel, and then policed the channel from sluice to village, closing illegal water exits along the way.

Irrigation workers also had tasks that were not connected with the fields, as on the occasion of village temple festivals. Until 1978, one or two persons from among them were assigned the task of announcing, from street to street, the temple festival in the Chithrai month (April–May) at the Kaliamman temple. At this and the other major temple festival in the Tamil month of Purattasi (September–October), they also helped with tasks connected with organizing the rituals and regulating the crowds when the rituals were on.

The working year of the irrigation workers begins with the opening of water on 1 June, and is complete with the second harvest, in early March. During the season when the surface-irrigated tracts did not receive water, from March to June, some of the workers contracted with individual owners to work on the fields, to

[43] The tasks of maintaining irrigation works and of repairing them (particularly when there were urgent repairs to be done) have been historically associated with forms of forced or compulsory labour, including compulsory labour that had legal sanction. The Madras Compulsory Labour Act of 1858, for example, was intended to be an 'Act to make lawful compulsory labour for the prevention of mischief by inundation and for enforcement of compulsory labour on certain works of irrigation in the Presidency of Fort St. George.' The Act's preamble declared that it was 'necessary for the common good to make it obligatory on persons of the labouring classes, when duly called on, to unite their labour to prevent . . . breaches or to repair them instantly', and that it was 'expedient to make legal provision for the enforcement of the duty, which by local custom is incumbent on village-communities, to furnish the labour required for the execution of certain works for the purpose of irrigation and drainage.' See, in this regard, the illustrative descriptions of forced labour in irrigation in Tirunelveli in Ludden (1985, pp. 142–4, 174–5, and *passim*); also S. Menon(1983, pp. 500 ff.) on forced labour on irrigation works in 19th cent. Thanjavur.

- keep guard over fields in which green manure or fodder crops were grown;
- keep guard over manure heaps laid out in the fields;
- do odd jobs, particularly involving hoeing work, in the village; and
- do manual tasks on surface-irrigated land, such as hoeing work or helping with basal manuring.

The work year for an irrigation worker under the common agreement was spread over about 280 calendar days. The work, which was distributed between guarding the standing crop, irrigation, and other tasks, was not of uniform intensity through the year. Crop-guarding, which took a large part of the work time, does not involve the kind of manual labour that irrigation and other tasks do; it requires that the worker guard the field through the night and be in the vicinity of the field and on call during the day. With the qualification that the intensity of the work over the year was not uniform, the average daily working time of a worker was about 10 hours per working day (350 eight-hour days a year); in the working year, irrigation and other tasks took about 54 per cent of work time and guarding the standing crop took about 46 per cent.

Before 1978, when the common agreement was in force, the agents allocated work between the workers.[44] Each irrigation worker was allotted a set of fields irrigated by a particular field channel, from 12 to 24 acres each, the extent varying according to the work-load that the nature of the area, particularly its topography, demanded. (Where the land has ups and downs and where there are fields that are above the level of the channel, there is relatively more work for the irrigation worker than where the land is flat.) If there were complaints that the load on an irrigation worker was too great in the *kodai* (June to October) season, he was given a new area in the *kalam* (October to March) season. For guarding the risen crop, the entire area was divided into zones, and these zones allocated between all but two of the irrigation workers, who did the rounds every night. The beats were changed every week.

The workers and agents met regularly, generally once every three or four days and at least once a week. At their meetings they discussed their work and problems that may have arisen: if a person had ruptured a field bund, encroached on another's field, tried to interfere with the irrigation, grazed his cattle on the fields, drained another's field soon after top-dressing was done, or tried to steal or tamper with another person's crop, the matter was raised at the meeting. If there were complaints by a landowner about the work of an individual irrigation worker, it was raised at the meeting by the agent. The workers also reported on the state of the crop to the agents (and lazy landlords

[44] The owners of canal-irrigated land who are resident in Gokilapuram do not control irrigation over the whole extent of the revenue village of Gokilapuram. The tasks of the workers who were appointed by common agreement previous to 1978, and who are directly employed by the owners of irrigated land today, are confined to the command area of five sluices (with direct command areas) with a total coverage of 297 acres of surface-irrigated land that is located closest to the village settlement. The responsibility for irrigation duties over the rest of the command area (which includes the whole command area of the tank) rests with landowners resident in Rayappanpatti whose canal-irrigated land holdings are located in the revenue jurisdiction of Gokilapuram village.

who hardly watch their crop considered this obligatory for them)—when it was time for transplantation, when weeding was overdue, when the fertilizer for top-dressing was not bought, when leaves in a field began to show signs of plant disease, or when a ripened crop remained uncut.

Remuneration

In 1977, the basic payment that cultivators made for irrigation and *kaval* on surface-irrigated land was 2 *kalam* (one *kalam*, or bag, is equal to 58 kilograms of dry paddy) per *sey* (2.4 acres) per crop, or about 48 kilograms of paddy per acre per crop, or 96 kilograms per acre per year. This amount was collected at the end of each crop from all cultivators, and to this fund was added whatever *inam* or extra payment that a landowner would add to the regular payment. The total collection was divided into 19 shares: one each for the workers and one single share for both the agents. In 1977, the irrigation workers got about 13 bags of paddy each per harvest, which is 26 bags (or 1,508 kilograms) of paddy per year. The market price of 26 bags of paddy in 1977 was Rs 1,560.

There had been changes, over the years since 1966, in the basic remuneration. Before 1966, irrigation work and guarding were done by different sets of people, only guards being appointed by common agreement. The combined wage for irrigation and guarding (where irrigation was done by hired labour) was about 66 kilograms of paddy per acre. In 1966, when the tasks were merged, workers lost out in that the payment for the combined task was about 56 kilograms per acre. Part of this payment was in sheaves of grain and seed paddy left over after sowing. A few years later, workers asked for wage measurement to be standardized, and, since the amount that could be threshed from a sheaf varied, for it to be substituted by a standard half-bag payment. The basic rate then rose to the equivalent of 72.5 kilograms per acre. The 1977 rates were introduced in 1973 or 1974, after the new varieties had completely altered the levels of production in the village.

To sum up the data on basic remuneration, there was an increase in the rate per acre paid to an irrigation worker from about 56 kilograms of paddy per acre per year in 1966 to about 96 kilograms per acre per year in 1977, an increase in earnings of about 17 per cent per worker from about 1,279 kilograms of paddy per worker per year to about 1,508 kilograms. Over this period, the amount paid to an irrigation worker as a proportion of the total product marginally declined: assuming production to have been 1.45 tonnes per acre in 1966 and 3 tonnes in 1977, the share paid to them was 4 per cent in 1966 and 3 per cent in 1977.

An irrigation worker got less per eight-hour day of work than, say, a cash-paid, daily-rated male worker.[45] In 1977, the cash equivalent of his basic annual remuneration, with additions as *inam*, was Rs 1,560. Setting this alongside the work schedule (and reiterating the qualification that the intensity

[45] See the discussion earlier in the chapter of comparative wage rates for daily-rated tasks and for long-term labour.

of work over the year is uneven), the remuneration per eight-hour day was Rs 4.47. The wage-rate for an eight-hour day's hoeing work in the village in 1977 was from Rs 6 to Rs 7.50.

There is, however, a trade-off here: although the rate per day is lower for the irrigation worker, the total earnings in a year are higher than for the average male casual labourer. A worker receiving Rs 7.50 a day would have had to receive 208 days of employment to earn as much as an irrigation worker in the year; as we have seen, the average number of days of agricultural work for an adult male in the village that year was 118. The earnings from 118 days of cash-paid, daily-rated employment would have been only 57 per cent of the annual earnings of an irrigation worker that year.

There were certain other elements in the remuneration of irrigation workers over this period. On payment of lease money to the owners of surface-irrigated land in the village, duck-keepers are permitted to let their ducks graze on the land before sowing. This money used to be divided between Gokilapuram and Rayappanpatti villages and part of the share kept in Gokilapuram was used by the agents:

- to pay an allowance to the workers when they had to take a bus and go upstream to do manual work at points on the canal near Rayappanpatti or at the head sluice;
- to pay, on occasion, for hoeing work on the channels: irrigation workers would not be paid at the normal daily rate for hoeing work, but deliberately less (they were paid Rs 5 against the regular Rs 7.50), the agents' explanation being that those who receive a total annual wage 'could not expect' to be paid at the regular daily wage rate for such tasks;
- to buy batteries and bulbs for the workers' flashlights and to buy whistles for their guarding work; and
- to buy a *dhoti*, shirt, and towel cloth for each worker at the Thai Pongal festival.

During the first crop, workers got 10 ducks' eggs each. Some of them also received, from individual owners, a small amount of grain during the preparation of seed-beds for the second crop.

There were two components of the total earnings that were taken away during this period. Irrigation workers were paid cash, Rs 1 in some cases and not more than Rs 5, by landholders for scattering fertilizer over the crop. This was an informal payment, and one that the agents stopped. For about 10 seasons after 1966, the workers were allowed to fish in the channels. This was later disallowed.

The agents said that in the early years of the common agreement, they could use group pressure to punish workers—mere exclusion from the group served as punishment. Fines and suspensions, however, quickly became regular forms of punishment after 1966. (An agent said: 'Earlier we could just say "sit aside for two days" and that itself was like a punishment; now we have to suspend workers or impose fines.') Punishment was imposed for violating the ban on fishing; for not coming to work, particularly during the irrigation turn, without letting the agent know; for sleeping over a guarding job; for permitting a field wall to crumble or cattle to graze or a person other than the owner to cut grass; for not preventing a person from tampering with the crop of another; and so on.

Some conclusions emerge from the preceding sections on the tasks and remuneration of irrigation workers and crop guards. The duties of irrigation workers, while centring on the task of water management in fields, encompassed a wide variety of other tasks. These included hoeing work on field bunds and channels, application of chemical fertilizers, work connected with the nursery stage of cultivation and, in some cases, with the construction of bunds and basal manuring. In some years, they went to the head sluice, diverted water into the channel, and prevented water from being drawn off illegally.

The irrigation worker must have a close knowledge of the irrigation system, of the field and channel layout of the village, and of the particular features of cultivation in each segment of the command area. He must know the water requirement of the paddy crop at each stage of its growth. He must be able to monitor the crop at every stage, to follow its progress, and detect symptoms of plant disease when they appear. There must be, in short, a comprehensiveness in his perception of paddy cultivation in the village.

There have been changes in the tasks of irrigation workers over time, such as changes in cultivation practices and water management that accompanied the introduction of new varieties. There have also been important changes, as we shall discuss further on, in the tasks of irrigation workers-cum-guards and in the method by which landowners recruit irrigation workers and allot them work. However, corresponding to the fact that certain major features of cropping patterns and surface-irrigation technology have remained intact over an extended period, it is also true that important elements of continuity characterize the skills required for tasks connected with irrigation and water management in the canal-irrigated tracts of the village.

The combination of irrigation tasks and guarding work represents a heavy work-load: an average of about 10 hours per day during the actual cultivation period of about 280 calendar days, about 54 per cent of the work time being spent on irrigation and other tasks and 46 per cent in guarding the standing crop. Annual earnings per worker increased by about 17 per cent between 1966 and 1977, although the share of the workers' wages in total product declined. Over the period, workers gained, to an extent, a standardized payment, with a larger proportion of wages paid in volumetric measures rather than in sheaves, whose yield of paddy, when threshed, varies. While the average wage rate per day for an irrigation worker was lower than for a daily-rated casual worker, total earnings were much higher than an average male worker could expect to earn, given the low average number of days of employment that a daily-rated hired labourer receives in the year.

While the work-load is heavy, the task is not such that labour time is uniformly distributed over the working year. For the individual worker, there is a discontinuity about the scheduling of work time over a work period; the nature of the task and the system of field channels and cultivation also make for squandering of labour time. To take an example: on a day on which there are two small (0.1 acres each) seed-beds to water at the end-reaches of a field channel, an irrigation worker may have to spend two hours up and down the channel, dredging the channel at several points between sluice and field, packing the channel walls, and blocking or regulating (with earth, stones, banana-tree trunks) the flow at every branch and field inlet, in order to get water to the seed-beds. And where plots are fragmented and farming operations not synchronized, if just one more plot is to be watered the next day, the entire procedure will, in all likelihood, have to be repeated. Fragmentation clearly makes for waste in the patterning and consequent utilization of field channels; workers themselves point out, for example, that the consolidation of seed-beds for the command area of a single channel at a single or a few specific points would save substantially on water and labour time.

The System after 1978: A Summary of Change

The system of recruitment and allocation of work described above was in effect from 1966 to the end of the agricultural year 1977–8.

Until 1966, with a three-year gap between 1960 and 1963, crop-guarding tasks in the fields were done by a group of workers under the terms of a common agreement (although the specific terms under which guarding was done had changed over time). Irrigation tasks were done by individual peasant cultivators for the plots that they cultivated and by agricultural labourers hired for the task of irrigating fields. In 1966, the system that has been described was instituted; at the suggestion of the head of the P. T. Rajan family, the system was made roughly similar to that of neighbouring Karunkattankulam, where the family has vast wet-land holdings and where irrigation and guarding tasks were done by the same workers. From 1966 to 1978, as we have seen, the system in Gokilapuram was extended to cover irrigation and guarding tasks.

From 1978, irrigation work and guarding have been separated. Irrigation workers now contract with individual cultivators for employment, and are no longer appointed by agents chosen by the committee that is formed at the meeting of landowners. Guarding work on the surface-irrigated tracts is still done by common agreement; two Maravar agents are appointed by the committee and the agents appoint six agricultural labourers from Maravar households to guard the standing paddy crop.

Even this system has not been functioning smoothly. The agents appointed in 1978 were not reappointed before the beginning of the next agricultural year—they continued to act as agents for five consecutive seasons without the common meeting being held. After that, crop-guarding in common in the canal-irrigated area did not exist at all for one crop; it was revived in the agricultural year 1981–2.

None of the old irrigation workers are crop guards under the present system. Of the 18 workers who were employed for irrigation and guarding tasks before 1978, 14 to 15 still do irrigation labour. There are no new entrants to the job; a single person who attempted to join the irrigation work-force after 1978—a person from Narayanathevanpatti, south of Gokilapuram—left because of his unfamiliarity with the irrigation pattern of the village.

While contracts are now between workers and cultivators (in some cases, two workers make their contracts together), these are made in such a manner that individual workers (or pairs of workers who have teamed together) irrigate plots of land that are contiguous and irrigated from the same sluice. Although contracts are now individual, it is still the case that no cultivator, even a poor peasant, does irrigation tasks with family labour: all of them employ irrigation workers for irrigation work.

The tasks of the irrigation worker no longer include guarding the standing crop. The number of acres irrigated per worker has risen, with 15 workers now irrigating the area formerly irrigated by 18 workers. There has been, as can be expected, a decline in the rate per acre paid to irrigation workers after the separation of the task of guarding the standing crop—they are now paid at the rate of about 72.5 kilograms of paddy per acre per year. Payments from the common fund are no longer made to them and they also do not have any tasks at the Chithrai festival in the local temple.

An important change for the workers is that they are now free to work at tasks other than those connected with surface-irrigated land—in the process of change irrigation workers have acquired, in part, the status of daily-paid casual labourers. When time permits, some of them (particularly those who are young or who have a relatively small area to irrigate) work at tasks such as hoeing on dry land and groundwater-irrigated land, ploughing, guarding tamarind and fruit groves, and guarding and irrigating groundwater-irrigated fields. Some even work in heavy manual operations in the tamarind groves.

In short: after 1978, contracts became individual and a common system across 300 acres was modified according to the command area of each sluice (that is, modified by *ayacut*); at the same time, irrigation tasks are still done exclusively by irrigation workers. While irrigation tasks were separated from guarding, the specific tasks of irrigation and the deployment of labour time in these tasks remain essentially the same, although traditional functions such as those in the Chithrai festival are no longer part of the job. Wages are only for irrigation and are collected by workers directly from their employers; some of the older elements in the remuneration are no longer there. As was not possible under the common agreement, irrigation workers now also work as daily wage workers in a variety of tasks in the village.

I shall briefly outline some of the factors associated with the dissolution of the system in 1978.

In a period of changing socio-economic conditions in the village, the old system, with its rigidities and dominated as it was by Maravar landlords, was becoming obsolete. Maravar owners were losing their traditional place of pre-eminence in the ownership of land, and landowners of other castes, no longer in the same subordinate position *vis-à-vis* the Maravars in the distribution of ownership of surface-irrigated land, had no similar stake in the continued existence of the system. After the opening of the PTR channel (which is an extension of the Palayamparavu channel: see Appendix 7) a new and unprecedented level of water supply began to be received by the village; there was a consequent loosening of the turn system for village irrigation (*murai*), and one technical rationale for a common system, the restricted availability of water, became less pressing. For their part, individual Maravar landlords were either unwilling to work as agents or were unable to work together. For a year, four agents were appointed instead of two, but even this could not continue. Nor would they agree to relinquish the task of being agents to persons of other castes in the village.

A reason frequently stated by Maravar landlords for allowing the old system to decline is that the irrigation workers no longer 'showed the same respect', and that they thought that the workers had to be 'taught a lesson'. One landlord and agent said that the workers had begun to run their own 'underground'—referring to occasional meetings between a group of irrigation workers without the agent, including one meeting to plan a demand for higher wages.

Irrigation workers in Gokilapuram have not, as a matter of fact, been unionized; nor have they formed or joined any regular organization of irrigation workers, as was the case with irrigation workers in some villages in the southern part of the Valley in the early 1970s. There have, however, been instances of irrigation workers in the village acting together spontaneously and as a group (see Appendix 8); in a village situation where the market for hired labour based on simple contracts between employers and workers has seen significant expansion in the recent period, the kind of subordination and code of conduct that landlords used to expect of the irrigation workers is quickly going out of date. Even as landowners, led by the landlords, entered a common agreement to organize irrigation and irrigation labour, irrigation labourers began to be aware of their common identity as workers and began to act on that awareness.

Irrigation Labour: Groundwater Irrigation

In contrast to surface-irrigated land, the nature of irrigation labour on groundwater-irrigated land is determined, in the first place, by the fact that the source of irrigation is privately owned and the utilization of water is controlled by the decisions of individual cultivators. The other important feature of groundwater-irrigated land in this regard is the fact that the crops grown vary from field to field; it also often happens that one well irrigates different segments of a single consolidated operational holding that are planted with different crops. Surface-irrigated land, by contrast, is wholly planted with paddy.

There are two field operations connected with irrigation on fields irrigated by wells. While these tasks demand certain skills associated with particular operations, they do not involve an overall understanding of the entire crop cycle, as irrigation work on paddy-planted, canal-irrigated land does. When the source of irrigation is a well, irrigation labour is more operation-specific. The first task is an operation that involves field preparation for irrigation: cutting and forming irrigation channels and plant beds. The size and pattern of channels and plant beds vary from crop to crop—they depend on the crop (or crops, where there is inter-cropping) grown, the size and spread of the plant, the spacing of plants in the field, the type of soil, and the slope of the field. This task is done by men working with hoes employed at time-rates. Even small cultivators generally hire workers for this task. In villages where groundwater-irrigated cultivation is advanced and is the dominant sector of cultivation and where the demand for labour for field preparation is high, fields are often given out on contract by those who do not work the fields themselves.

The second operation is, of course, that of irrigating the field itself. This is done when the electricity is on and involves dredging and clearing the channels while water flows. Some landlords hire workers on a permanent or semi-permanent basis for this task, paying the worker monthly wages; often such workers look after the pump room (many make it a part-time residence), perform irrigation tasks, guard the field and crop, and supervise labour employed for other operations on the field. In the case of peasant families working land irrigated by wells, whose members are themselves engaged in the major manual operations on the land, irrigation tasks and guarding the crop in most cases (particularly among poorer households) consume the highest share of male family labour on groundwater-irrigated land.

It is to be emphasized that in contrast to surface-irrigated land, there are no systems of irrigating or guarding land in common on groundwater-irrigated land. Here tasks are done exclusively by hired labour in the case of households whose members do not work on the land (such as those of the landlords), or by family labour, or, in some cases, by a combination of both hired and family labour.

Some Traditional Tasks

There are different types of traditional tasks in the village. Some are associated with artisan work: the Achari caste provides woodworkers, goldsmiths, and blacksmiths, the Potmaker caste moulds earthenware. There are traditional service functions, such as those performed by the village barbers, washer-people, and medicine men. Other services include preparing corpses for cremation, done by a Chakkiliyar family, and beating drums, done by Parayar families. There are also tasks, as we have seen, associated with irrigation and traditional tasks performed by Maravar irrigation workers. There are non-labouring traditional functions that are associated with particular castes: the traditional village headman, for instance, comes from a particular family among the Maravars, and he has certain prerogatives and 'duties' at, to take an example, village festivals such as the one celebrated at the Kaliamman temple every April.

This section deals briefly with certain menial and agricultural labour functions; it does not attempt to be exhaustive, but to illustrate certain broad features of these tasks. Two points can be noted at the outset. First, traditional tasks are almost always caste-specific and, secondly, menial traditional tasks are almost always performed by workers of the scheduled castes, particularly the two most oppressed among them, the Parayar and Chakkiliyar castes.

Chakkiliyars are the most oppressed caste of all in the village; their women traditionally clear dung and pile it into farmyard heaps and men from the caste are employed by the panchayat as sweepers and latrine-cleaners. People from the caste, particularly women, are considered expert at work on the threshing-floor. Only Chakkiliyar women are employed for threshing dry-land coarse grain. This is a task that provides little reward: with the stagnation and decline that has occurred in the production of unirrigated cereal crops, there is little grain to thresh and threshing coarse grain is now a minor operation. In the case of paddy, it was traditional for Chakkiliyar women to do tasks on the threshing-floor other than actual threshing. They were, and are, considered the most deft at winnowing and removing the chaff and dust and mud that sticks to the paddy and at removing stones from paddy and from rice.

In earlier years, workers from Chakkiliyar and Parayar families were not taken on for harvest work, a feature of the work system that lasted particularly long in respect of the Chakkiliyar caste. The members of these castes had no regular way of earning paddy during the harvest season, and rights were granted them to sweep the roads and collect the grain that fell on the roads when sheaves were carried from field to threshing-floor. The custom is evidently an old one, and by all accounts these rights of certain families to sweep existed before the Periyar system brought water in the 1890s. The right to sweep roads, paths, bunds, and canal banks was apportioned between Parayar and Chakkiliyar families; families were given specific stretches of pathways and roads to sweep, and the right to sweep them has been inherited as property by their descendants. In the case of one family, the right to sweep—'from the peepul tree to the canal bridge'—was inherited by a Parayar agricultural labourer from his elder sister. He had no children, and, around 1977, he sold the right to another Parayar household for a down-payment of Rs 40.

The average pickings are about one bag (58 kilograms of dry paddy) per season: they range from half a bag to two bags. Pickings have fallen in recent years. First, the new varieties do not shed as much grain; secondly, grain drops on the ground only when carried as head-loads, and owners now take tractor trailers to points on the road near paddy-fields and transport sheaves to threshing-floors. This change has particularly affected those whose rights extend on main roads.

It is also women and children of these castes that do most of the gleaning during the harvest season. Gleaning, however, is on the decline. Tough and open-eyed, most owners today stop women and children from picking what they can off field-floors and stand guard against harvest labourers colluding with the gleaners in the business of collecting grain.[46]

There are some noteworthy features of tasks such as these. I have mentioned two: the fact that they are generally caste-specific and that the most lowly of the tasks are performed by scheduled caste workers. Their other important feature is that they are peripheral to actual production tasks. In the case of the sweeping rights, the traditional task represents a remnant of the past that is now a minor survival mechanism for the poorest people of the most depressed castes in the village.

Some General Features of Agricultural Employment in the Cumbum Valley

At this point, certain general features of the labour process in agriculture in Gokilapuram can be considered.

Constraints on the supply of labour were not a major feature of the market for hired labour in Gokilapuram. Different interrelated aspects of the labour market contributed to this situation. Aggregate labour supply far exceeded the aggregate demand for labour: witness the fact that a landless agricultural labourer received employment, on the average, for only 140 days a year. In the peak seasons, forms of labour hire that have adapted to seasonality are available to employers: examples of such forms are the employment of large groups of contract workers and of groups of migrant labourers that have contacts with landowners within the village. The mobility of the labour force and new, flexible forms of labour hire have made for a work-force that is readily available to employers at peak seasons. In nearby Gudalur, where the demand for labour on paddy land and groundwater-irrigated land coincided at one point in the year, the peak-season market for labour was tighter than in Gokilapuram. Here

[46] A situation that bears a conspicuous resemblance to the one described by the Hammonds: 'There was another respect in which the resources of a labouring family were diminished towards the end of the [eighteenth] century, and this too was a loss that the rich thought trifling. From time immemorial the labourer had sent his wife and children into the fields to glean or leaze after the harvest. The profits of gleaning, under the old, unimproved system of agriculture, were very considerable . . . With the introduction of large farming these customary rights were in danger. It was a nuisance for the farmer to have his fenced fields suddenly invaded by bands of women and children. The ears to be picked up were now few and far between, and there was a risk that the labourers, husbands and fathers of the gleaners, might wink at small thefts from the sheaves' (1913, pp. 107–8).

recent changes in cropping patterns (with a shift to planting coconut on groundwater-irrigated land) have reduced peak-season demand. Employers in Gudalur also make extensive use of migrant labourers during the groundnut harvest and of different types of piece-rated wages to speed up work during peak seasons.

It has been suggested in the Indian literature that an important means by which employers ensure a supply of hired labourers in the peak season is by means of an interseasonal labour-tying arrangement, that is, by offering workers privileged employment in the slack season or by means of loans advanced by employers to workers (Bardhan and Rudra 1981; P. K. Bardhan 1984*a*, ch. 5).[47] In her important survey of the literature in the field, Kalpana Bardhan sums up the argument thus: 'It seems that employers use the slack market to recruit for the peak operations when the market will get tighter. A labourer gets employed in the slack period not primarily by bidding down current wage further, but rather by committing concessional supply for part or whole of the peak period' (K. Bardhan 1983, p. 43). It is certainly true of Gokilapuram that landlord-employers act to retain and strengthen their control and authority over the labour force, and that it is in their interest to be able to compete for and employ, at different times of the year, those workers who are known to be the best. And though there is no supply constraint, landlords still have to undertake, generally through employees or underlings, the task of efficiently deploying workers on different fields at different times of the year, particularly during peak seasons. At the same time, I do not know of such explicit interseasonal labour-tying arrangements in Gokilapuram, by means of which individual workers enter into specific arrangements to work for lower wages in the peak season in order to ensure that they are employed in the slack season. The information from the village also indicates that when employers in Gokilapuram give loans, including interest-free loans, their motivation in doing so is unlikely to be the need to ensure a supply of labour in the peak season. On the contrary, workers place themselves at the beck and call of specific employers to ensure that they receive employment of some sort whenever it is available. For many agricultural labourers in Gokilapuram an important means of employment and sustenance in the slack season is the performance of labour services; for some of them labour services can be the major form of employment at this time.

In an important contribution to the literature, Utsa Patnaik has proposed that with the adoption of green-revolution technology, 'methods of labour control begin to change in the direction of longer-term contracts which carry, built into them, debt-bondage of a new type' (Patnaik 1985*b*, p. 28). A point of sharp contrast between the labour market in the Cumbum Valley and the situation described by Patnaik is the fact that in the Valley in the contemporary

[47] It must be noted that Bardhan has taken care to qualify this position; he also argues that there is an inverse relation between the occurrence of labour-tying arrangements and levels of unemployment (P. K. Bardhan 1979*a*).

period, it is *casual*, generally *short-term* labour that is the most prevalent form of labour hire, whether workers are employed at time-rates or piece-rates, whether they are employed singly, in small groups, or in contract 'gangs'. This is in spite of the fact that workers at certain special tasks who are employed on long-term contracts and farm servants are also part of the labour force. As far as this region is concerned, there has not been any general tendency for methods of control in agricultural tasks to change in the direction of long-term contracts or of long-term contracts based on relations of debt bondage between hired labourers and employers. Although 'green revolution' technology does raise overall labour demand and requires a more exacting deployment of labour ('more stringent requirements with respect to the time distribution of labour inputs', as Utsa Patnaik writes (Patnaik 1985*b*, p. 28)), it does not necessarily call for a proliferation of long-term contracts. Whether or not employers require a labour force employed on contracts that last through the crop season and longer, and whether, consequently, there is a general shift in methods of labour control towards long-term contracts, depend on the nature of the crops that are grown and the way in which the market for hired labour has evolved in a region.

Take as an example a landlord-employer who has 10 acres in Gokilapuram planted with paddy and some acres of lift-irrigated land planted with banana and vegetable crops as well as some tamarind and fruit trees. Given the labour input requirements of these crops, in terms of the quantity of labour power required and the acute seasonality of labour demand, there will be days of the year when the employer has to employ more than 100 workers on the various plots of land, as well as days on which very few workers are needed. It would be grossly uneconomical for this owner-employer, at the present level of technology, to have the work on this farm done by workers who are on long-term contracts. In any case, in this area of abundant labour supply, labour is readily available (or 'on tap', to use the phrase that appears in Utsa Patnaik's writings) without ensuring a labour supply through long-term contracts.[48]

[48] The importance of casual labour in the agricultural labour force and its predominance in different regions of the country have been widely noted in the Indian literature: see e.g. Kandaswamy (1964); Djurfeldt and Lindberg (1976); Dasgupta (1977); Parthasarathy (1977); Mencher (1978); Chattopadhyay (1979); Nagaraj (1981); Parthasarathy and Pothana (1983); Hatti and Raagard (1985); C. R. Reddy (1985); Drèze *et al*. (1986); Drèze and Mukherjee (1987); Gough (1987); B. Harriss (1987); van Schendel and Faraizi (1984). An exception to this phenomenon, and a useful corrective to over-generalization, comes from the data collected by Sheila Bhalla in Haryana in the early 1970s (S. Bhalla 1976). Bhalla divides Haryana into three agricultural zones, based on the degree of adoption of the new technology in agriculture. In her region A, which covers Karnal district and adjacent areas of abundant irrigation, 57% of male agricultural labourers were on long-term contracts. Bhalla's careful description of the 'making of a particular kind of rural proletariat' (the phrase is from Byres 1981, p. 437) has been an important contribution to the study of employer–employee relationships in Indian agriculture; the features of the production system that gave rise to or were associated with the widespread existence of this form of labour hire in region A in the early 1970s are still to be studied.

Were the new levels of labour demand generated by new production methods in agriculture from the 1950s to the 1970s met by employing increasing numbers of bonded labourers? The data from this study area show that they were not. There are still contributions to the literature that insufficiently recognize the fact that many types of dependent labour, representing different degrees of socio-economic dependence, are represented in the grey zone that lies between bondage and freedom. Tom Brass criticizes John Harriss's study of Nesal village in North Arcot district, Tamil Nadu, and Scarlett Epstein's resurvey, conducted in the 1970s, of two villages in Karnataka that she had surveyed in the 1950s for failing to recognize the existence of bonded labour (Brass 1986*a*). Brass asserts that in Nesal and in the irrigated ('wet') village studied by Epstein, unknown to the scholars who actually studied the villages, bonded labour increased as a result of irrigation and the expansion of the new technology in agriculture. The reasons for this increase, according to Brass, were largely the same in the two regions. According to him, in Harriss's study area,

> Those who benefited most from the development of the productive forces were the rich peasants who compensated for the high costs of technical inputs . . . by cutting back on wage costs which remained under their firm political control, that is by employing cheap bonded labour instead of the relatively more expensive free wage labour. (Brass 1986*a*, p. 56)

Following the same line of argument, Brass writes that Scarlett Epstein's restudy 'revealed' that

> the installation of canal irrigation in the 'wet' village had stimulated the development of a capitalist agriculture, and the demand for labour doubled accordingly. As in the case of the area studied by Harriss, this demand was met not by the employment of free wage labour, but rather by the utilisation of cheaper bonded labour, the latter increasing by about 243 per cent over the 1955–1970 period. (Brass 1986*a*, p. 56)

There is no serious empirical substantiation for this remarkable assertion, particularly of its reference to 'northern Tamil Nadu'; it is simply not correct to say that the increased demand for labour in parts of 'northern Tamil Nadu' where there was an extension of irrigation and new inputs was met by employing bonded labourers.[49] And, with reference to the discussion in the previous chapter, it is also entirely possible that there are villages in the regions of North Arcot district where high-yielding varieties and new irrigation facilities have been introduced in which there are no bonded labourers at particular points in time. At the same time, even in these villages, the question of whether the labour force is completely free remains, of course, an open one. An assumption, explicit or implicit, that hired labour is *either* bonded *or*

[49] Nor is it correct that wages are determined completely outside the market. If wage costs were under the employers' 'firm political control' (and that this was all that was needed to lower them), why didn't they cut back on wage costs *before* the introduction of the new technology?

completely free wage labour is a hindrance to any useful taxonomy or other analysis of hired labour, free and unfree.

To labour the point before moving on. The introduction of new farming techniques was accompanied by a large-scale expansion in labour demand in the Cumbum Valley. This expansion was not met by a disproportionately high increase in the employment of workers on long-term contracts relative to the employment of workers on short-term (daily-rated or piece-rated) contracts, nor was it met by a large-scale increase in the employment of bonded labourers in production tasks.

10

Resurvey 1986

In June 1986, I undertook a brief sample survey of landless agricultural labour households in Gokilapuram (see Appendix 1). The sample, which was a random sample chosen from a list of agricultural labour households in the village, covered about 11 per cent of the agricultural labour force as recorded in the village returns of the Census of India 1981. The schedule that was used was similar to the schedule used for the census-type survey of 1977. For each household, data were collected on the age, occupation, and educational status of each household member; on agricultural and non-agricultural employment and earnings (including earnings from self-employment and other sources); and on labour services. Data were not collected, as they were in 1977, on assets owned by households or on household indebtedness, as such data had been collected as part of another sample survey conducted in the village in 1985.[1] The paragraphs that follow report, on the basis of results from the two surveys, on some of the changes that took place among landless agricultural labourers in Gokilapuram between 1977 and 1986.

From 1977 to 1986 there was an unambiguous decline in the average number of days of employment available to a landless agricultural labourer. There was a decline in the average number of days of hired employment in agricultural as well as in non-agricultural tasks. (The number of days of self-employment in non-agricultural tasks is not being considered here.) The decline in the number of calendar days of employment in agricultural tasks was of the order of 10 per cent and the decline in the number of days of agricultural plus non-agricultural employment was of the order of 15 per cent (see Tables 10.1 to 10.3).

This change can usefully be seen in terms of changes in the supply of labour as well as in the demand for it. If we were to assume that the demand for labour remained unchanged, a decline of 10 to 15 per cent in employment is roughly commensurate with the increase in the supply of labour over a period of nine years. Nevertheless, if the labour absorption data for 1977 and 1986 are disaggregated—particularly if they are disaggregated by crop and by gender—it becomes clear that changes in the pattern of demand for labour are very important in understanding changes in labour absorption over the period between 1977 and 1986.

The most important change in cropping patterns during this period took place on groundwater-irrigated land. In 1977, groundwater-irrigated cultivation in the vicinity of the village was more diverse than in 1986; groundwater-

[1] See Swaminathan (1986 and 1989) for the results of the 1985 survey.

Table 10.1 *Annual number of calendar days of employment per adult landless agricultural labourer, all castes, Gokilapuram, 1976–1977, 1985–1986*

Crop/operation	Men		Women		All	
	1976–7	1985–6	1976–7	1985–6	1976–7	1985–6
Paddy:						
Transplanting	1	n	23	22	13	11
Weeding	n	n	21	20	12	10
Harvesting	11	10	13	6	12	8
Threshing	8	12	5	7	6	10
Other paddy operations	19	19	3	1	10	10
All paddy operations	39	41	65	56	53	49
Coarse grain	8	2	16	5	12	3
Irrigated non-cereals	15	6	20	12	18	9
Groundnut	3	2	3	2	3	2
Tamarind	17	23	9	7	13	15
Others	4	9	n	n	2	5
All crops	86	83	113	82	101	83
Crop guarding	22	18	0	0	10	9
TOTAL	108	101	113	82	111	92

Notes: Each cell is the quotient of the total number of days of work in the particular crop (or crop operation) divided by the number of agricultural workers (those who work solely in agriculture and those who do both agricultural and non-agricultural labour) in the group. The denominator excludes workers in landless agricultural labour families who work only at non-agricultural tasks.

n = negligible.

Sources: Survey data, 1977; sample survey of landless agricultural labour households, Gokilapuram, June 1986.

irrigated fields, as we have seen in earlier chapters, were planted with irrigated coarse grain, vegetables, cotton, banana, and other crops. By 1986, the shares of coconut and banana in groundwater-irrigated cultivation increased at the expense of cotton, which was no longer grown in the immediate vicinity of Gokilapuram, as well as some vegetable and irrigated coarse-cereal crops.

Banana is a crop that absorbs more labour than, for instance, irrigated coarse cereals. At the same time, labour absorption on banana fields is substantially weighted towards men (see Chapter 5). Coconut in the Cumbum Valley used to be grown in traditional coconut groves, on the banks of canals and of the river,

Table 10.2 *Calendar days of employment in each crop (or crop operation) as a percentage of calendar days of annual employment, adult landless agricultural labourers, all castes, Gokilapuram, 1976–1977, 1985–1986*

Crop/operation	Men		Women		All	
	1976–7	1985–6	1976–7	1985–6	1976–7	1985–6
Paddy:						
Transplanting	1	n	20	27	13	13
Weeding	0	n	19	25	12	12
Harvesting	12	12	11	7	12	10
Threshing	10	14	4	9	6	11
Other paddy operations	23	23	3	1	10	12
All paddy operations	46	49	57	69	53	58
Coarse grain	10	2	14	6	12	4
Irrigated non-cereals	17	7	18	14	18	11
Groundnut	3	3	3	2	3	2
Tamarind	20	28	8	9	12	19
Others	4	11	n	n	2	6
TOTAL	100	100	100	100	100	100

Note: n = negligible.

Sources: Survey data, 1977; sample survey of landless agricultural labour households, Gokilapuram, June 1986.

Table 10.3 *Number of days of employment in the year, adult landless agricultural labourers, Gokilapuram, 1976–1977, 1985–1986*

Caste	Men		Women		All	
	1976–7	1985–6	1976–7	1985–6	1976–7	1985–6
Scheduled	135	169	184	119	161	144
Non-scheduled	103	88	133	112	118	99
All	118	122	158	115	139	119

Note: Each cell is the quotient of the following fraction: the sum of all days of hired employment in all crop operations including guarding and all wage-paid non-agricultural employment divided by all agricultural and non-agricultural workers in landless agricultural labour households.

Sources: Survey data, 1977; sample survey of landless agricultural labour households, Gokilapuram, June 1986.

and on field walls. The expansion of coconut cultivation in the Valley from the late 1970s onwards has not been the result of increased planting in these areas, but of planting coconut trees on groundwater-irrigated land, that is, on land that had hitherto not been used for coconut cultivation. (Coconut on groundwater-irrigated land, which comes to fruition in four years, is intercropped with other crops for the first two years. After that, it is grown to the exclusion of other crops.) In the Valley, coconut is cultivated for the fruit of the tree; that is to say, coconut cultivation has not generated employment in processing other products of the coconut tree, and there were no large-scale efforts to employ workers in coir-making or weaving coconut leaves into mats and other products. A major feature of coconut cultivation is that it absorbs little labour once the tree has come to fruition, and a shift to coconut cultivation on groundwater-irrigated land means a steep decline in labour absorption per hectare. A few workers are employed for harvesting the crop and for field maintenance and the few who are employed are predominantly men.

The decline in the cultivation of irrigated non-cereal crops other than banana and coconut has meant a decline in the employment of hired labour in agriculture. In the case of men, the decline was partly offset by new employment in tamarind-harvesting. Between 1977 and 1986, the tamarind traders-cum-landlords of Gokilapuram village consolidated their grip on the tamarind economy of the whole Valley. Tamarind trees in the Valley, as elsewhere, belong to private individuals and institutions, to institutions of local government (such as the panchayat and panchayat union) and to other departments of government (the Forest Department and the Public Works Department, for instance). Tamarind merchants from Gokilapuram contract for the usufruct of these trees and, over the years, they have brought tamarind trees in every corner of the Valley under their control. These trees are harvested manually, and the number of men workers (who climb the trees to beat down the fruit) in a group is generally higher than the number of women workers, who gather the fruit for transportation or storage. The tamarind merchants of Gokilapuram take workers, particularly men, from the village to different parts of the Valley for the harvest. The fruit are finally brought back to Gokilapuram, where they are seeded and de-fibred. They are then packed and sold to merchants at the market in Theni.

The tamarind economy is an example of how changes in the product market (in this case, the development of a commercialized market for a commodity) has led to the assembling of a new labour force for the cultivation of the commodity. The new and unprecedented levels of employment in tamarind cultivation are not the consequence of new and unprecedented levels of *production*, but have resulted from the fact that the produce of many trees is being brought, for the first time, to the market.

In the case of labour hired for non-agricultural tasks, no new avenues of steady employment were created over the nine-year period between 1977 and 1986. In 1985–6, a short-lived source of non-agricultural employment was created when the main channel from the Periyar river to the village was lined to prevent seepage and water losses. The job mainly involved hoeing work, which is done by men and, as it happened, this was a major source of non-agricultural employment for scheduled caste agricultural labourer men.

To summarize: average employment, measured in calendar days of agricultural and non-agricultural hired labour per worker, declined by about 15 per cent over the nine-year period. Disaggregating this, the effect of an increase in the supply of labour on the number of days of employment in agriculture for men seems to have been offset by an increase in the employment of men for harvesting tamarind. The employment of women in agriculture suffered a double blow, on account of an increase in the number of women workers and an absolute decline in the demand for labour in lift-irrigated agriculture. This decline was a result of the shift in cropping patterns away from cotton, vegetables, and other crops towards coconut and banana cultivation. In the case of non-agricultural labour, while work on irrigation channels provided non-agricultural employment for men in the reference year (1985–6), no such new source of employment, even short-lived, existed for women. As a result, the decline in the number of days of agricultural employment for a woman worker from a landless agricultural labour household in 1986 compared to 1977 was of the order of 27 per cent, as was also the decline in the number of days of agricultural and non-agricultural employment. The steepest decline in the average number of days of employment in 1986 compared to 1977 was among scheduled caste women, among whom the number of calendar days of employment per capita declined by about 35 per cent (Table 10.3). This was followed by the decline in the number of days of employment among non-scheduled caste women, which was of the order of 16 per cent, and then by non-scheduled caste men, among whom the number of calendar days of per capita annual employment declined by about 15 per cent. The only subgroup of workers among whom there was an increase in the number of days of per capita annual employment was scheduled caste men; the increase was of the order of 25 per cent (Table 10.3).

Turning to changes in wage rates, there was an *increase* in the real wage-rates paid for cash-paid, daily-rated operations in 1986 compared to 1977. The increase was by about 18 per cent for men and about 11 per cent for women. The gap between wages paid to men and wages paid to women for cash-paid, daily-rated tasks was *wider* in 1986 than in 1977; in 1986, the ratio between male and female wage rates was between 2.5 and 2.6. The amount received as wages in 1986 by an individual worker (man or woman) for a day's harvesting and the average amount received by an individual woman worker for a day's transplanting remained the same or was slightly less than the corresponding amount in 1977. Once again, women lost out relative to men as a result of changes that took place between 1977 and 1986. While the wage rate for cash-paid, daily-rated operations rose, the share of these operations in women's employment declined; by contrast, the share of transplanting in women's employment increased, and transplanting is a task performed by women in which average earnings stagnated or declined.

The period between the census-type survey of households in 1977 and the sample survey of landless agricultural labour households in 1986 was a period in which a major government-run anti-poverty programme, the Integrated Rural Development Programme (IRDP), was implemented in Gokilapuram and the rest of the country. The impact of the IRDP on poverty in Gokilapuram has been the subject of special studies for which field-work was carried out in 1985 (Swaminathan 1986, 1989). The major objectives of the IRDP have been summarized as follows:

(i) reaching households in the *target group*;
(ii) creating, by means of a loan-cum-subsidy, a *permanent productive* asset with the beneficiary households; and
(iii) generating income from the asset on a *sustainable basis* in order that the household be *raised above the poverty line*.

(Swaminathan 1986, pp. 108–9)

In the Gokilapuram sample in 1985, 50 per cent of beneficiaries received loans-cum-subsidies for the purchase of milch cattle, 17.5 per cent for the purchase of sheep and goats, 17.5 per cent for the purchase of bullocks and carts, and 15 per cent for small shops and other forms of self-employment (called ISB—industry and small business—loans). Leakages of IRDP assistance, measured in terms of the proportion of beneficiaries who were outside the target group, were 10 per cent in the case of milch cattle loans, 14 per cent in the case of sheep loans, 43 per cent in the case of bullock and cart loans, and 50 per cent in the case of ISB loans.

In 1985, it was calculated that for an agricultural labour household of average size with an average number of workers per household receiving 150 days of hired employment in the year and with an average ratio, within the household, of male to female workers, the gap between household earnings from wage labour and the official poverty line was Rs 2,489. On the average, earnings from loans-cum-subsidies for milch cattle brought in only about 26 per cent of this amount, earnings from loans-cum-subsidies for sheep-rearing about 46 per cent of the amount, and loans-cum-subsidies for bullocks and carts, about 53 per cent.

There have been many reasons for the IRDP providing beneficiary households with lower-than-projected incomes. First, initial costs can be very high: in Gokilapuram, for example, the costs of acquiring the necessary certificates, of opening a bank account, and transport and other search costs amounted to over 50 per cent of the subsidy for acquiring milch cattle and 35 per cent of the subsidy for acquiring sheep.[2] Although the IRDP was primarily a milch-cattle

[2] In the MIDS (1980) evaluation of the IRDP, initial costs were estimated to average 65% of the subsidy for small farmers and 49% of the subsidy for agricultural labourers. In Konur village near Dindigul (Swaminathan 1986), initial costs in the acquisition of milch cattle were estimated at 54% of the subsidy.

scheme, no effort was made to upgrade local cattle breeds. In addition, beneficiaries from poor households could not afford to give the animals concentrate feeds, and levels of lactation were very low, with an average of 3.7 litres per day in the first month of lactation and 3.3 litres per day in the ninth month of lactation. There was inadequate pasture for sheep and goats acquired under the IRDP and veterinary services were very poor.

The meagre benefits that were available under the IRDP were, for many households, short-lived. When a loan-cum-subsidy is given, it provides a welcome injection of liquidity in a poor household. But in many cases, the IRDP asset was not a permanent income-bearing asset. By 1985, 25 per cent of the milch-cattle loan beneficiaries had lost their animals through sale or death. As for sheep loans, within a year and a half of acquisition, 26 per cent of the original stock acquired had died and 57 per cent had been sold. In 1986, when I did my sample survey of landless agricultural labourers, not a single household in the sample that had been a beneficiary of IRDP schemes for the purchase of cattle or sheep still had an animal. The milch cattle had been lost through sale or death and the sheep had all been lost through death, as a result of a livestock disease that had spread in the region. Despite the fact that insurance premiums had been paid, no IRDP sheep loan beneficiary had received insurance payments. When livestock insurance was introduced in the village as a part of the IRDP, it was a new idea for the village poor, and the failure to settle insurance claims despite the payment of premiums was a particularly cruel epitaph to the IRDP for many poor households. A member of a household that had lost its sheep said: 'They told us to pay the [premium] money regularly so that if the animals fell sick and died, we would not lose all the money that we had spent. We thought: "This is a very good thing," and paid the money regularly. But when the sheep died, the money they promised us was not to be seen. We asked the people from the bank, we asked the veterinarian, we asked the Panchayat Union Chairmaṇ, and we stood in front of the Panchayat President's house and gave him a petition, but we did not get any money. We were cheated, it is a shameful business.'

With respect to the indebtedness situation among agricultural labour households, there has been a sharp increase in the amount borrowed from banks as a proportion of total indebtedness, from 17 per cent in 1977 to about 80 per cent in 1985 (*ibid.*, Table 7.8), and a corresponding decline in borrowings from outside the formal sector. This reflects, of course, the fact that the largest debt that a landless agricultural labour household incurred was as a result of borrowing in order to buy the IRDP asset. At the same time, debt–asset ratios increased among agricultural labourers, from an average of 0.19 in 1977 to 0.49 in 1985 (*ibid.*, Table 7.12).

The net effect of the fall in women's employment (despite the increase in wages in one set of operations) and the failure of the IRDP to cover the poverty gap has been that 72 per cent of households in the 1986 sample survey of landless agricultural labour households had household per capita earnings (these are total earnings, including wages, earnings from self-employment, and from other sources of income) that were below the officially specified poverty line. This is marginally higher than the corresponding figure from the 1977 census-type survey of households in Gokilapuram, which was 70 per cent. The value of the Sen index for landless labour households that are below the poverty line in terms of household per capita earnings was 0.38 at both surveys. The value of the Gini coefficient for the distribution of assets between all households increased between 1977 and 1985 from 0.71 to 0.81 (Swaminathan 1988, p. 78, table 2.12).[3] There was no substantial change in the number of households reporting labour services between the surveys of 1977 and 1986: in 1977 the proportion of households was about 42 per cent and in 1986 it was about 38 per cent.

So are there any improvements at all in the material conditions of life of agricultural labourers in the village between 1977 and 1986? If there are, they are in the figures for school enrolment: thanks to the noon meal scheme,[4] introduced in the State in 1982, aggregate school enrolment went up. I do not have enrolment data for 1977, but observation, conversations and the 1986 data leave little room for doubt on this (see Table 10.4). However, the data show a very sharp difference in the access to school education of boys and girls. *All* the boys and 81 per cent of the girls in the age group 5 to 9 years in the sample were enrolled in primary schools. In the age group 10 to 14 years, 80 per cent of boys were enrolled in schools, while the proportion of girls in the same age-group who enrolled in schools plummeted to 31 per cent. The rate of dropping out was clearly very high, particularly among girls. In addition, there are children who never really dropped in: many children, particularly girls, who were already too 'old' when the noon meal scheme was introduced

Table 10.4 *Children enrolled in schools as a percentage of all children in landless agricultural labour households, by age group, Gokilapuram, 1986*

Age group	Enrolment rate		
	Boys	Girls	All
5–9	100	81	90
10–14	80	31	61
5–14	88	59	73

Source: Sample survey of landless agricultural labour households, Gokilapuram, June 1986.

[3] These measures have been weighted for purposes of the 1985 sample by using the 1977 socio-economic stratification of households.

[4] The full title of the scheme was 'Chief Minister's Nutritious Noon Meal Programme'.

probably did not benefit from it, since it was already 'too late', or perceived to be too late, for them to join school anew. There were examples, in 1986, of households with children aged 17 to 18 years who were illiterate and had younger siblings between 5 and 12 years of age who could read and were enrolled in schools.

Another corollary of the increase in enrolment in schools was a decline in work participation rates among children; the decline, predictably, was more marked among boys than girls (see Tables 10.5 and 10.6). The number of boys (boys and girls defined as children below the age of 15) from landless agricultural labour households who were in the work-force declined from 13 to 7 per cent between the 1977 and 1986 surveys. The corresponding decline among girls was from 22 to 20 per cent. From 1977 to 1986 the number of working children as a proportion of all workers declined from 9 to 5 per cent among boys and only marginally, from 13 to 12 per cent, among girls. In addition to a decline in the work participation rates among children, there was

Table 10.5 *Working children as a percentage of all children in landless agricultural labour households, Gokilapuram, 1977, 1986*

Year	Boys	Girls	All
1977	13	22	17
1986	7	20	13

Note: 'Children' refers to persons below the age of 15.

Source: Survey data, 1977; sample survey of landless agricultural labour households, Gokilapuram, June 1986.

Table 10.6 *Working children as a percentage of all workers in landless agricultural labour households, Gokilapuram, 1977, 1986*

Year	Boys	Girls	All
1977	9	13	11
1986	5	12	8

Note: 'Children' refers to persons below the age of 15.

Source: Survey data, 1977; sample survey of landless agricultural labour households, Gokilapuram, June 1986.

also a decline in child labour as a proportion of the total number of days that agricultural labourers spent in hired labour. Among landless agricultural labour households, the days of child labour in agriculture as a proportion of the total number of days of agricultural labour (including guarding) declined from 2 per cent to 0.5 per cent from 1977 to 1986.

To summarize: in terms of earnings, and the removal of poverty, there were no basic changes in the conditions of life of agricultural labourers in Gokilapuram between 1977 and 1986. The number of days of annual agricultural and non-agricultural employment per woman worker at the survey in 1986 was 27 per cent less than the corresponding figure in 1977. There was change for the better in respect of aggregate school enrolment; rates of enrolment were, however, much lower among girls than among boys. There was a decline in the rate of work participation among children and in child labour between 1977 and 1986; once again the decline was more marked among boys than among girls.

11
Summary and Conclusions

Economic development has traditionally been measured in terms of economic growth, particularly as reflected in the growth of incomes. More recently, the economic development of a society has been judged by the extent to which the basic needs of the people have been provided. There is, however, another dimension to the development process: it concerns the freedom of employment of members of the labour force, their freedom to dispose of a resource that they each possess, their labour power. This is the major area of inquiry of this book. In particular, this book has dealt with the impact of production conditions and of the material conditions of life of agricultural labourers in a rural region on the freedom of those agricultural labourers to sell their labour power.

Such an investigation is warranted for different reasons. The degree of freedom of the individual worker to sell her labour power to the employer of her choice is an indicator of the direction of historical change, and of progressive socio-economic transition. As is generally recognized, a cardinal feature of the system of wage labour under capitalism, and one that distinguishes it from pre-capitalist systems of production (save the most primitive), is that the direct producer under capitalism (that is, the wage labourer) is free to work for the employer of her choice. That is to say, she is not compelled by different types of other-than-economic coercion to work for a particular employer. An assessment of the degree to which hired labourers in a given region have or do not have this freedom is therefore one index of the historical stage of socio-economic organization of production in that region. Wage labour, conceived in the abstract, is classically supplied by workers who are propertyless, by those who have no share in the ownership of the means of production. The freedom or unfreedom of the labour force also has a crucial bearing on politics, on the strategy and tactics of bringing about political change.

The question of the freedom of the labour force is of intrinsic importance: it is a good thing in itself to move away from thraldom, and to establish that basic self-dignity and freedom from servitude that the freedom to choose employers implies. It is also of importance instrumentally: a worker who is free to choose where she works, for whom she works, and (even if to a limited extent) the kind of work she will do has, as is obvious, a much larger opportunity set than a worker who is unfree to do any of these.

The particular aspect of freedom and unfreedom on which this work focuses, namely, the freedom of the agricultural labourer to sell his or her labour power, has profound implications for the lives and labour of millions of human beings. (In India—to give an idea of the numbers involved—a projection from figures from the Census of 1981 puts the agricultural labour force at no less than 68 million people in 1990.) In order to give the concepts of free and unfree labour empirical flesh and blood, and to make this investigation a manageable one, it had, clearly enough, to be confined to a particular place and time. I chose to live and work in Gokilapuram village in the Cumbum Valley in Tamil Nadu, an area where there had been a sharp increase in the number and proportion of agricultural labourers in the labour force in recent decades, where the new agricultural technology and new farming practices had taken root and where there had been a substantial development of the economic infrastructure. On the basis of data from the village and the Valley collected between 1977 and 1986, I shall try and answer the questions raised at the beginning of the book.

Are hired labourers in agriculture in the Cumbum Valley propertyless? The vast majority of agricultural and non-agricultural labourers in Gokilapuram and the Cumbum Valley own neither land nor other means of production. At the same time, a significant feature of agricultural change in the region is that other sections of the population—including wide sections of the peasantry—are drawn into the market for hired labour as sellers of labour power. It is also of relevance to the current status of agricultural labourers that the creation of a landless labour force, bereft of ownership of the means of production, long pre-dates the creation of a free wage labour force in Indian agriculture.

How free are workers in the Valley to work for employers of their choice? Theoretically the range of possibilities extends from complete bondage to complete freedom. In practice, in Gokilapuram, the empirical evidence showed that by the definition of bondage proposed in Chapter 8, there was only *one* bonded labourer in the village. The answer to the obvious question concerning the situation at the other end of the spectrum cannot be given with the same degree of quantitative confidence, since unfreedom often takes subtle forms, forms that are not always perceptible immediately. It is also true that individual agricultural labourers may move between different positions on a freedom–unfreedom scale. Nevertheless, in the context of the rather unwarranted assertions of some scholars in the field that wage labour has hardly emerged, that in truth workers are completely unfree to choose their employers, the raw data arrest attention: no less than 58 per cent of landless agricultural labour households, questioned in detail about their freedom to choose employers,[1] asserted that they were indeed free to do so.

Nor is that perception implausible. In the context of the introduction of new agricultural technology and new farming practices, employers needed workers on a new and, in the circumstances, unprecedented scale. The new levels of labour power that were employed were a major factor in the socialization of agricultural production, in the casualization of the labour force, and the impersonalization of the relationship between hired labourers and employers.

[1] See ch. 8 for the questions on freedom and unfreedom.

Manifestly, it is not possible for a modernizing farming system such as that of the Cumbum Valley, where crucial crop operations are highly employment-intensive, to be conducted solely on the basis of relations of personal dependence between the employer and the individual agricultural labourer. Take a hypothetical case, that of a landlord-employer who owns 20 acres of surface-irrigated land on which paddy is grown, 15 acres of groundwater-irrigated land on which cotton, vegetables, and fruit are grown, a few acres of unirrigated land, and groves with mango, coconut, and tamarind trees. Such a landlord would need to hire thousands of person-days of labour power a year. With these levels of demand, and with the existence of multiple sources of increased demand for labour in the village, a farm such as this cannot be sustained by a system of labour hire that presupposes the personal dependence of the one hired upon the one who hires. Farming here has to be based on newer forms of hire, inevitably on a more impersonal relationship between employer and hired labourer and consequently on the greater anonymity of the individual worker.

The factors that contribute to the extension of the freedom to choose employers do not, of course, all derive from conditions of supply and demand in the labour market, or solely from what are conventionally considered to be 'economic' circumstances. Factors outside the market have a profound impact on the development of wage labour. Among the factors that have played a role in the Cumbum Valley are: new forms of labour hire; new opportunities for migration and labour mobility; mass actions of agricultural workers; rural literacy and education; the development of mass-transport systems; the role of legislation; and the existence of different institutions of civil society.

How are the others in the survey to be characterized, those agricultural labour households that were neither bonded nor completely free to choose their employers? While it is possible, in the abstract, to differentiate between free wage labour and unfree labour, it is an altogether less tractable problem to separate freedom and unfreedom in the daily lives of workers and to separate the impact of different sets of influences on the freedom and unfreedom of the individual worker. The unfreedom of workers who were neither bonded nor completely free to choose their employers took different forms, their freedom to choose employers was circumscribed in different ways and in different degrees. The most common manifestation of this kind of unfreedom was what has been called the right of first call of employers over workers. Under this arrangement, discussed at length in Chapter 8, a worker looks in at a particular employer's house every morning to check whether agricultural or non-agricultural wage work is available. If it is, the worker is employed and paid a regular day's wage. If it is not, the worker seeks work elsewhere. However, if wage work is not available with the first employer but the employer gives the worker another job—for example, a domestic task that involves manual labour—the worker does the task even if it means forgoing wage labour elsewhere and forfeiting a day's wage.

Closely associated with the different types of constraints on the worker's freedom to choose employers is the other important manifestation of unfreedom, the system of labour services. Labour services are tasks performed by workers for employers at rates well below the market wage rate—in the Gokilapuram data, at rates that were as low as one-fifteenth of a day's wage or, on occasion, for nothing at all. A worker who performs labour services for an employer does not *sell* his or her labour power to the employer of his or her choice: he or she *gives it away* to the employer free or almost free. A further, and crucial, feature of the system of labour services is that it presupposes the personal dependence of the worker on the employer.

Empirically, the performance of labour service was not restricted to a particular section of the labour force: men and women, persons of all castes in the village and of different ages—from children to the elderly—performed labour services. Almost all the employers who demanded and received labour services were landlords. (There were three non-landlords belonging to this category in the village data; one of them was a rich peasant and another a land-owning businessperson.)

The data also show that labour services in agriculture were less prevalent than in non-agricultural and domestic services. Purely agricultural labour services were, if anything, on the decline. In agriculture, work in most tasks, particularly employment-intensive tasks such as field preparation, transplanting paddy, weeding, harvesting or threshing grain, was not done by means of labour service, but by workers who were paid wages. Agricultural tasks that were done by means of labour services were typically tasks that, although necessary for the crop, require only a person or two and a short time to complete. There is little job-specification involved in tasks that are done through labour service, and the tasks themselves generally have to do with overall field maintenance and with monitoring the growth of the crop. By comparison, the system of labour service comes into full play in the sphere of non-agricultural and domestic labour. As has been discussed earlier in the book, the image of a rich landlord who gets all his errands run and odd-jobs done, his house cleaned and his cattle tended, and his wood hewn and water drawn by servants to whom he pays an arbitrary pittance in cash or kind or, on occasion, even nothing, remains true to life in Gokilapuram to this day.

Why do workers agree to perform labour services? Why do they agree to give an employer a right of first call over their labour power and submit to the kind of servitude that such a relationship entails? There is no *legal* coercion for them to do so, and in Gokilapuram, generally speaking, workers are not forced to perform labour services because landlords threaten them with physical violence. In many cases, as we have seen, a worker who does a range of labour services for a landlord can refuse to work at anything but a regular day's wage when he or she works at a production task (and, indeed, may stop working for the landlord if the landlord tries to get away with less than full payment). The most important reason is, of course, that workers are compelled by poverty and unemployment to seek different ways of ensuring that they get wage employment whenever it is available and of ensuring subsistence at times when there is no employment to be had. While poverty and unemployment are the most important (and often the only) reasons for a worker to compromise her freedom to sell her labour power to the employer of her choice, the specific forms that labour service takes—the particular employer for whom a worker labours, the kinds of jobs the worker does, the degree of servility involved and so on—are influenced by a wide range of factors. In Gokilapuram, these factors include: the caste and gender of the worker; traditional relationships of loyalty and subservience between working families and certain employer households; illiteracy, indebtedness, and the assignment, by tradition, of specific village tasks and functions to particular castes, families, and individuals.

For many workers, then, non-wage employment in the form of labour service is a means of placing a booking for wage employment when it is available, of gaining tenure with an employer. Workers *present* their labour power free or for a paltry amount to an employer in order to be able to *sell* their labour power for a wage when wage employment is to be had with the employer, and in order to stave off complete household disaster during the long lean season. In G. A. Cohen's insightful formulation, 'the worker here agrees to be a bit of a serf in order to qualify for proletarian status.'[2]

Do continuing constraints on workers' freedom to sell their labour power represent an extension into the present of pre-capitalist relations? Some scholars have argued that such unfreedom represents essentially *capitalist* exploitation, because restrictions on the freedom to sell labour power (including the phenomenon of bonded labour) are essentially arrangements whereby a capitalist entrepreneur cuts costs in order to enhance profits. Pre-capitalist remnants in production relations are thus explained away as capitalist reality misperceived. Such either–or interpretations may have their uses, but not for the purpose of illuminating the diverse and intricate complexity of current reality. For it is clear that traditional, pre-capitalist forms of subordination can be and are utilized, sometimes directly and sometimes in transmuted form, to cut costs by an employer in agriculture who otherwise employs wage labour for production tasks. Indeed, the central problem in the study of employer–employee relations in village India is that of analysing the specific ways in which different traditional forms of subordination are appended to, or integrated with, wage labour. The data from Gokilapuram, as we have seen, show that almost every feature of the physiognomy of unfreedom among agricultural labourers today derives organically from the past.[3]

It seems clear that different forms of unfreedom—reflected in different types of constraints on the free choice of employers as well as in the system of labour service—will not just wither away with the mere passage of time. The factors that provide a material basis for different types of unfreedom in

[2] Personal communication, 27 July 1988.

[3] The expression borrows from Bloch (1966, p. 247).

Gokilapuram are deep-rooted, and it is evident that the growth that has occurred in the productive forces does not in the natural course of events necessarily or completely overcome these different types of unfreedom.

As is evident from the foregoing, the freedom and unfreedom of agricultural labourers are crucially influenced by agricultural growth and production conditions in agriculture and by the material conditions of their lives. It was for this reason that this study probed with special interest issues concerned with growth, production and the conditions of agricultural labourers' lives in the region that was studied. In the event, this investigation also served to illustrate, in microcosm, certain themes that recur in analyses of India's agrarian development, namely, growth, uneven growth and the proposition that growth *per se* does not guarantee equitable distribution. While the specific combination of factors that has made the Cumbum Valley something of a pace-setter for the agriculture of south India is unique, a delineation of these factors helps to identify some of the building blocks of agricultural growth in other, and wider, regions.

Agricultural Growth: Salient Features

The Cumbum Valley has a stable and relatively reliable irrigation system. Surface-irrigated land (*nanjai*) in the Valley is in the first and best-served segment of the command area of the Periyar system of irrigation. Lift-irrigation has advanced, particularly as a consequence of the construction and energization of wells. (The spread of rural electrification has been a major factor in the extension of well-irrigation in Tamil Nadu.)

The Valley has an expanding sector of commercial cultivation. On surface-irrigated land there is a stable and specialized sector of high-yielding paddy cultivation, and yields in Gokilapuram village, which varied between 4 and 5.4 tonnes of rice per hectare in 1976–7 and in 1981–2, compare favourably with those in Japan and Taiwan in the 1970s. There has been a remarkable pattern of crop diversification on land irrigated by wells (*thottam*), where advanced cultivation is investment-intensive, risk-exposed, and often price-responsive. An example of the crop diversification that was achieved in the late 1970s came from the village of Kamayakavundanpatti, where there were more than 25 major crops grown over an agricultural year. Among the other features of cultivation in the Cumbum Valley are the adoption, in certain crops and on certain farms, of advanced agricultural techniques involving scientific farming practices, and an increase, in these areas, of the productivity of labour. In common with agricultural regions in India that are rice based or non-wheat based, mechanization is limited, and the labour process is characterized by a low organic composition of capital.

There has been an exceptionally large expansion of the market for hired labour. The specific changes in patterns of land use that took place in the

Valley made for an increase in labour absorption, particularly on surface-irrigated land and on groundwater-irrigated land. Many of the tasks that are necessary for the cultivation of the crops grown involve employment-intensive forms of hire. Another feature of the market for hired labour in the region is that wide sections of the village population participate in it as hired labourers, including large numbers from peasant households. Landlords are the principal employers of hired labour in the village, with the peasantry as secondary employers.

Government-run schemes for the dissemination of agricultural information (known as 'agricultural extension' schemes) are, in practice, selectively extended—they are of most benefit to the rich and literate. Nevertheless these schemes have been relatively effective in the Valley in so far as they have been a major factor in the high levels of agricultural information that are prevalent among the most successful 'innovative farmers' of the region. In addition, very important aspects of the vanguard character of the agrarian regime of the Valley are the production experience and skills of its work-force, the capacity of manual workers to adapt to diverse forms of labour, and the creation, alongside the large contingent of casual labourers, of a smaller corps of workers who are employed at tasks that require special skills.

A widespread and organized rural credit network; the development of the transport infrastructure, of a network of motorable roads, and lorry and passenger transport; and the development of a trade and commerce network for the agricultural and agro-based produce of the region are other noteworthy features of the agrarian regime. A diversification of the occupation structure of the village population reflects the influence of the new commerce on agriculture.

Agricultural growth in the Cumbum Valley has been uneven, having progressed in certain sectors and farms and lagged behind in others. Unirrigated agriculture continues to dominate the sown area of the Valley, the spread of new farming techniques is unequal across different classes of the cultivating population and across farms, and there are severe constraints, technical and socio-economic, on the unfettered growth of agricultural productivity and of the productivity of labour. These constraints are particularly evident in dry-land farming, but they can also be seen across the agriculture of the region as a whole. The agrarian regime is also marked by the accumulation and reproduction of capital among significant strata, particularly in the hands of landlords (including those who combine cultivation with other forms of business activity), sections of the rich peasantry, and among merchants who handle the produce of the region.

In Gokilapuram, despite new developments in the land market and changes in the ownership structure in the contemporary period, the class of landlords has continued to own most of the land and other assets in the village; the Gini coefficients in this regard show astonishingly high levels of concentration of ownership. In 1977, the Gini coefficient for the distribution of the extent of ownership holdings among all households (including the landless) in the village was 0.86, and the corresponding figure for operational holdings was 0.83. The data show a staggering concentration of the ownership of the means of production—and of the sources of future accumulation—in the hands of the major landlords. By way of example, in 1977, major landlord households, which comprised 1.8 per cent of all households, owned 52 per cent of the crop land, 67 per cent of the trees and groves, 60 per cent of the plantation land, and 72 per cent of the agricultural machinery and other non-land agricultural assets held by all households in the village. The Gini coefficient for the distribution of the value of crop land among households in the village in 1977 was 0.93. It was 0.97 for agricultural assets other than crop land; 0.97 for trees and groves; 0.98 for plantation land; 0.89 for draught animals; 0.87 for other livestock; 0.78 for non-agricultural assets; and 0.87 for all assets combined. A sample survey in the village in 1985 indicated that this concentration of ownership of assets in the village increased: the Gini coefficient for the distribution of assets for households in the sample (appropriately weighted) rose from 0.77 in 1977 to 0.81 in 1985.

Alongside the overall process of polarization, small production on economically unviable plots of land is a tenacious feature of cultivation. The vast majority of households actually engaged in cultivation, particularly the middle and lower sections of the peasantry and agricultural labourers with operational holdings, carry on production on very small plots of land. To take an example, the average size of a surface-irrigated operational holding among poor and middle peasant households that had operational holdings of surface-irrigated land was 0.64 acres. The average size of a groundwater-irrigated operational holding among poor and middle peasant households that had such holdings was 1.74 acres.

Although tenancy no longer plays the role that it did, say, 50 years ago, tenancy remains a significant secondary form of land tenure in the village. Of the relevant totals, the leased-in proportion of groundwater-irrigated, surface-irrigated, and dry-land operational holdings was 25 per cent, 22 per cent, and 12 per cent respectively. Leases were not registered as required by law, and rents were far above the statutory limit (which in Tamil Nadu was a 75 per cent share or its equivalent for the tenant and 25 per cent for the lessor). Forms of rack-rent still persist. Rent systems are complex, with different forms of fixed rent on surface-irrigated land and of share-cropping on unirrigated and groundwater-irrigated land.

Changes in Labour Absorption

As in other parts of India, there was an increase in labour absorption in paddy cultivation resulting from the introduction of high-yielding varieties of paddy in the Valley. Relative to labour absorption in South-East and East Asia, employment per hectare in this high-yielding region was extremely high, varying between 350 person days and 519 person days per hectare. This was much higher than the levels (of less than 200 person days per hectare) envisaged in the literature on high-yielding rice cultivation in Asia. The result is a paddy cultivation system that is characterized by high yields per hectare and very low yields per person-day of labour relative to East and South-East Asian standards.

A very sharp rise in labour absorption accompanied changes from unirrigated agriculture to the irrigated cultivation of non-cereal crops. Such crops are typically grown on land that is irrigated by wells. When an unirrigated field on which coarse grain is grown is irrigated and planted with, for example, tomato or other vegetables, cotton, or banana, the number of days of employment per hectare can increase by a factor of up to 12 to 16.

While new farming practices generally served to increase labour absorption in agriculture, there was a decline in labour absorption in bullock-supported operations. These operations are ploughing, carting, and the second threshing in paddy cultivation on surface-irrigated land; and ploughing, carting, and threshing as well as bullock-powered irrigation (now replaced by electric motor pumps) on tracts irrigated by wells.

In the years after our first survey in 1977, changes in land use led to rapid changes in patterns of labour absorption and in the number of days of hired labour available to the labour force in the village. Women's employment in the village and Valley was severely affected by a shift in cropping patterns from a range of irrigated non-cereal crops to coconut in the years between 1977 and 1986.

Increase in the Labour Force

There was, in the State as a whole, a sharp increase in the proportion of agricultural labourers in the rural work-force between 1961 and 1981, from 22 per cent to 40 per cent. This was a general increase (the proportion has increased in *all* taluks of the State); the process was also marked by higher rates of growth in areas where base-year incidence was low, resulting in a closing of the gap between taluks in this respect over time. The increase in the proportion of agricultural labour appears to be strongly linked to the spread of irrigation, particularly by wells, in the State; it was particularly high in areas where there was a substantial area under well irrigation and was still higher in areas where a large sector of well irrigation coincided with a large canal-irrigated sector.

In the village, the pushing forward of the productive forces in agriculture in advanced sectors of cultivation, and the interaction between modernizing farming systems and a structure of property ownership that is grossly inequitable have been important factors in a remarkable process of differentiation over a single generation. This process is manifested in the creation of a new section of workers who have lost their land or are unable to pursue traditional tasks and trades. Fully 58 per cent of landless agricultural labour households in 1977 had first-generation ascendants who were *not* agricultural labourers; 47 per cent were either peasant or artisan households. The years have also witnessed a loss of land by households that are now agricultural labourers: of the landless agricultural labourer households, about 60 per cent had ownership or operational holdings in the family in an earlier period, and the data indicate that the pace of land-loss increased in the 1960s and early 1970s. Agricultural labour is also the most caste-heterogeneous class, the occupational category that draws on the widest range of castes in the village and in which the largest proportion of almost every caste in the village is concentrated.

Agricultural labourers in the village are, for the most part, a propertyless mass in respect of ownership of the means of production. (In the village, there were 257 landless agricultural labour households, or 40 per cent of households, and 11 major landlord households, or 2 per cent of households. The combined assets of the 257 landless agricultural labour households added up to less than 60 per cent of the average assets of a *single* major landlord household.) The few assets that agricultural labour households did own were concentrated in non-agricultural, mainly household possessions.

Features of the Annual Labour Schedule of Agricultural Labour

Mass unemployment. On average an agricultural labourer in Gokilapuram was employed for only 140 days at agricultural and non-agricultural tasks in 1976–7. This average, however, conceals very wide variation within the labour force; owing to a combination of factors, there is significant inequality in the number of days of work gained by each worker. In 1985–6, the average number of days of employment available to an agricultural labourer was about 119, or 15 per cent *less* than in 1976–7. The decline was particularly steep—about 27 per cent—among women workers. This fall was a consequence of changes in the supply of labour as well as of a shift towards coconut cultivation and away from more labour-absorbent irrigated non-cereal crops.

Seasonality. Unemployment is concentrated in a lean period that lasts three to four months and is also dispersed between crop operations over the rest of the year, with male employment having been generally more sporadic than female at the time of the first survey. The crop cycle in paddy cultivation sets the pace for seasonality; the major influences that counteract seasonal peaks are, first, groundwater-irrigated cultivation, which goes on through the year (unlike surface-irrigated cultivation, which is timed for the release of water into the river), and, secondly, employment opportunities at non-agricultural tasks.

Mobility. The labour force of the village, while being settled in that it is resident in the village, is mobile, a characteristic that it appears to share with agricultural labourers in other parts of the State. Fields on which daily work is done are scattered over a 10-kilometre radius around the village; and there is seasonal migration (for example, to harvest groundnut in southern villages of the Valley) as well as long-term migration to the cardamom, tea, and coffee plantations in Kerala and Tamil Nadu. Despite the hardships of migrant life, migration is a catalyst of proletarian qualities in the labour force: a migrant worker is relatively free of the traditional bonds of the village and is also a source of news and information about conditions of life and work outside the village. A migrant's remittances can be important sources of monetization of the household income of the family that remains in the village.

The diversity of wage systems. A wide variety of wage systems has crystallized in different tasks, in agriculture and outside it. These differ, task by task, in the form of wage (different types of time-rates and piece-rates), in the mode of payment (kind, cash, cash plus kind), in the methods of recruitment of workers, in the intensity and duration of the working day, and in whether the tasks are gender-specific (as most tasks other than harvesting are).

Agricultural Labour: Material Conditions of Life

Wage rates. Wage rates are low in relation to subsistence needs. They are also low in relation to levels of payments at certain unskilled tasks in the urban organized sector: for example, the wage rate of an agricultural labourer in 1976–7 was about 27 per cent of the wage rate of an unskilled textile worker. If the rice equivalent of a day's agricultural wage is taken as a comparative measure, wage levels in Gokilapuram are far below those in East Asian and some South-East Asian countries.

Changes in wage rates. It is clear that the demand-inducing effects of the new technology do not ensure a rise in wage rates. The wage payment records of a landlord family, covering the years 1948 to 1976, the period during which the major changes in agricultural technology and farming practices took place, show that real wages either stagnated or declined; that they fluctuated sharply, particularly after the 1960s; and that the gap between the grossly unequal rates of payment to men and women actually widened. The amount of rice that could be purchased with a day's wage at cash-paid, daily-rated operations was 3.02 kilograms in 1948 and 3.06 in 1976; for women it was 1.34 kilograms in 1948 and 1.31 kilograms in 1976. There is no evidence from official data or individual studies of a secular trend of rising wage rates for any part of Tamil Nadu till 1980–1. (Cross-sectional comparisons are instructive here. In the late 1970s, the amount of rice that could be purchased with a day's agricultural wage in Taiwan and Japan was between 15 and 20 kilograms.)

From 1977 to 1986, there was some change in the trend in wage rates. In the village, real wage rates for cash-paid, daily-rated operations were higher in 1985–6 than in 1976–7 by 18 per cent for men and 11 per cent for women. At the same time, the gap between wage rates paid to men and women was wider in 1985–6 than in 1976–7. In addition, the amount received per day as wages by an individual worker (that is, the share of the piece-rate that the individual worker received) at harvesting and transplanting stagnated or declined between 1976–7 and 1985–6. (Over the same period, the crop operations for which wages stagnated or declined became a larger proportion of women's aggregate annual employment.)

Poverty. Taken together, these factors have resulted in very high levels of poverty: in 1977, about 70 per cent of all landless agricultural labour households were below India's official poverty line. Households below the poverty line were more dependent on agriculture (in that a higher share of their total earnings came from agricultural wages) than those above, which had a larger share of earnings from other sources. In 1986, the proportion of landless agricultural labour households below this poverty line was 72 per cent.

Indebtedness. Poverty on this scale is combined with deep indebtedness: of all landless agricultural labour households 63 per cent were in debt; the average debt–asset ratio among the indebted was 28 per cent, with about 25 per cent of households struggling under a debt–asset ratio of over 50 per cent. Debt is generally incurred to cover day-to-day expenses of the household and on rituals and festivals and loans are mostly taken at usurious rates of interest: at the survey the modal rate of interest at which loans were taken was between 40 and 100 per cent per annum. By 1986, as a result of the spread of loans disbursed under the Integrated Rural Development Programme (IRDP) in the village, the major share of loans taken by landless agricultural labour households was taken from commercial banks. Nevertheless, in a panel of agricultural labour households, the average debt–asset ratio actually *increased* from 0.19 in 1977 to 0.49 in 1985.

The Cumbum Valley, in short, is an area of relatively high agricultural growth. New crop varieties have been introduced, yields per hectare of irrigated crops are high, and farming practices have been modernized. There have been changes in irrigation patterns and improvements in the transport system and a relatively widespread rural credit network has been established. The levels of agricultural information among those who take decisions regarding crop management are, by the standards of the State, very high indeed. This agricultural growth has enhanced the demand for hired labour and has contributed to the growth of wage labour. At the same time, such growth as has occurred cannot be a sufficient condition for ensuring that a worker is free of extra-economic, non-contractual constraints on the free sale of his labour power to the employer of his choice.

A whole set of factors, as we have seen, works towards the commoditization of labour power, towards impersonalizing the relationship between worker and employer and establishing the freedom of the worker to sell her labour power to the employer of her choice. At the same time, many factors continue to work against the dissolution of constraints on the free sale of labour power. To begin with, unfreedom continues because workers cannot just opt out, walk out of the village and find a job elsewhere in a place and occupation (in a town and factory, for instance) where they will be free.

The failure of industry to absorb the vast numbers of the agrarian unemployed as well as the interaction between the phenomenon of a surplus population and the pre-capitalist aspects of the village community cannot but have a profound significance for the socio-economic status of the manual worker in the village. To recapitulate: in Tamil Nadu as a whole, 75 per cent of the work-force was, by the Census of 1981, rural. The agricultural part of the work-force (i.e. the Census categories of cultivator and agricultural labourer together) constituted 61 per cent of the total work-force (that is, of the Census category of 'main worker') and 79 per cent of the rural work-force. The main feature of occupational change has not been a shift *away* from agriculture, but a restructuring of the work-force *within* agriculture, through a process by which the number and proportion of manual labourers increased at the expense of other sections of the rural population. Industrial employment is not an alternative that is available to the increasing numbers of manual workers in the village, and so for most of them the only option is to live and work in the too-familiar village, often consigned to narrow, pre-ordained places in its socio-economic hierarchy. A labour force that is far in excess of the means of employment is trapped within the village; it is little wonder that the scale of poverty and unemployment, amidst growth, remains himalayan.

Combined with poverty and unemployment are other factors that work against the dissolution of unfreedom in employer–employee relations: the social and economic dominance of a small group of landlords; caste discrimination and the practice of untouchability; severe discrimination against women; indebtedness and usury; widespread illiteracy and the terribly slow rate of spread of primary and secondary education among the poor, and especially among women and people of the scheduled castes; and, in general, the slow penetration of scientific culture in the daily lives of the people. The continued operation of these factors contributes to the imperfect development of wage labour and to the survival of labour service for longer than at first glance seems compatible with the general level of development of the productive forces.

What categorical conclusion, then, is warranted by the evidence from Gokilapuram village and the Cumbum Valley? Without doubt: the system of wage labour is established and entrenched, and is dominant in most production tasks. Also without doubt: unfreedom and the system of labour service have not withered away or been consigned to the past; on the contrary, they play an important role in the daily lives of agricultural labourers and in production relations. In their socio-economic being, contemporary agricultural labourers recapitulate, as it were, many characteristics of different historical stages through which agricultural labour passed on the road to becoming proletarian.

Appendix 1

Notes on Methods of Collection of Primary Data

Census-Type Survey of Households, 1977

Survey Method and Survey Period

Four persons did a pilot survey and pre-tested the questionnaire in April 1977. Sixteen persons worked for about two weeks from May 1977; a smaller number stayed on to fill in and correct schedules till the end of June. Four went back regularly to correct discrepancies that had become apparent in the course of computation.

The original objective of the survey was to undertake a study of agrarian relations as a whole in Gokilapuram village and, as a consequence, the information gathered at the survey was more elaborate than it might have been had our area of interest been narrowed down to agricultural labour at the outset. In the event, however, the comprehensiveness of the data that we gathered has been of value in helping us locate the class of agricultural labourers in the larger socio-economic structure of the village.

The survey was a census-type survey of households, with one questionnaire used for each household. In this 10-page schedule, the major items of information elicited from households in the survey related to: family size, caste, age structure, and educational level and occupation of each household member; land owned and operated by the household; tenancy for fixed rent and share; production and costs of production on the household operational holding, including data on irrigation, major inputs, means of production owned or hired, livestock, and employment of family and hired labour by crop and agricultural operation; marketing of agricultural produce; income from business activity other than agriculture; labouring out at agricultural and non-agricultural tasks and wages; agricultural and non-agricultural assets; indebtedness; and aspects of consumption.

An effort was made to identify the actual head of each household, and the main respondent was generally the best-informed member of the household. Generally, however, most of the members of the household were present when the interviewers (we generally went to households in pairs although in the later stages, when experience had been gained, we worked singly) came round to the household for interviews; this was particularly useful when we recorded the labour schedules of individual members.

As investigators, we worked from early morning till late in the evening and often late into the night; two persons checked the questionnaires and two took detailed case studies of the life histories of individual respondents chosen in the course of the survey. The duration of each interview varied: an individual who had a single source of non-agricultural income as the sole source of total earnings took a minimum of 45 minutes, while peasant households with an operational holding as well as agricultural and non-agricultural labouring out took a few hours. In the case of one big landlord household, where the interview was frequently interrupted, and the head of household took a great deal of time in answering each question, the questionnaire took a total of about seven hours to complete.

There were, of course, problems of concealment, particularly by richer households, of sources and levels of income and of assets and their value. In cases where a discrepancy appeared to interviewers between the obvious material status of a household and the data that the head of household had provided in the interview, and in cases where questionnaire data were internally inconsistent, we checked the sources and levels of income of the household and the kind of assets it owned with other persons who were well informed about the household, and then went back to the household to reinterview it in respect of the suspect items of data and correct, as best as possible, the questionnaire. (In cases where it occurred, one item of data that was particularly difficult to ascertain with accuracy was the level of income of a household from money-lending.) We also checked back the entire body of data on land holdings of households with the village accountant (*karnam*) and some others—old and experienced agricultural labourers were particularly useful in this task.

Reference Period

For items of data that relate to the working year, the reference period of the survey is the agricultural year 1976–7; for items of data that relate to a point in time—land holdings, asset holdings, debt, family size, age structure—the reference period is the date of the interview, generally between May and June, 1977. The reference year, 1976–7, was a relatively poor year for agriculture in Tamil Nadu as a whole; for a variety of reasons, the main one being the relative reliability of irrigation in the Cumbum Valley, it was not a bad agricultural year in the irrigated tracts of the Valley.

Case Studies and Other Primary Data

After the survey of May–June 1977 and up to 1986, I worked on case studies of individual agricultural labourers and others in the village; on a series of studies of employer–employee relationships in certain crop operations, among which were studies of contract labour in paddy harvesting and transplanting, irrigation workers and crop guards on surface-irrigated land, seasonal migration at the time of the groundnut harvest and migration to plantations, and labour in vineyards and at traditional village tasks; and on collecting village-, block-, taluk-, and district-level secondary data. I must note that considering the time that I spent in the village and Valley (as well as in the neighbouring taluks and the cardamom, tea, and coffee plantations to the east and west of the Valley), my work was not as productive of hard data as it should have been; a great deal of time was spent trying to learn about agriculture and the region by studying and working with the agricultural labourer families of Gokilapuram.

Sample Survey, 1986

This was a sample survey of 43 landless agricultural labourer households, covering the caste of each household; the age, sex, occupation, and educational level of each member of the household; the deployment of annual labour time (reference year: 1985–6) by each member of the household; labour service; and earnings, agricultural and non-agricultural, of each household.

Collection of Data on the Deployment of Labour Time

At the surveys of 1977 and 1986, while filling in information on agricultural labour, we first took the names of each person in the household who had worked at any agricultural tasks over the reference year. We then recorded, for every crop season, crop by crop, operation by operation, the labour of each worker in the household, that is, for each operation we recorded the number of calendar days worked by each person, the hours of work each day, and the earnings for that day (with separate entries for the earnings of farm servants or those such as irrigation workers and crop guards who were paid by the month or season). Where there appeared to be discrepancies in the data, we checked back with the household (and sometimes with helpful neighbours), and we checked labour schedules with a few selected households at the end of the next agricultural year.

In the household questionnaire, for every working member of the household we also collected similar information on each manual task outside crop production, on salaries, on artisans' earnings and earnings from other such traditional callings, and net earnings and income from sidelines and other sources.

Between 1978 and 1984, I also did case studies of each agricultural operation (some of these are discussed in detail in Chapter 9).

In addition, from December 1978 to December 1979, a young harijan agricultural labourer, I. Balusami, who is a friend and is literate, maintained labour diaries for four Pallar landless agricultural labourers: himself, his wife B. Sundari, and a related couple, A. Madasami and M. Kaliyammal, who were their neighbours. Every day, he wrote in whether each person received work, the particular task at which he or she worked (the crop, operation, type of land), the remuneration each person received, and the hours of work. (This was our first try at a labour diary and, in a subsequent format, there was space for entering the employer's name and occupation as well.) Balusami kept the diary for a whole year, and it has been a valuable reference point for checking out different kinds of data on the labouring year in the village.

APPENDIX 2

CASE STUDY: FARMING PRACTICES IN RICE CULTIVATION ON SURFACE-IRRIGATED LAND

I take an example of changing farming practices from a case study of surface-irrigated (*nanjai*) farming in the household of one major landlord in Gokilapuram village.

SNR, who heads the landlord household that has the largest holdings of surface-irrigated land among households resident in the village (22 acres in 1977 and 30 acres in 1981), has been supervising paddy cultivation on the family land holdings since 1950. In the 1940s, they grew two crops of paddy, *parunnel* (in the *kodai* season) and *samba* (in the *kalam* season), with a yield per crop of about 483 to 580 kilograms per acre. A part of the land was leased out, and the family had nine tenants (*kuthagaikkarar*, paying a fixed rent) working their surface-irrigated land. By 1950, paddy yields had risen to about 967 kilograms per acre. This was not because of the application of any special fertilizer or pesticide, but because, with the price of grain rising, landlords saw to it that each operation was performed with greater care, and was no longer haphazard. Manure was spread evenly, bunds built more firmly, weeding was more thorough; this itself helped to increase yields on surface-irrigated land.

In 1955, sulphate produced in Sindri began to be sold in the Valley, and SNR used it for paddy cultivation. In 1965, he began to apply urea (imported from Japan) to surface-irrigated land. By 1960, paddy yield had risen to about 1,160 kilograms per acre, and by 1966 to about 1,450 kilograms per acre. Meanwhile, from 1950 onwards, the family had begun to evict the tenants who worked the land and, by 1960, they had 'not more than six' tenants.

SNR remarks that it could not be said that, in 1965, a very big change had come about in paddy cultivation on surface-irrigated land. The application of urea and sulphate, combined with more carefully conducted farming operations, had made a difference to surface-irrigated cultivation and to yields, but the crop pattern and the pattern of employment of labour had not changed, and the major change in paddy cultivation for this landlord household was yet to come.

In 1967, the *parunnel–samba* crop pattern was broken; SNR began to have ADT–21, an improved variety of paddy, planted. This made some difference to yields, which went up to 1,740 kilograms per acre. Again, SNR said that he would not say that 1967 was the year of big change in agriculture. It was only when new seeds were accompanied by new fertilizers, pesticides, and large-scale government loans for agriculture that a big change came about in cultivation. The real development was after 1968, and particularly after 1970, when there was all-round development in surface-irrigated cultivation and, with it, in groundwater-irrigated cultivation.

After 1968, SNR began to take agricultural loans, and took a close interest in publications and broadcasts on agriculture and in the information provided by Agricultural Extension Officers. By, 1970, the yield from surface-irrigated land was about 2,417 kilograms per acre, with new varieties being planted in both seasons.

Regular use of pesticides began in 1970. Since then, SNR has taken care to find out whether pests are developing immunities to particular pesticides, and tries to use substitutes when he feels that particular brands are not effective. From the 1971–2 season, he began to use nitrogen–phosphorus–potassium fertilizers in addition to sulphate and urea and, in 1976, bought a Massey-Ferguson tractor with a cage-wheel, tiller, disc plough, and trailer. Between 1967 and 1970, SNR also evicted all the tenants from his land, except one, who has a leased-in holding of 0.6 acres and who has, in addition, to work at supervising the cultivation of the household as a whole. The economies of an establishment run with hired labour and on an expanded scale had become apparent. 'It was wasteful to let out the land on fixed rent (*kuthagai*),' SNR said. 'We would lose a lot of income if we allowed tenants to take half the produce.'

In 1981–2, on the better tracts of SNR's surface-irrigated holdings, the first harvest yielded about 3,095 kilograms of paddy per acre (or 7,645 kilograms per hectare).

APPENDIX 3

CASE STUDY: KARUPPAYYA ACHARI, BLACKSMITH

Karuppayya Achari, blacksmith, runs one of the two forges in Gokilapuram village. He is 55 years old. He is a thin man, with a thick grey moustache, few teeth in his mouth, and thick grey-and-white hair pushed back over his head. He lives with his wife and five children in one room of a two-roomed hut, and carries on his trade in the 10-by-12-foot courtyard of the hut.

Karuppayya Achari is acutely aware that the blacksmith's trade has fallen on hard times, and that if the family continues to depend on what he earns from his trade, they must face deepening poverty. He has given some thought to the causes of the decline in his trade and spoke with clarity of how the large-scale production of agricultural implements, increased production from the land, increased mechanization of agriculture, and the loss of land by the cultivating peasantry have resulted in the gradual erosion of his trade.

As far as he can remember, Karuppayya Achari's family has not had land. His father was reputedly the best blacksmith in the village, and he himself began to learn the trade, he said, 'from the day I was born'. At that time, there was only one forge in the village, under the tamarind tree where the rice-pounding co-operative society now is. As a boy, he was sent to school, but left after a short time—he is still illiterate—and from the age of 10 he never left his father at the forge. By the age of 20, the quality of Karuppayya Achari's work was good, and he began to take up independent tasks.

The blacksmith does two main types of work: on bullock carts and on ploughs. For the bullock cart, he makes the axle, the linchpin, the cylinder that fits into the centre of the wheel, metal washers for the wheel, and metal hoops that fit round the wheel. In his childhood and now, work on bullock carts has been paid for in cash (see Table A3.1). The blacksmith also makes the iron-tipped end of the plough and attaches it to the plough.

He is paid 6 *marakkal* of grain in the small *marakkal* per year from each household that gives him their plough to work. The payment is not necessarily in paddy—'except earth, we accept any grain that is given to us'. The amount paid to blacksmiths for work on ploughs—6 *marakkal* per plough—has been the same from his childhood.

Table A3.1 *Comparative payments to the village blacksmith, 1947, 1977*

	1947 (Rs As.P.)	1977 (Rs)
Axle and linchpin	3.0.0	25
Metal hoop	5.0	10
Cylinder	3.8.0	10
Washers (for 8 washers)	n/a	60

Note: n/a = not available.

In Karuppayya's childhood, he and his father used to make sickles, hoes, choppers, and other implements at the forge. Since then, however, large-scale production has completely displaced him from this field of activity. First, years ago, a company of the Tata group began to make implements and sell them in Cumbum. Later, in about 1965, a factory in Vatthalakundu began to manufacture implements. Since that time, people have not come to him to buy implements. Karuppayya Achari explained that he could not afford to make implements and stock them. Years ago, people would be willing to wait 5 to 10 days. But today they buy implements when the need arises in their surface-irrigated, groundwater-irrigated, or plantation lands. They want the implement immediately, or else agricultural operations—particularly work on the estates—will be delayed and their profits will fall.

Karuppayya Achari gets regular work from the households of only four major landlords. He also gets some work from peasant families. Most of the people in the village have shifted to the other blacksmith, who runs a bigger establishment, where work is done faster than Karuppayya can. Tradition dies hard: the blacksmith also does some free labour for the big landlords all year round. He sharpens sickles and knives, and repairs hoes, choppers, sickles, and other implements for the big landlord houses without receiving payment.

The blacksmith's trade is declining. Every year, Karuppayya Achari says, has been worse for him than the last.

Question. How many years is it since your trade began to decline?

Answer. How many years is it since these new seeds came and the tractors came? It is since then that I have done badly; from that day our trade has been ruined. Every year, in the month of Thai, we would be sure of work on at least four newly constructed bullock carts, and would get to do repair work on carts throughout the year. But now, where four carts would take manure to the fields, one tractor with a trailer will do the work. And those without a tractor will save money on a cart by hiring a tractor and trailer.

Question. How has work on ploughs been affected?

Answer. From the day they began to get more paddy per *kuzhi*, the work on ploughs has gone down. Earlier, if a landlord had 5 *sey* of land, he would give 10 tenants half a *sey* each to plough for a fixed rent. Each tenant would keep his own plough. But when the landlords saw that the number of bags of paddy that they could get per *sey* kept on increasing, they did not want to leave land to the tenant and give him half the number of bags. So landlords took back their land and used tractors to plough it. Others also took their land and hired tractors to plough it. The reason for the tenants being sent out was that the harvest brought more bags of paddy. Also, the big farmers began to use tractors. The landlords also thought that the tenants would use the land-to-the-tiller law to fight. So where there were 10 ploughs for us earlier, now there are only one or two. Where we would get 60 *marakkal* of grain, we now get only 6 or 12. And, as for the tenants, they must do wage work, or labour on the estates.

Question. Is there any work on the tractor for a blacksmith?

Answer. NS and VK (who are major landlords) used to ask us to make the prongs for tractor tillers. They paid three rupees per prong if they provided the iron and five rupees if we had to buy the iron. But it is difficult to manage with that work, since the coal and iron cost two rupees and the bus fare to Theni to buy the iron costs three rupees. The money that is paid is just enough to pay costs. If I charge two rupees for the work, I would have to charge seven rupees for the prong, and for seven to eight rupees a spare can be bought from the company. That is why a tractor owner does not come to us for work on his tractor. He buys the part for seven or eight rupees in Madurai. He can buy it in a shop without delay, and does not have to wait.

Costs have gone up for Karuppayya Achari. Thirty years ago, coal cost Rs 5 per bag. Now it costs him Rs 17. He now uses low-quality coal instead of the higher-quality coal that he used to, and pays 10 paise per measure for it. Iron used to cost 5 annas per *veesai*. Now it costs between Rs 4 and Rs 10 per kilogram, depending on the quality.

The remuneration of the blacksmith in monetary terms was Rs 995 last year. The old remuneration of 6 small *marakkal* of grain per plough remains unchanged. The blacksmiths have gone from house to house, asking for an increase, or at least to receive 6 *marakkal* in the 4-measure *marakkal*, but everyone has refused, and told them that it was difficult enough for them to pay 6 small *marakkal*.

Question. Can you not join together and demand 6 big *marakkal* instead of 6 small *marakkal*?

Answer. There are too few of us to do that, and anyway I am not sure if the other blacksmiths would join me. The landowners would go to the next village and get blacksmiths there to do the job until we agreed to accept the 3-measure *marakkal*. We cannot live here if we refuse to work at the old rate. Even to get the 6 *marakkal* is a great effort. In some houses we are told that there was no harvest in the dry land and they will pay us next year—to wait one year and take 12 *marakkal* instead of 6. You talk of higher rates, but even to get these 6 *marakkal*, we must carry a sack and wander from house to house.

Question. What do you do when you ask for the 6 *marakkal* and it is not given?

Answer. We take it that one month's gruel is gone.

Two persons in the household work as hired labourers: Karuppayya Achari's wife Rajathiammal (45) and his son Mani (15). Women from the family have worked for wages in agriculture since Karuppayya Achari's mother's time. Rajathiammal did agricultural labour for 54 days last year. She works only in the lighter operations, such as transplantation and weeding, and not in the operations which call for heavier labour, or those 'where there are many men on the job', such as harvesting and threshing. She also worked at seeding tamarind for more than two months (64 days) last year.

For the first time, a male in the blacksmith's family is working as a hired labourer in agricultural operations and in non-agricultural tasks. His son, Mani, has been working for a woman's wage for the past two years. Last year Mani harvested paddy for 20 days, and worked 36 days in construction work. 'This is work that I never had to do,' Karuppayya Achari said.

About the future Karuppayya Achari said: 'I do not like the way things are going. When I see the way things are going, I think to myself: "Why did I grow up in this trade?"' His fears now are for the children, particularly his son, Mani: 'We could not send him to learn some new trade. To study, a person must have money. We had no money, so he learnt the blacksmith's trade. Now we are sending him to do wage labour, so that he can earn some money when there is no work at the forge.'

May 1977

APPENDIX 4

DEFINITIONS OF ASSET CATEGORIES

Crop land refers to all surface-irrigated (*nanjai*), unirrigated (*punjai*), and groundwater-irrigated (*thottam*) ownership holdings.

Trees and groves are of two types: regular tamarind, coconut, and fruit groves, valued by the unit of land, and individual trees, valued at rates per tree.

Plantation land, in the case of households in Gokilapuram, includes only land in the eastern Erasakkanayakkanur and High Wavy hills and in the south-east, where the price of land varied between Rs 4,000 and Rs 5,000 per acre.

When valuing trees and groves (*thope*) and plantation land, average prices were obtained from interviews with a number of well-informed respondents in the village. In the case of crop land, the value per acre of highly productive, surface-irrigated land fell within a fairly well-defined range, and we were able to confirm whether the data entered in the questionnaires fell within known village-level limits. A combination of factors has pushed the value of surface-irrigated land, whose extent is limited by the command area of the Periyar system, to extraordinarily high levels—an acre of land cost from Rs 45,000 to Rs 60,000 in 1977, and about Rs 100,000 per acre in 1985. These factors include the high concentration (in a relatively small region) of wealth in the hands of a small section and the investment of part of this wealth in the (highly competitive) purchase of a resource that is essentially a limited one, land in the Periyar tract. This wealth was gained from new profits in surface-irrigated and groundwater-irrigated cultivation; from inflated paddy prices in the first half of the 1970s, particularly when the smuggling of rice across the border was rife and very lucrative; from tree produce; and from usury and business activities outside agriculture. Another very important source of income, particularly in the second half of the 1970s, was cardamom. The price of cardamom on the world market peaked in the years that immediately followed the destruction of the Guatemalan cardamom crop after the earthquake of 1975. The price of surface-irrigated land is not entirely transitory, and prices of land in the better tracts have levelled off at the present range since about 1974, and a number of transactions in surface-irrigated land occurred at these price levels during the second half of the 1970s.

In the case of unirrigated and groundwater-irrigated land, the range of prices per acre was wide, depending on varied conditions of fertility, topography, location, and water availability.

Other agricultural assets refers to agricultural implements and machinery and to means of production other than land. The category includes ploughs, animal-driven puddlers, hand sprayers, power sprayers, electric and diesel motor pumps, carts, tractors, and tractor accessories.

It is in the nature of this category of assets that differences between households in the number and type of assets are marked by large and progressively bigger leaps in the value of asset holdings. A large number of households has no such implements at all. A household that has a minimal asset holding in this category owns only a ploughshare and yoke—together costing about Rs 35 in 1977. At a next stage a household may own an electric pump set and perhaps a power sprayer or a bullock-driven puddler, the combined value of which was about Rs 5,600 in the reference year (electric pump set: Rs 4,500; power sprayer: Rs 950; puddler: Rs 150), a vastly higher amount than the value of assets of the one-plough household. At another level, a few households own tractors in addition to ploughs, sprayers, pumps, carts, and other assets, and the price of a tractor and tractor accessories alone was about Rs 100,000 in 1977.

Draught animals is a category comprised of animals actually used for ploughing and carting. There was wide variation in the quality and market value of these animals: from Rs 100 to Rs 900 in the reference year, 1977, for a weak and ill-fed pair of an inferior breed to Rs 2,500 to Rs 3,000 for the hefty, sure-footed, docile, and well-matched Kangeyam animals owned by richer landlords.

The number of draught animals in the village has, over a long period, been on the decline, and a large section of households has no draught animal at all. Some of the factors that have contributed to this decline are the loss of land by many peasant and tenant-peasant households; a decline in the practice of exchange labour, particularly in the ploughing operation (although not its extinction: see Chapter 5); and the extension of tractorization. Tractorization has also reduced the opportunities for agricultural labourers to labour out with their own draught animals for wages on the land of others.

Other livestock includes all milch cattle, poultry, goats, and washermen's donkeys. The Cumbum Valley is not a region where animal husbandry has traditionally developed as a specialized branch of agriculture; there has been no specialized rearing, for example, of draught animals, milch cattle, sheep and goats, or poultry. Changes have occurred with regard to milch cattle, however, and even in 1977, differential interest-rate schemes, forerunners of the IRDP, had begun.

Non-agricultural assets consists of a detailed checklist of assets, physical and financial, not covered by other categories. The composition of these assets showed great variation over households. At one extreme, the poorest had hardly any assets at all, living in a rented or mortgaged hut, with a few worn-out earthen utensils that could hardly be valued. At the other extreme, the richest had entries against every group of assets, physical and financial, on our list—from jewellery and electronic sound equipment to company shares and bonds in state-sponsored investment schemes.

There were problems of concealment in this category. Some things, of course, cannot be concealed—a large house, an automobile, or a large piece of non-agricultural land. Some households have little to conceal: in the case of a poor family living in a one-roomed hut, most household assets are apparent to the eye. There are, however, assets that cannot be accurately assessed in an interview or even by checking with knowledgeable neighbours: for example, the value of gold and jewellery, of financial assets, furniture, and household appurtenances, particularly in a many-roomed residence.

APPENDIX 5

FROM PEASANT TO PROLETARIAN: A CASE STUDY OF P. CHINNASUBBAYYAN

P. Chinnasubbayyan (born 1940) is a working man of Gokilapuram. He grew up in a Pallar scheduled caste middle peasant household in the village, working on the family land, and as a farm servant and casual daily labourer. In the late 1960s and through the 1970s he earned a living mainly as a poor peasant and as a hired labourer. In 1979 he lost his operational holdings and became a landless agricultural labourer in the village and he migrated later that year to Chittampara in Idukki district, Kerala, where he worked as a watchman in a cardamom estate owned by a plantation owner-cum-landowner from Cumbum.

Chinnasubbayyan's parents, Pasuvan and Periyamma, had two sons and a daughter. Pasuvan was a middle peasant who owned about 5 acres of unirrigated land in the eastern part of the village and leased in two plots of surface-irrigated land of a combined extent of 2.2 acres for a fixed rent in kind and one 0.9-acre plot of groundwater-irrigated land on share tenancy. The extent of the first plot of surface-irrigated land was 1.2 acres, leased in by Pasuvan for about 30 years till 1966 (the lease ended with Pasuvan's death) from Shanmughasundara Chettiar, a small landlord, who later sold the land. Pasuvan paid him 20 bags (1.16 tonnes) of paddy per crop as rent. The second plot of surface-irrigated land was a 1-acre plot, and was leased in from Rangasami Pillai of Thevaram, which is about 20 kilometres away. (Rangasami received it as part of the scheme to award agricultural land to freedom fighters.) Pasuvan paid the owner 928 kilograms of paddy per crop, and the lease continued, with Chinnasubbayyan's elder brother as tenant, when we first met Chinnasubbayyan in 1977. The plot held on share tenancy was leased in by the household, first by Pasuvan and later by his elder son, for 40 years until 1977, when the tenant was evicted by the owner.

Pasuvan had a cart, 2 bullocks, and a plough, 9 or 10 milch cows, and about 20 chickens. The small surplus that the family earned helped him save a few thousand rupees to build a house in 1958. They built the house together, Pasuvan, Periyamma, and their children, hiring only a single *kothanar* (village mason) to help them.

Chinnasubbayyan began to work at an early age. When he was seven, he took the family's goats and cows out to graze; at 12, he began to help his mother with the *kavalai*, the leather water-bag in which water was drawn from the well with the help of cattle. He also helped in irrigating the fields and hoeing on field bunds and he helped women workers weed the crops. At 15, his father taught him to plough and, by then, he was familiar with all the major operations on the household holdings. Apart from this, he supplemented family earnings by labouring out, particularly with the draught animals. Every crop season, he ploughed for wages with the animals for about 20 days, worked as a hired labourer with the animals at irrigation from the well for about 10 days, and hitched the animals to the cart and drove it on hire for about 10 days. For each of these tasks, he was paid Rs 6 a day, and he earned about Rs 240 per paddy season (there are two paddy seasons in a year).

In 1961, when he was 21, Chinnasubbayyan began to work with a landlord as a *pannaiyal* (farm servant), working for 6 months at a wage of Rs 60 a month. For two years after that, he worked for a landlord who had the government contract for building houses for harijans. Chinnasubbayyan worked at construction himself, and supervised other workers. They worked for a year at constructing harijan housing in Gokilapuram and another year at Rayappanpatti, and, at the end of it, Chinnasubbayyan received an award for his labour. He was paid Rs 125 a month and an allowance of 50 paise a day over the two-year period. The landlord also gave him clothes. During that period, his father and brother worked the land and took care of the house. Apart from sharing in the expenses of his father's household, where he lived and ate, Chinnasubbayyan was able to turn over Rs 150 that he had saved from the two years of construction labour to his father.

As soon as the work in Rayappanpatti was over, the landlord 'married me off'. His family spent Rs 1,000 on the wedding, and Chinnasubbayyan returned to household cultivation where, on surface-irrigated land, they grew two crops of paddy, on groundwater-irrigated land, two coarse-grain crops (*cholam* and *ragi*) with pulses intercropped and, on dry land, *ragi*, *samai*, and groundnut.

Pasuvan died in 1966, a hard-working man who left his sons no debts. The years that followed his father's death were decisive for Chinnasubbayyan; it was a period of instability and of a shift to semi-proletarian status and, by 1979, to the position of a hired labourer.

Head of a Poor Peasant Household: 1977

In 1977, Chinnasubbayyan was the head of an eight-member household consisting of himself (37), his wife Mariamma (30), his mother (about 53), their two daughters, Murugeswari (5) and Mariamma (1), and three sons, Muniyandi (11), Chinnamuniyandi (9), and Murugan (7). This was a relatively large household: the average size of household in the village in 1977 was 4.9 and, for households of the Pallar caste, 4.8. The adults in the household had never been to school and were illiterate: Chinnasubbayyan could barely—big, scrawled letters—write his name. In 1977 only his second son, the 9-year-old Chinnamuniyandi, was being sent to school. The children were not yet old enough to work at major manual tasks, and Chinnasubbayyan, his wife, and mother worked at the family holdings of dry land and leased-in surface-irrigated land and they did agricultural and non-agricultural tasks for wages.

Chinnasubbayyan still owned some dry land, 1.2 acres in the east of the village, on which he grew a single (and meagre) crop of coarse cereal intercropped with pulses. After the death of Pasuvan, two of the landlords who had leased land to him extended the leases with Chinnasubbayyan's elder brother Periyasubbayyan as tenant. After Chinnasubbayyan took in surface-irrigated land on lease himself in 1972, the two brothers decided to live and cultivate their land separately.

Chinnasubbayyan leased in 0.67 acres of surface-irrigated land from Murthy Naicker, a rich landlord-cum-merchant and occasional money-lender of Vaikkalpatti, a village to the north-east of Gokilapuram. The land had been mortgaged to Murthy Naicker by a member of a Maravar landlord family in Gokilapuram and Murthy Naicker leased it on fixed rent to Chinnasubbayyan during the period of mortgage. Chinnasubbayyan paid a high rent for the land, 15 bags (of 58 kilograms each) of paddy per crop, which worked out to about 45 per cent of gross output in a good season, apart from having to meet the costs of all inputs.

Chinnasubbayyan is a very competent farm worker. In 1977, his family had been using new varieties of paddy for almost 10 years. He saw the burgeoning harvests of the big landlords who had begun to plant new varieties and began to plant them himself. 'We have faith in the new methods,' Chinnasubbayyan said, 'but compared to the landlords we do not have the opportunity to use them fully.' Poor peasants could not afford to hire labour as freely as the landlords, and landlords did not have to hesitate, as a poor cultivator did, before deciding to spend money on the fertilizer, pesticide, and the tractor time that they thought necessary for the crop.

There was no real hesitation, however, among poor cultivators about introducing new varieties, since they produced much more paddy. Of course, there was some talk from those who did not like change, 'talk intended to frighten people', Chinnasubbayyan said. Rumour had it, and Chinnasubbayyan laughed as he reminisced, that 'cattle would die and goats would vomit if they ate the straw from the new paddy, and that human beings would get heart attacks from eating the rice. In Chinnamanur a threshing bullock died, and it was said that it was from eating the straw. I went to Chinnamanur, I saw the animal dead, but this did not keep us from introducing the new varieties—we readily accepted them.' Production was much higher and, once the new rice entered the market and people's cooking-pots, the rumours died out very quickly.

The most serious blow to household paddy cultivation was during the drought, when the second crops in the 1973–4 and 1974–5 seasons were badly affected. 'We had only straw left and poor and broken paddy and husk to pay as wages,' Chinnasubbayyan said. 'I had had no previous experience like this in my life.'

In addition to working on the household operational holding, the three adult members of the household worked as hired labourers. Mariamma and Periyamma together worked 194 days in the 1976–7 agricultural year as hired agricultural labourers on paddy, coarse cereal, cotton, groundnut, banana, and vegetable fields and spent another 60 days working for piece-rates at removing seed and fibre from tamarind. Chinnasubbayyan worked as a hired labourer for 62 days at tasks for which he took the household draught animals with him, for 3 days at harvesting groundnut, and for 5 days at threshing coarse cereal.

The system of exchange labour, though declining, still existed and, in the case of this household, mainly on dry land. Labour was exchanged between relatives and members of the same caste, mainly at the time of weeding and ploughing, 'when there is a need to drive more than one plough through the field.'

Chinnasubbayyan did perform labour services for the landlord-lessor, 2 to 3 days per crop, guarding the sheaves at Murthy Naicker's threshing-floor. He did not do any domestic labour services for the landlord. He would have had to do more had the landlord lived in the same village, but one reason Murthy Naicker, an absentee, had taken Chinnasubbayyan on as a tenant was that Chinnasubbayyan lived in the village where the land was located, and the labour services that the landlord could demand were correspondingly less.

Chinnasubbayyan also attempted to earn some money from sideline business activities. When the banana crop in the village ripened, he and Ramasami, the head of the Pallar *kothu* (group of contract labourers, see Chapter 9), went from field to field, looking for bananas that had been knocked down by the wind before ripening enough for harvest and which could not be marketed with the regular crop. They bought these bananas and 'ripened' them by smoking them in a makeshift oven, made of earth and dead leaves. Every time they had a consignment large enough, they took the bananas—by bus and by hitching rides on lorries for small payments—across the border to Kerala, where they sold them in Vandiperiyar and other markets in the hill region. Chinnasubbayyan earned about Rs 900 that year, and was planning to look for other damaged fruit to ripen and sell in markets close to the border where plantation workers spent their wages.

He had made another, unsuccessful, attempt to enhance household earnings. Twenty-one months before we met him in May 1977, he had bought a milch buffalo on a loan-cum-subsidy scheme from a commercial bank. The household got about 5 litres of milk a day from the animal, and they paid their dues (calculated at a rate of Rs 1.60 per litre) to the village milk co-operative society. Four months after the buffalo calved, the calf died. The milch animal went dry and the family sold it for Rs 300. Out of the existing liability of Rs 1,500, Rs 100 was deducted. There was 'no way of repaying all that we owe now,' and Chinnasubbayyan said that he expected a 'notice' from the bank at any time.

The failure of the paddy crop in two seasons and the bank loan had landed the household in considerable debt. They still had to pay about Rs 1,000 to the bank and Rs 600 to a private lender; the second loan, taken in 1976, had on it a rate of interest of 30 per cent per annum.

When he spoke to us in 1977, Chinnasubbayyan spoke of the fact that over the previous 20 years, joint families were breaking up, and of the growing importance of the cash nexus in village life. He spoke of the landlord who had been his employer in the early 1960s, who was 'interested in harijan welfare'; there was 'less benevolence now'. As for present employers, 'they don't give meals with wages now; they don't even give a worker buttermilk to drink, because they want to sell the milk.'

Loss of Land and Migration: 1979, 1981

Crisis hit Chinnasubbayyan's household in 1979, and he was forced to leave Gokilapuram and migrate to the Cardamom Hills in Kerala for employment. The land he had leased in was land that had been mortgaged to the lessor; the mortgage was redeemed and the land sold. The new owner evicted Chinnasubbayyan without compensation. The household had debts of Rs 4,000. Unable to repay a loan of Rs 1,000 to one of the lenders, Chinnasubbayyan mortgaged the household ownership holding of dry land to him. After losing the surface-irrigated land that he leased in and the dry land that he owned and had mortgaged, Chinnasubbayyan sold the plough animals. He had bought the bullocks in 1977 for Rs 1,050, and he sold them for Rs 600. He sold a 5-year-old bullock cart, bought for Rs 500, for Rs 350. The banana-vending that he had been doing brought in only a small and desultory income, and only for a few months in a year.

At this time, when he, his wife, and mother were struggling to support the family, his wife's cousin, from the village of Veerapandi, who worked as an accountant at a cardamom estate owned by a landlord of Cumbum, asked Chinnasubbayyan if he wanted to work on the estate as a watchman. Chinnasubbayyan knew nothing of cardamom cultivation, but he took the job and left Gokilapuram for the hill town of Chittampara, which is in Idukki district in Kerala, and about 20 kilometres from the border town of Kumili.

At the estate on working days, from 8 a.m. till past 3 p.m., Chinnasubbayyan went with the estate workers to the hillside fields, where he did small jobs around the field. After their work was done, he did his rounds of the estate. From August to November–December, when the crop ripens, he and two other watchmen-guards, armed with choppers, patrolled their estate-wide beats from 6.30 p.m. till past midnight. On Sundays as well, he had to be present to guard the estate. Chinnasubbayyan also had to do odd jobs and run errands for the administrative staff of the estate and for the owner and his family when they came up from the plains.

Years of sustained trade union activity in Kerala have made for estate wages that are higher than wages for agricultural labour in the village. Chinnasubbayyan earned Rs 10 per day for 6 days a week (Rs 60 paid every Saturday), on Tamil New Year's day he was given Rs 240, and Rs 50 each on the festivals of Pongal and Deepavali—about Rs 3,460 a year in all. Once in two months he was given two days off and he went to the village.

Chinnasubbayyan sent home approximately Rs 100 a month. He had savings of about Rs 500 that were kept with the owner: he did not have an account, but was 'sure' that the owner would not cheat him. In 1981, his son also had a job on the estate; his wife would have come too, but there were no vacancies in the work-force.

Chinnasubbayyan's wage, however, was less than that of other estate workers, who earned Rs 10.61 per day and Rs 1.30 bonus for every day they worked. They belonged to the trade union; Chinnasubbayyan, fearful for his job, had not joined either the CITU (the trade union organization of the Communist Party of India (Marxist)) or the INTUC (the trade union organization of the Congress party), although both had asked him to join. He did not intend to join the trade union, even though staying away meant a lower wage (joining the union, he feared, may have meant no wage at all): 'It is the owner who is feeding me rice; can I betray the person who brought me here, confident of my loyalty?' He would not stand in the way of the union, Chinnasubbayyan said, but he would not join it.

Differences in conditions of work in Gokilapuram and in the cardamom plantations of the Kerala hills had made a deep impression on Chinnasubbayyan. 'Here a worker is a worker, whether man or woman, and whatever the caste. For a day's work, man, woman, harijan, Chettiar, all receive the same wage.' There was no segregation of castes—workers of different castes lived in similar quarters and beside one another, they drew water from the same wells and taps, the tea-shops did not have separate glasses in which to serve tea to harijans, and there was frequent intermarriage between men and women of different castes and between people from Tamil Nadu and Kerala. 'In the village, for at least eight hours' work—and it may be more—my wife is given Rs 3. Here for a day's work a worker, man or woman, gets Rs 10.61 and Rs 1.30 bonus, and those who spray pesticides work an hour less and get a rupee more.' The unions were responsible for the difference, Chinnasubbayyan said. 'Because of the union, workers are more aware of their rights, they do not cringe in front of the owner, and the owners cannot behave as they like towards the workers.'

His own dream for the future, when he saved 'about 5,000 rupees', paid off his debts, and redeemed his mortgage, was to return to the village as a peasant cultivating the land.

By June 1986, Chinnasubbayyan—cowherd, middle-peasant, farm servant, construction worker, agricultural labourer, poor tenant-peasant, self-employed itinerant fruit vendor, and cardamom plantation worker, all in the space of his 46 years—was back in Gokilapuram as an agricultural labourer.

May 1977 to June 1986

Appendix 6

Case-Study: G. Selvam, Farm Servant

G. Selvam is a young upstanding agricultural labourer, who has bonded himself as a *pannaiyal* (permanent farm servant) out of economic necessity. Selvam's family is of the Parayar scheduled caste.

A usurious loan of Rs 100 taken over six years ago from CT, a petty usurer, led directly to his present condition as farm servant. At that time, the loan was taken for subsistence needs and was perceived as a temporary expedient. On account of a 120 per cent interest rate, the loan of Rs 100 became a liability of Rs 220 over a year. The usurer pressed Selvam, then 31 years old, to sell his house in order to repay the loan. Selvam, refusing to abandon the family house site, went around asking for a way to work off his debt. The opportunity presented itself in the form of the landlord SCC. This landlord, who was looking for a young and strong farm servant, was willing to advance the money to clear the debt, provided Selvam attached himself as a farm servant for a remuneration of Rs 65 per month plus one sheet and a dhoti, a shirt, and a *thundu* (towel-cloth) a year.

Selvam took an advance of Rs 100 for going to work as a farm servant, and used that to clear just under half his debt. Then, after the first month of work, he took a loan of Rs 120 to clear the debt.

Since then, that is, for six years, Selvam has been working for well over 13 hours a day. He worked at the salary of Rs 65 a month for four years. Two years ago, when paddy prices soared to nearly Rs 150 per 58-kilogram bag, the farm servants in the village asked their employers for a raise. Selvam was given a wage rise that was long overdue: in 1977 he was paid Rs 110 a month.

SCC, like some other big landlords in the village, has found it much to his advantage to hire a farm servant in this way. He has advanced small sums of money to Selvam over the years, sums always taken 'temporarily', but with no real chance of the debtor repaying the debt and getting out of his present condition. Selvam makes it clear that he is not paid anything near the remuneration he should be getting for this work. 'There is no choice,' he says, 'I can't leave my *mudalali* (employer, landlord) unless I can clear my debt of Rs 300. I would certainly like to leave.'

In his childhood, Selvam was not as badly off as he is today. His father, Gabriel, was a poor or lower-middle peasant, cultivating surface-irrigated land that had been leased in from landlords of the village. His mother worked as a hired labourer.

His father sent Selvam to school. Selvam studied in the Mission School in Gokilapuram, down the road from where he lives today, up to the fifth class. He completed the sixth class at the NMR School in Gokilapuram. Selvam was a good student and his father sent him to Vatthalakundu, where Selvam studied in the seventh and eighth classes, finishing when he was 16 years old. He stayed at a hostel at Vatthalakundu, and his father sent him Rs 20 a month. Selvam returned to the village after finishing the eighth class. He can read and write Tamil and can still read some English.

Selvam's father tilled 6 *kuzhi* (3.6 acres) of surface-irrigated land belonging to landlord ST on *kuthagai* (fixed rent) for 20 years. He also worked 1.5 acres of groundwater-irrigated land belonging to SP, a landlord of Uthamapalayam. Selvam also worked on the land leased in by the family. Selvam married when he was 20 years old. His wife, Alphonse, is from a family of tenant cultivators from Pudupatti in Uthamapalayam taluk.

In about 1967, there was a sharp decline in the agriculture conducted by the family. The surface-irrigated land had poor soil and bad drainage, and standing water after the rain affected the crop. The land was manured only for one crop, one of the reasons for the yield being poor. The groundwater-irrigated land had a well with plenty of water and had a pulley and rope to draw water with. But household cultivation was in decline, the family began to incur debt, and the age-old symbol of a peasant family heading towards destitution became apparent: the cattle owned by the family became thin and weak.

About the same time, eviction took place on a large scale in the village. Earlier, the paddy crops were the traditional *parunnel* and *samba* varieties; later, with the introduction of new seeds, yields were higher, and agriculture became more profitable for the landlords. There was also fear among them, Selvam said, that the tenants would assert their right to cultivate the land. Landlords brought pressure on the tenants to leave the land. Rents were raised, rack-rents were imposed on tenants. In some cases, landlords brought great pressure on tenants to leave, offering them small amounts of money to do so. The peasants were disunited, Selvam said, and, out of fear, accepted the money and left the land.

Agricultural Wage Work

Selvam made it clear that the big landlords can always bring pressure on agricultural labourers; as for agricultural labourers, there is little scope for their advance today. Opportunities for employment in some tasks have gone down since the arrival of tractors, Selvam pointed out. The tractor has robbed those with ploughs and bullocks of ploughing work, and those with carts and bullocks of work in basal manuring. Even during threshing, the tractor has deprived agricultural labourers of employment. While earlier, there would be four days' work at the threshing-floor and four bullocks would be needed to trample the grain for the second threshing, now the tractor can be driven over the sheaves to complete the task in less than an hour.

Unemployment is high and wages are poor, and women's wages have actually been brought down, from Rs 3 per day to Rs 2.50 per day. There was agitation by Communists five years ago in southern villages of the Valley, Selvam recalls, when men workers won an increase in their grain wages, from 4 measures per day to 5 measures, and women an increase to 3 measures for threshing.

While wages in the village do not vary directly with the caste of the worker, discrimination against the scheduled castes, the overwhelming majority of whom are landless labourers, is deeply entrenched. Most of the farm servants in the village are harijans.

Selvam has an extremely busy working year. During the 1976–7 agricultural year, he worked on both crops of paddy on surface-irrigated land, in different operations in the cultivation of irrigated *cholam*, *ragi*, tomato, and banana on groundwater-irrigated land, and in *samai* on unirrigated land. In addition to agricultural work every day, he did domestic tasks at the landlord's house. Selvam's wife Alphonse laboured at agricultural operations for 57 days in 1976–7, and earned Rs 158.70 as wages. She seeded and cleaned tamarind for 3 days, for which she earned Rs 7.50.

Selvam's wage is lower than other farm servants in the village, some of whom are paid Rs 130 and Rs 140 by their landlord employers. But when Selvam asks for a wage increase, the landlord arrogantly insists: 'I have already increased your wages. You used to get Rs 65, now you get 110.' The standard that SCC uses, Selvam says, is not the wage paid to other farm servants in the village, but the Rs 65 pittance that Selvam himself used to receive before he was paid Rs 110 a month.

When Selvam is unwell, the landlord may give Selvam a small amount of money to buy medicine, but he will not pay for treatment for Alphonse or the children when they fall ill.

As for his hut (the retention of his house site was an immediate reason for Selvam bonding himself to the landlord), Selvam cannot afford to erect a complete hut. When he gets a small amount of money, he adds a row of bricks to the hut. Some months ago, he bought a door-frame. Now there is a door-frame, fixed with mud into a few rows of country-made brick, with no wall around it or roof above. The single room in the hut is 8 feet by 6 feet. 'We do not know when this hut will be built,' Alphonse said. 'We lay a few bricks, and it may be a few months before we can add more bricks. It may take years to complete the hut in this manner.'

With all this, Selvam must also face rudeness and shouts from the landlord, the loud arrogance of SCC as taskmaster of 'his' farm servant.

'Life is difficult for me,' Selvam said. He sees his children at 5 a.m. (when he leaves the house), when the older children may be just beginning to wake up. It is seldom, and only in the slack season, that he comes home before they are asleep for the night. And the hours in between are filled with arduous, back-breaking labour that covers every task that an agricultural labourer in the village can perform.

I have seen Selvam, early in the morning, heaving farmyard manure on to a cart and driving the cart to the paddy-fields, unloading it in neat piles across the field. I have seen him, bare-chested, barefoot, and dressed in an old and torn lungi, with a soiled cloth on his head to protect him from the sun, walking with a hoe across his shoulder over the ridge east of the village on the road leading to SCC's groundwater-irrigated field in neighbouring Anamalaiyanpatti. I have seen him at the field, cutting and clearing water channels before the electricity comes on and irrigation water rushes up; and in the banana field, hacking at the tough young shoots that grow by the banana trees, cutting dry and withered leaves off the tree, and straightening tree trunks with wooden props that have been chopped and shaped by him. I have seen him, in the evenings, working in SCC's house, watering, feeding, and washing the cattle or chopping a tree trunk for firewood for SCC's kitchen. I have seen him at the bus-stop in Uthamapalayam, straining to lift a motor pump, newly repaired, on to the carrier of a Gokilapuram-bound bus. Standing under a bus-shelter in Gokilapuram after 11 o'clock at night, I have heard the sound of a cart coming down the road from Anamalaiyanpatti. It is Selvam, urging the bullocks through pouring rain back to the house of SCC. I have seen him after 11.30 at night, rain smudging the red mud that clings to his face and body, entering his hut and drinking gruel and pickle by the weak light of an oil lamp, and preparing to catch some sleep before the next day of labour begins.

Selvam states clearly that it is not 'loyalty' that keeps him with the landlord. It is his debt, and the difficulty of finding alternative employment if he were to leave. 'I have heard of the union of agricultural labourers in East Thanjavur. A union is needed, unity of agricultural labourers is needed. If there is a union in Gokilapuram, I will join it.'

May 1977

APPENDIX 7

THE PERIYAR IRRIGATION SYSTEM IN GOKILAPURAM

This appendix describes the Periyar irrigation system in Gokilapuram. A right-bank channel of the system, the Palayamparavu Channel, feeds the surface-irrigation network in the village.

The southern tip of Gokilapuram village is about 4 kilometres south of the village settlement; a long stretch of the wet land of the village is contiguous to the village settlements of Anaipatti and Rayappanpatti villages (see map). There is no wet land in the revenue jurisdictions of those villages and the surface-irrigated holdings of landowners in them are located within Gokilapuram revenue village. Up to point A on the map, there are five left-bank sluices with direct command areas (*ayacuts*). Up to B, the channel runs through Gokilapuram and contiguous to Rayappanpatti and Anaipatti. Between A and B there are two left-bank sluices. From B, one left-bank branch channel goes on to Thamaraikulam tank, whose command area of about 543 acres is located in Uthamapalayam village (PWD Uthamapalayam 1981). Before reaching Thamaraikulam, there are five left-bank sluices whose command area is in Gokilapuram.

To return to the main channel, there are six sluices at C and D. Four have direct command areas, one channel leads to Kalaneerkulam tank, whose command area of about 105 acres is in Gokilapuram, and another to Karunkattankulam tank, which is the terminal point of our channel and whose command area, in Karunkattankulam village, is about 755 acres (Uthamapalayam Taluk, Revenue Records, 1980–1). In the upper reaches of the channel to Karunkattankulam, a left-bank sluice at E opens out into Gokilapuram village.

After 1978, a new channel, the PTR Channel,[1] beginning from Karunkattankulam, was commissioned. In order to feed this channel, which draws mainly on Periyar water, more water than before is released into Karunkattankulam and, consequently, into the Palayamparavu Channel.

To sum up, there are 17 sluices with direct command areas in Gokilapuram village, which irrigate about 580 acres in all. The extent irrigated by individual sluices varies widely, from 3 acres through one of the sluices at D, to 180 acres through the sluice at E. The indirect command area, through two sluices in Kalaneerkulam, is 105 acres. The command area consists of land that is almost entirely cropped with paddy—the returns for 1980–1 show a gross cropped area under paddy of 1,369 acres (net sown area 684.5 acres) and no other crop on surface-irrigated land.

Before the Periyar system was inaugurated in the 1890s, irrigated land in the village was located in the command area of the tank, and in the command area of the three present-day sluices, at C, E, and F. Of these, only in the land fed by the tank could two irrigated crops be raised; even in good years the other land irrigated by the Vairavanar–

[1] Named after P. T. Rajan, major landlord and Justice Party Minister, whose family has vast wet-land holdings all along the Periyar tracts. Their holdings—of surface-irrigated, lift-irrigated, and dry land, and of groves and plantation land—are located in different parts of the Valley and district; the family seat is at Uthamapalayam (see also ch. 9, n. 40).

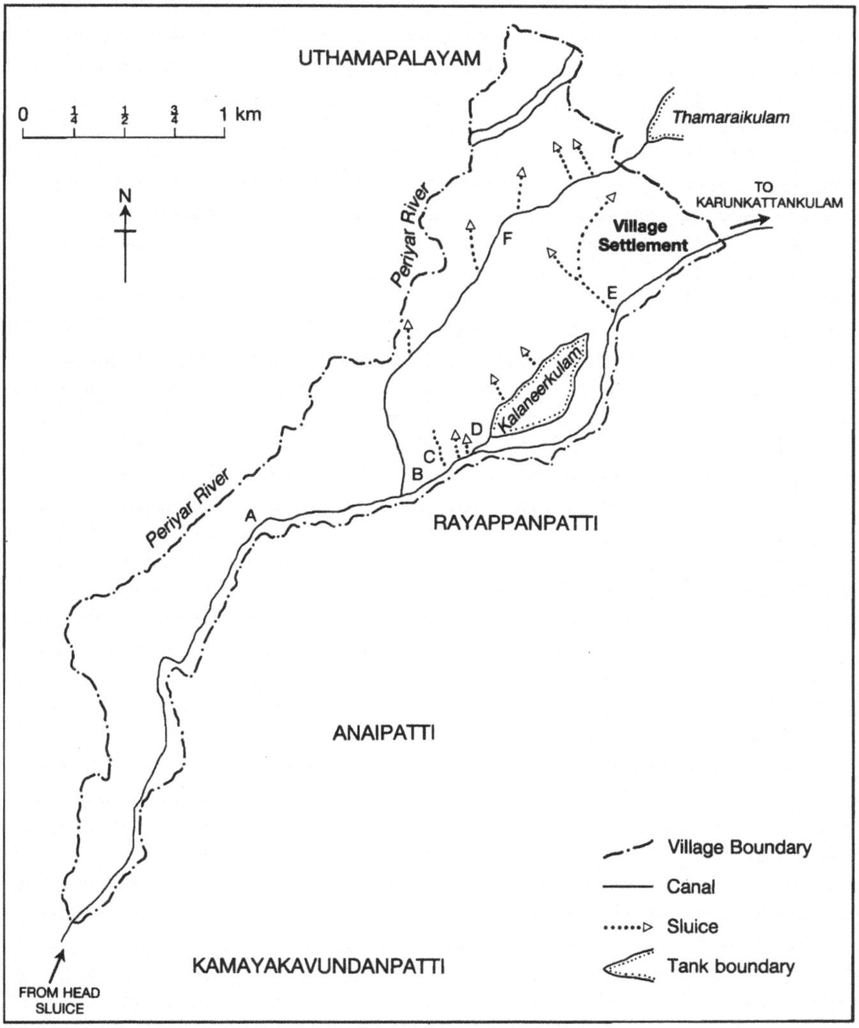

Map A7.1

Suruliyar–Varattar system, which was in existence before Periyar water came to the village, had only a single annual irrigated crop.

The release of Periyar water stimulated an important process of intensification of irrigation on tracts that were formerly poorly irrigated. In an area where water supply was irregular and scanty, and over large portions of which water was available only for a single crop, it was now abundant and available for two crops. Some extension of cultivated land also took place, although there are not sufficient data to establish the precise acreage.

The revenue returns for 1980–1 show that 685 acres were under canal and tank irrigation in the village. Since the resettlement, anicuts have been repaired and the channel has been fitted with a head sluice and with subsluices along its course. There has also been an extension of the irrigated area. In the 1960s, a stretch of fields and fallow land in the tract between the river and the canal south of A was brought, *en bloc*, under canal irrigation. Land was levelled and the canals were breached to irrigate the fields. Prominent landlords of Rayappanpatti and Gokilapuram have holdings in this area, and they were responsible for its forced conversion to irrigated land. They also made continual efforts at considerable expense to convert the land into registered wet land, to record it as part of the regular command area of the Palayamparavu Channel. These efforts have partly paid off—irrigation has been regularized in that the penal rate on irregular water utilization has been minimized.

Water is distributed along the channel by a turn system. There are, first, turns between the three villages that receive water, namely, Uthamapalayam (through Thamaraikulam), Karunkattankulam, and Gokilapuram itself. These villages receive water by rotation, for 48 hours each. (In a year when there is no shortage, Karunkattankulam and Thamaraikulam, by agreement with wet-land owners from Gokilapuram, exceed their 48-hour limit somewhat in the first few turns, since preparation of seed-beds in these villages generally begins before Gokilapuram.)

There is also an intra-village turn system (*murai*) between the 17 sluices and the tank with two sluices. The volume of water allowed through each vent is a function of the extent of command area of each vent. Sluices are either shuttered (these are generally open for a specified period) or fitted with open pipes. Most of these pipes are open continuously for 48 hours, the flow being restricted only by the diameter of the pipe. Fig. A7.1 illustrates the *murai* within the village over the 48-hour village turn. It should be noted that although the area south of A does not have the same status of registered wet land in the village records as other areas, it receives water equally in the village *murai*; the owners in this area have sufficient influence to ensure that.

In a year of scarcity, access to irrigation water depends largely on two factors: the nearness of a field to the head reaches of a channel or sluice, and the status of a landowner in relation to other owners in the command area and his ability to block other outlets and take water to his field. In an area which also has the feature that the land holdings of powerful landlords are scattered across the command area, it is not surprising that the strictness with which the *murai* is enforced varies inversely with the availability of irrigation water. We have been told that the present turn system was put into effect 'about seven years after the PWD took charge of the canals. Before that, in years of scarcity, only those whose lands were in the upper reaches or those who could threaten others and take water to their fields could irrigate their crops.'[2] It is noteworthy that the year mentioned (about 1930) is the period following the two years 1928 and 1929, 'which were years of very severe drought in the [Periyar] Delta—perhaps the two worst years in the history of the Delta' (Hall 1930, p. 58).

Two crop years in the 1970s, 1973 and 1975, were years of shortage: while official records concerning the losses in these years were not available, persons in the village estimate that, in 1975, 375 to 400 acres in the village were affected. During this period there was a more rigorous enforcement of the *murai*: inadequacy demanded more careful husbandry of the scarce resource. This did not, however, make for a more equitable distribution of water between all landowners in the command area. In 1975, the

[2] Interview, 1981, with Raphael Udayar, then 94 years old, of Rayappanpatti, corroborated by others. Raphael Udayar's father made the southernmost breach in the canal wall in 1907; his son, tenaciously litigious, regularized their claim to canal water in 1921.

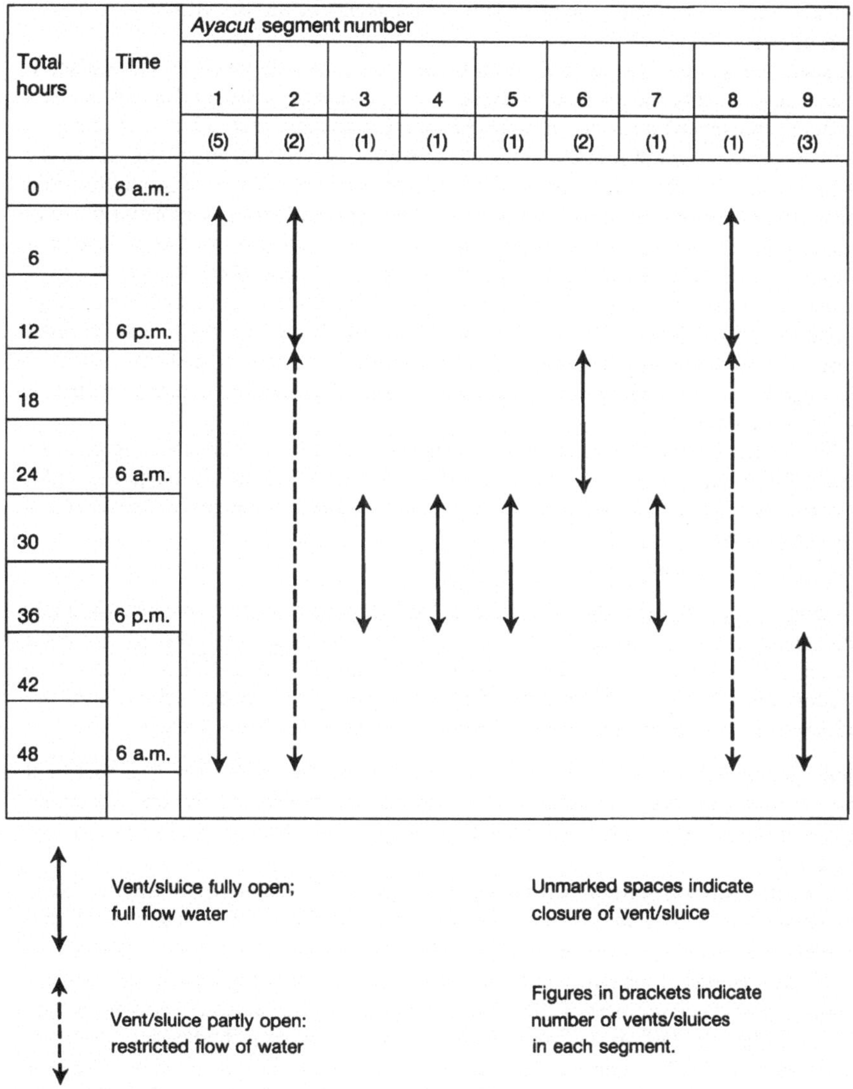

Fig. A7.1

restriction on tapping groundwater in the Periyar command area was lifted, and bore-wells fitted with diesel motors were permitted to be installed for that year. All those who installed bore-wells in the 300-acre area abutting the village whose names were ascertainable were prominent landlords.

Conversely, in years of plenty, the *murai* loosens. Since 1978, the year the PTR Channel was commissioned, there has been an increase in the volume of water that flows through the canal. During the first crop in 1981–2, for instance, while the 48-hour turns between villages was maintained, the rules of the *murai* within the village were hardly put into effect.

A specific feature of surface irrigation in the Cumbum Valley and, in particular, in the command area of the Palayamparavu Channel, is that the command area is not flat, but slopes downward from the contour channel towards the river. Thus it can be seen on the map that all the sluices along the channel after the channel enters the village are left-bank vents. Given this slope, and given that water management in paddy-fields requires that fields be flat, the division of the total command area into plots for farming are a function of the subdivision of land into individual holdings as well as of the line of contour. As a consequence of the slope of the command area, there is no problem of drainage from paddy-fields in Gokilapuram, and this feature of farming in the village contributes to the fact that cultivators stick with paddy. There is also a substantial return flow of irrigation water from the fields to the river; this is in the form of surface run-off from outlets leading from field channels and of groundwater that seeps below from channels and fields.

The system of surface irrigation in the village and in the Periyar system is, in general, a channel-to-field system and not a field-to-field system. While this is the general rule, the extent of field-to-field irrigation has increased because of changes in the pattern of ownership of land.

In 1937, it was noted that:

> Owing to the extension of cultivation on hill slopes consequent on the wholesale cutting of trees—the natural water courses, which some years ago brought little silt . . . now flow down laden with heavy coarse sand and pebbles and fill up the channels within a few hours . . . This state of affairs is most pronounced in the case of the Uthamuthu *and Palayamparavu channels* . . . (Chave 1937; my italics)

While specific data on the extent of silting today are not available, it is clear that field channels and tank beds get silted every crop season, and that there are substantial seepage losses. A great deal of labour time is spent on clearing field channels (see Chapter 9).

There are also evaporation losses from tank surfaces. Tanks in the Cumbum Valley are shallow, with a large waterspread. The fact that tank floors are at a higher level than the fields is a check against dead storage and against waterlogging, but evaporation is high. In at least one of the tanks near the village, trees that provided shade and checked evaporation over a large part of the waterspread till 25 years ago have now been cut.

In the 1930s, the Officer on Special Duty in Connection with the Periyar System of Irrigation, J. F. Hall, suggested that the possibility of altering the present system of channels into a properly regulated system be investigated and that the costs and benefits of such alteration be made the subject of a special study in order that the duty on irrigation water in the Valley might be increased and that, in general, the use of irrigation water in the Valley be optimized (Hall 1930, pp. 46 ff., particularly para. 39).[3] It appears that a more rational system of water use and of the use of labour power

[3] It is clear that irrigation statistics that deal with irrigation solely in terms of area irrigated are very inadequate; accurate analyses of irrigation require data also on the *volume* of irrigation water required and available and the *period* for which water is available. Although I do not have such data, it has been noted in official documents that the rate of consumption of irrigation water in the Cumbum Valley is higher than in other parts of the Periyar system, and that there is a much lower duty on water in the Valley than elsewhere in the system (Hall 1930, pp. 46 ff.).

in surface-irrigation tasks in the village would have to include a system of shuttered and lined channels (including field channels) that were regularly dredged and remodelled on the ground to allow the most efficient use of water. It would also have to include a change in farming practices: the consolidation of seed-beds at the head reaches of channels, for instance, was mentioned even in the early literature[4] as more efficient than the present practice of dispersing seed-beds—and consequently taking water—to every field in the command area of a sluice. Changes such as these, however, are likely to be more than the present system of ownership and control of land can bear.

[4] By a Mr Dowley, Executive Engineer, PWD, at Madurai, cited in Hall (1930, pp. 36 ff.).

APPENDIX 8

CASE STUDIES OF IRRIGATION WORKERS

These are accounts of two episodes when the irrigation workers (*neerpaichi*) joined together, in defence of the rights of a tenant in one case, and for payment in good quality grain in the other.

K. Servar, headman of the Parayar caste group in the village, is a respected member of the village community. He is experienced in traditional veterinary medicine and goes to great pains to take care of sick and disabled animals—he is willing to answer calls at any time of the day or night, even when the owner of the animal does not have money or grain for immediate payment. From 1954 to 1976, Servar was a tenant-at-will on hereditary trust property controlled by a landlord family of Rayappanpatti.

In 1976, a new trustee took over on the death of the old. Without letting Servar know, the new trustee had the land ploughed. Servar appealed to the landlord to permit him to continue for at least another year—the previous year had been a bad one and he had incurred debts of more than Rs 2,000. Servar completed the nursery ploughing and preparation of the seed-bed. One day when he was out of the village, the landlord had the seed-bed ploughed over. Servar complained to others in the village against the landlord and also filed a complaint against him at the police station.

At this point, the irrigation workers intervened, and told the agent that Servar ought to be permitted to continue for another year. (Even the irrigation agents of that year recall that it was because of the intervention of Rayar Karuppana Thevar, the oldest and most experienced of the irrigation workers, and the other irrigation workers that Servar's tenancy was saved for a year.) For each operation, the landlord sent his workers; they were prevented from entering the field by the irrigation workers. Weeding was done by two sets of people: Servar's own workers and workers employed by the landlord. When the crop was cut and threshed, the landlord still claimed the crop. Rayar and other irrigation workers kept guard over the crop, not allowing the landlord's men to move it until a compromise was worked out. Ultimately, Servar cultivated the land for the whole year free of rent and was paid Rs 2,500 as 'compensation' for the loss of the 23-year tenancy. It is also recorded that Servar, out of gratitude to the irrigation workers, added to the extra payment (*inam*) that went with the payment for irrigation work that season.

At the end of each harvest, two irrigation workers, on behalf of all the irrigation workers, would go from house to house collecting the paddy wage due to them from cultivators. In one season, when they had to collect grain after the harvest, they went as usual to the house of RSN, the landlord who has the largest surface-irrigated land holdings among households in the village. When the grain was being measured out they found, to their dismay, that RSN was paying them in very poor paddy taken from the second threshing and full of husk and broken grain. Paddy such as this yields very little whole rice, and these were wages being paid by one of the richest landlords in the village—his reputation for being a miser was certainly being proved. Neither of the irrigation workers said anything, however. They took the grain to the other workers, showed it to them and told them who had measured out such poor grain, and then together worked out a plan for the next day.

The next morning, the two bags in which RSN's grain was filled were taken to a place near the Vinayaka temple in the centre of the village and kept there with two irrigation workers stationed beside them. They took turns by the bags, which stayed there until the evening. The place they chose is the busiest crossroad in the village—it is also the village bus-stop—and is a point which almost every person in the village passes at least once a day. And when people passed on that day, they invariably stopped by the sacks and asked the irrigation workers, 'What are you doing with these two sacks of husk and broken grain in this place?' And the workers would answer, 'Oh these?—this is the paddy that RSN measured for us as wages this season.' From morning till evening, people passed the crossroad; each asked the same question and got the same reply, and each person went away laughing at the stinginess of the man.

From the next season, RSN paid the irrigation workers in whole grain.

Unionization of Irrigation Workers in the Cumbum Valley

In 1971–2, during a period of general militancy among agricultural labourers in the Valley, activists of the All India Kisan Sabha and the Communist Party of India (Marxist) in the Valley called a meeting of irrigation workers at a place near Gudalur in the south of the Valley. The meeting was attended by irrigation workers from Gudalur, Cumbum, Narayanathevanpatti, and Kamayakavundanpatti villages. They later formed a union, the Cumbum Area Irrigation Workers' Union, which was active through the first half of the 1970s.

Their major demands concerned an increase in wages, standardization of measures in which grain wages were paid, separate payment for tasks other than those directly concerned with irrigation, payment in paddy of good quality, and security of employment.

Wage increase and standardization of measures. In the regular volumetric measure, 4 measures (*padi*) comprise one *marakkal*, and 12 *marakkal* comprise a bag, which is equivalent to 58 kilograms of dry paddy. A regular bag thus consists of 48 measures. A *coolie marakkal*, which employers use specifically for wage payment, however, contains only 2.5 to 3 measures, and one bag of 12 *coolie marakkal* contains only 30 to 36 measures.[1] The union demanded that irrigation workers be paid 2 bags per *sey* (2.4 acres) in the regular *marakkal* and not the *coolie marakkal.*

Separate payment for tasks other than irrigation. The union demanded separate wage payment for all tasks other than those concerned specifically with irrigation; for instance, a payment of Rs 8 per *sey* for top-dressing, similar wages for applying pesticides, and a regular daily wage for hoeing work such as construction and maintenance of field bunds.

Paddy quality. The union demanded that paddy wages be paid in grain obtained from the first threshing and not, as was the practice, from the second threshing.

[1] In certain areas, rent is collected in a *kuthagai* (rent) *marakkal*. This measure, predictably, holds more than the regular *marakkal*—about 5 *padi*.

Protection against dismissal. Most irrigation workers are men with many years of irrigation labour behind them. The union demanded that landowners should not be allowed to hold out the threat of dismissal against them.

The first demand, relating to payment in the regular *marakkal*, was won in most of the villages in which the union was active; the other demands were not gained in most villages.

While canvassing membership for the union, its activists exposed a case of bondage. Suruli, an agricultural worker of Cumbum, took a loan of poor-quality millet ('that not even an animal would eat') from GB, medium landlord, Congress Party man, and secretary of a local charity hostel. Suruli was made to repay the debt by working on GB's surface-irrigated land. His debt began to accumulate, and, by the time he was contacted by activists of the union, Suruli had not been paid for seven harvests. The union demanded of the landlord that Suruli be paid. Their demand was eventually successful: they had him pay 105 measures of paddy to Suruli as back wages.

GLOSSARY

Many of the words in this glossary can have more than a single meaning. In general, I have given only the meaning that is relevant to the context of this book.

Volumetric measures

4 *padi* (or 'Madras measures') = 1 *marakkal*
12 *marakkal* = 1 *kalam*
1 *kalam* is the equivalent of 58 kilograms of dry paddy
1 *coolie* (wage) *marakkal* = 3 *padi*
1 *kuthagai* (rent) *marakkal* = 5 *padi*

Land measures

1 *kuzhi* = 0.6 acres
4 *kuzhi* = 1 *sey* = 2.4 acres
1 *sey* is almost 1 hectare, which is the equivalent of 2.47 acres

Paddy seasons

Kodai	June to October
Kalam	October to March

Forms of rent

kuthagai	fixed rent; hence *kuthagaikkarar*: tenant who pays a fixed rent
varam	share-cropping; hence *varamdar*: share-cropper

Months of the Tamil calendar

Chithrai	April–May
Vaikasi	May–June
Ani	June–July
Adi	July–August
Avani	August–September
Purattasi	September–October
Aippasi	October–November
Karthikai	November–December

Markazhi	December–January
Thai	January–February
Masi	February–March
Panguni	March–April

Botanical names of some cereal crops

Tamil name	Common English name	Botanical name
Cholam	Sorghum; great millet	*Sorghum vulgare*
Cumbu	Spiked millet; pearl millet; bulrush millet	*Pennisetum typhoides*
Ragi	Finger millet	*Eleusine coracana*
Samai	Little millet	*Panicum miliare*
Varagu	Kodo millet	*Paspalum scrobiculatum*

Miscellaneous

Anicut, *anaikattu*	Masonry barrage
Ayacut	Command area
Beedi	Smoking-tobacco rolled in *tendu* leaf (*tendu* is *Diospyros melanoxycon*)
Block	Administrative subdivision in the district development administration, below the district and division and comprising contiguous villages
Cheri	The section or settlement of the village where harijans reside
Chit fund	A rotating saving-cum-credit association
Deepavali	Festival held on the day of the new moon in the Tamil month of Aippasi, celebrated with firecrackers, sweets, and new clothes, and by decorating homes with lights
Dhoti	Cloth used as lower garment by men
Dravida Kazhagam (DK)	Leading organization of the non-Brahman social-reform movement in Tamil Nadu
Harijan	Name first used by Mohandas Gandhi for castes condemned to 'untouchable' status by Hindu tradition; synonymous with 'scheduled caste'
Jajmani	Network of patron–client relations
Jati	Caste or subcaste
Kavalkarar	Crop guard
Lungi	Lower garment worn by men

Mirasdari	Pre-colonial proprietary right in land, which was inheritable but inalienable
Murai	Village-level irrigation turn system
Neerpaichi	Irrigation worker
Panchayat	Institutions of local government at the village level. In the present context, distinct from the village panchayat as a group of male elders who adjudicated in intra-caste and inter-caste issues
Pongal	Festival associated with the grain harvest, held in the beginning of the Tamil month of Thai
Scheduled castes	Castes specified under the provisions of the Indian Constitution for purposes of affirmative action; synonymous with 'harijans'
Taluk	Administrative subdivision in the revenue administration, below a district, and comprising a collection of contiguous villages
Thalayari	Village watchman
Thope	Trees and groves
Thotti	Village sweeper and cleaner
Udayar	Upper-caste, non-Brahman *jati*; dominant caste in Rayappanpatti village, which borders Gokilapuram
Vokkaliga	Upper-caste, non-Brahman, Kannada-speaking *jati*; dominant caste in several villages of the Valley
Zamindari	Statutory intermediary property right in land, encompassing the right to collect rent

BIBLIOGRAPHY

AERC (1961), *Vadamalaipuram Village*, Village Survey No. 48, AERC, University of Madras.

—— (1962), *Dusi*, Village Survey No. 55, AERC, University of Madras.

—— (1964), *Keeranatham Village*, AERC, University of Madras.

—— (1969*a*), *Resurvey of Aralikottai: A Dry Village in Ramanathapuram District*, AERC, University of Madras.

—— (1969*b*), *Resurvey of a Dry Village in the Fertile Thanjavur District*, AERC, University of Madras.

—— (undated *a*), *Madigai*, Village Survey No. 16, AERC, University of Madras.

—— (undated *b*), *Rajagambiram*, Village Survey No. 40, AERC, University of Madras.

—— (undated *c*), *Sengipatti Village*, Village Survey No. 42, AERC, University of Madras.

—— (undated *d*), *Aralikottai Village*, Village Survey No. 49, AERC, University of Madras.

Agarwal, Bina (1981), 'Agricultural Mechanisation and Labour Use: A Disaggregated Approach', *International Labour Review*, 120(1), pp. 115–27.

Agricultural College and Research Institute (undated), *A Study of the Impact of River Valley Project on the Socio Economic Structure of the Regions Benefited: Parambikulam Aliyar Project Region: A Bench Mark Survey*, Agricultural College and Research Institute, Agricultural Economics and Rural Sociology Section, Coimbatore, Tamil Nadu.

Ahluwalia, Montek S. (1978), 'Rural Poverty in India: 1956–57 to 1973–74', Staff Working Paper No. 279, World Bank, Washington, May.

Ahmad, Zubeida (1984), 'Rural Women and their Work: Dependence and Alternatives for Change', *International Labour Review*, 123(1) (Jan.–Feb.), 71–86.

Alagh, Y. K., Bhalla, G. S., and Bhaduri, Amit, 'Agricultural Growth and Manpower Absorption in India' in Pranab K. Bardhan *et al.* (1978).

Archer, Leonie J. (ed.) (1988), *Slavery and Other Forms of Unfree Labour*, Routledge, London.

Arnold, David (1979), 'Dacoity and Rural Crime in Madras, 1860–1940', *Journal of Peasant Studies*, 6(2) (Jan.), 140–67.

Athreya, Venkatesh B. (1984), 'Vadamalaipuram: A Resurvey', Working Paper No. 50, MIDS, Madras, Aug.

—— (1985), 'Gangaikondan 1916–1984: Change and Stability', Working Paper No. 56, MIDS, Madras, June.

Bardhan, Kalpana (1977), 'Rural Employment, Wages and Labour Markets in India: A Survey of Research', *EPW*, 25 June, 2 and 9 July.

—— (1983), 'Economic Growth, Poverty and Rural Labour Markets in India: A Survey of Research', Working Paper No. 57, World Employment Programme, Rural Employment Policy Research Programme, ILO, Geneva.

—— (1984), 'Work Patterns and Social Differentiation: Rural Women of West Bengal', in Binswanger and Rosenzweig (eds.) (1984).

Bardhan, Pranab K. (1970), 'The Green Revolution and Agricultural Labourers', *EPW*, Annual Number.

—— (1973), 'Variations in Agricultural Wages: A Note', *EPW*, 26 May.

—— (1978), 'On Measuring Rural Unemployment', *Journal of Development Studies* (Apr.), 342–52.

—— (1979*a*), 'Agricultural Development and Land Tenancy in a Peasant Economy: A Theoretical and Empirical Analysis', *American Journal of Agricultural Economics*, 61(1), 48–57.

—— (1979*b*), 'Labour Supply Functions in a Poor Agrarian Economy', *American Economic Review*, 69 (Mar.), 73–83.

—— (1979*c*), 'Wages and Unemployment in a Poor Agrarian Economy: A Theoretical and Empirical Analysis', *Journal of Political Economy*, 87 (June), 479–500.

—— (1984*a*), *Land, Labour and Rural Poverty: Essays in Development Economics*, Columbia University Press, New York.

—— (1984*b*), 'Determinants of Supply and Demand for Labour in a Poor Agrarian Economy: An Analysis of Household Survey Data from Rural West Bengal', in Binswanger and Rosenzweig (eds.) (1984).

—— and Rudra, Ashok (1978), 'Interlinkage of Land, Labour and Credit Relations: An Analysis of Village Survey Data in East India', *EPW*, Annual Number, Feb.

—— —— (1980), 'Terms and Conditions of Sharecropping Contracts: An Analysis of Village Survey Data in India', *Journal of Development Studies*, 16(3) (Apr.), 287–302.

—— —— (1981), 'Terms and Conditions of Labour Contracts in Agriculture: Results of a Survey in West Bengal', *Oxford Bulletin of Economics and Statistics*, 43(1), (Feb.), 89–111.

—— —— (1986), 'Labour Mobility and the Boundaries of the Village Moral Economy', *Journal of Peasant Studies*, Apr.

—— Vaidyanathan, A., Alagh, Y. K., Bhalla, G. S., and Bhaduri, A. (1978), *Labour Absorption in Indian Agriculture*, ILO-ARTEP, Bangkok.

Barkat-e-Khuda (1980), 'Division of Labour in Rural Bangladesh', *Bangladesh Development Studies*, 8(4), (Autumn), 107–12.

Barker, R., and Cordova, V. (1976), 'Labour Utilisation in Rice Production', Resource Paper No. 5, International Rice Research Institute, Los Baños, Dec.

—— Herdt, Robert W., and Rose, Beth (1985), *The Rice Economy of Asia*, Resources for the Future/International Rice Research Institute, Washington.

Bartsch, W. H. (1977), *Employment and Technology Choice in Asian Agriculture*, ILO, Geneva, Praeger, New York.

Basant, Rakesh (1981), 'Attached and Casual Labour Wage Rates', *EPW*, 3 Mar.

—— (1985), 'Agricultural Technology and Employment in India: A Survey of Recent Research', Working Paper No. 1, Gujarat Institute of Area Planning, Ahmedabad.

Bautista, Germelino (1977), 'Socio-Economic Conditions of the Landless Rice Workers in the Philippines: The Landless of Barrio Sta. Lucia as a Case in Point', in Hirashima (ed.) (1977).

Béteille, Andre (1980), 'The Indian Village: Past and Present', in E. J. Hobsbawm, Witold Kula, Ashok Mitra, K. N. Raj, and Ignacy Sachs (eds.), *Peasants in History: Essays in Honour of Daniel Thorner*, Sameeksha Trust, Oxford University Press, Calcutta.

Bhaduri, Amit, Rahman, Hussain Zillur, and Arn, Ann-Lisbet (1986), 'Persistence and Polarisation: A Study in the Dynamics of Agrarian Contradictions', *Journal of Peasant Studies*, 13(3) (Apr.), 82–9.

Bhalla, G. S. (1978), 'The Impact of the Green Revolution on Rural Structural Change', Policy Research Programme, United Nations Centre for Regional Development, Nagoya, Oct.

Bhalla, Sheila (1976), 'New Relations of Production in Haryana Agriculture', *EPW*, Review of Agriculture, 11(13) (Mar.), A23-30.

—— (1979), 'Real Wage Rates of Agricultural Labourers in Punjab, 1961–77: A Preliminary Analysis', *EPW*, 30 June.

—— (1987), 'Trends in Employment in Indian Agriculture, Land and Asset Distribution', *IJAE*, 42(4) (Oct.–Dec.), 537–60.

Binswanger, Hans P., and Rosenzweig, Mark R. (1984), 'Contractual Arrangements, Employment, and Wages in Rural Labour Markets: A Critical Review', in Binswanger and Rosenzweig (eds.) (1984).

—— —— (eds.) (1984), *Contractual Arrangements, Employment, and Wages in Rural Labour Markets in Asia*, Yale University Press, New Haven.

—— *et al.* (1984), 'Common Features and Contrasts in Labour Relations in the Semi-Arid Regions of India', in Binswanger and Rosenzweig (eds.) (1984).

Blackburn, S. H. (1978), 'The Kallars: A Tamil "Criminal Tribe" Reconsidered', *South Asia*, 1.

Bloch, Marc (1966), *French Rural History: An Essay on its Basic Characteristics*, University of California Press, Berkeley and Los Angeles.

Boag, G. T. (1915), 'Proposals for Resettlement of Madurai District', *Resettlement Report*, Madurai District, BP.117 (Sett.), 29 Apr. (Pres.).

Boserup, Ester (1970), *Women's Role in Economic Development*, Allen and Unwin, London.

—— (1976), 'Traditional Division of Work between the Sexes, a Source of Inequality: Women in the Labour Market', Research Series No. 21, International Institute for Labour Studies, Geneva.

Bouton, Marshall M. (1985), *Agrarian Radicalism in South India*, Princeton University Press, Princeton.

Boyce, James K. (1989), 'Population Growth and Real Wages of Agricultural Labourers in Bangladesh', *Journal of Development Studies*, 25(4) (July), 467–89.

Brass, Tom (1986*a*), Review of Patnaik and Dingwaney (eds.) (1985), *Journal of Peasant Studies*, 14(1) (Oct.), 120–36.

—— (1986*b*), 'Unfree Labour and Capitalist Restructuring in the Agrarian Sector: Peru and India', *Journal of Peasant Studies*, 14(1) (Oct.), 50–77.

Bray, Francesca (1986), *The Rice Economies: Technology and Development in Asian Societies*, Blackwell, Oxford.

Breman, Jan (1974), *Patronage and Exploitation: Changing Agrarian Relations in South Gujarat, India*, University of California Press, Berkeley.

—— (1978), 'Seasonal Migration and Cooperative Capitalism: The Crushing of Cane and of Labour by the Sugar Factories of Bardoli, South Gujarat', pt. 1, *Journal of Peasant Studies*, 6(1) (Oct.), 41–70.

—— (1979), 'Seasonal Migration and Cooperative Capitalism: The Crushing of Cane and of Labour by the Sugar Factories of Bardoli, South Gujarat', pt. 2, *Journal of Peasant Studies*, 6(2) (Jan.), 168–209.

Breman, Jan (1985), *Of Peasants, Migrants and Paupers: Rural Labour Circulation and Capitalist Production in West India*, Oxford University Press, Delhi.

Byres, T. J. (1981), 'The New Technology, Class Formation and Class Action in the Indian Countryside', *Journal of Peasant Studies*, 8(4) (July), 405–54.

Census of India 1961 (1964), i, *India*, pt. 2*a*(i), *General Population Tables*, Delhi.

—— (1965), ix, *Tamil Nadu*, pt. 10, *District Census Handbook*, ii, *District and Townwise Primary Census Abstract*, Madras, for various districts.

Census of India 1971 (1974*a*), ser. 1, *India, Miscellaneous Studies*, Paper 1, *Report on Resurvey of Economic Questions: Some Results*, Delhi.

—— (1974*b*), ser. 19, *Tamil Nadu*, pt. 10*b*, *District Census Handbook, Village and Townwise Primary Census Abstract*, Madras.

—— (1975), ser. 1, *India*, pt. 2*a*(i), *General Population Tables*, Delhi.

Census of India 1981 (1983), ser. 1, *India*, pt. 2*b*(i), *Primary Census Abstract, General Population Tables*, Delhi.

—— (1984), ser. 1, *India*, pt. 2 Special, *Report and Tables based on 5 percent Sample Data*, Controller of Publications, Delhi.

—— (undated), ser. 20, *Tamil Nadu, Provisional Population Totals*, Madras.

Chambers, Robert, and Harriss, John (1977), 'Comparing Twelve South Indian Villages: In Search of Practical Theory', in B. H. Farmer (ed.), *Green Revolution? Technology and Change in Rice Growing Areas of Tamil Nadu and Sri Lanka*, Macmillan, Delhi.

Chatterji, Ruchira (1984), *Marginalisation and the Induction of Women into Wage Labour: The Case of Indian Agriculture*, Working Paper No. 32, World Employment Programme, Rural Employment Policy Research Programme, Mar., ILO, Geneva.

Chattopadhyay, Manabendu (1979), 'Agricultural Labourers of Birbhum', *IJAE*, 34(3) (July–Sept.), 42–50.

Chave, E. H. (1937), *A Monograph of the Periyar System*, Thallakulam, Madurai.

Chopra, Suneet (1985), 'Bondage in a Green Revolution Area: A Study of Brick Kiln Workers in Muzaffarnagar District', in Patnaik and Dingwaney (eds.) (1985).

Chung, Young-Il (1981), 'Transformation of Hired Labour and Cooperative Labour Organisations in Korean Agriculture', *The Developing Economies*, 19(4) (Dec.), 398–423.

Clay, Edward J. (1976), 'Institutional Change and Agricultural Wages in Bangladesh', *Bangladesh Development Studies*, 4(4) (Oct.), 423–40.

Cohen, G. A. (1978), *Karl Marx's Theory of History*, Princeton University Press, Princeton.

—— (1985), 'Are Workers Forced to Sell their Labour Power?', *Philosophy and Public Affairs*, 14(1) (Winter), 99–105.

—— (1986), 'The Structure of Proletarian Unfreedom', in John Roemer (ed.), *Analytical Marxism*, Cambridge University Press.

—— (1987), 'Are Disadvantaged Workers who Take Hazardous Jobs Forced to Take Hazardous Jobs?', in G. Ezorsky (ed.), *Moral Rights in the Workplace*, State University of New York Press.

Collins, E. J. T. (1969), 'Harvest Technology and Labour Supply in Britain, 1790–1870', *Economic History Review*, Dec., pp. 453–73.

Connell, John (1973), *Labour Utilisation: A Bibliography of Village Studies*, Institute of Development Studies, Brighton, Sussex, Dec.

—— and Lipton, Michael (1977), *Assessing Village Labour Situations in Developing Countries,* Institute of Development Studies, Village Studies Programme, Oxford University Press, Delhi.

Darling, Malcolm (1954), 'Report on Labour Conditions in Agriculture in Pakistan', Government of Pakistan, Ministry of Labour, published by the Manager of Publications, Karachi.

Dasgupta, Biplab (1977), 'New Technology and the Agricultural Labourers in India', in Hirashima (ed.) (1977).

Dhaliwal, R. S., and Grewal, S. S. (1982), 'Temporal and Spatial Variations in Wages of Agricultural Labourers in Punjab', *Economic Affairs,* Jan.–Mar.

Dixon, Ruth B. (1978), *Rural Women at Work: Strategies for Development in South Asia,* Resources for the Future, Johns Hopkins University Press, Baltimore.

Djurfeldt, Goran, and Lindberg, Staffan (1976), *Behind Poverty: The Social Formation in a Tamil Village,* Scandinavian Institute of Asian Studies Monograph Series No. 22, Curzon Press, Oxford and IBH, New Delhi.

Drèze, Jean, and Mukherjee, Anindita (1987), *Labour Contracts in Rural India: Theories and Evidence,* STICERD Development Research Programme Discussion Paper No. 7, London School of Economics, Feb.

—— Leruth, Luc and Mukherjee, Anindita (1986), 'Rural Labour Markets in India: Theories and Evidence', paper presented at the Eighth World Congress of the International Economic Association, New Delhi, Dec.

Dumont, Louis (1986), *A South Indian Subcaste: Social Organisation and Religion of the Pramalai Kallar,* trans. M. Moffatt, L. and A. Morton; rev. by L. Dumont and A. Stern, ed. M. Moffatt, French Studies in South Asian Culture and Society, i, Oxford University Press, Delhi.

Elster, Jon (1985), *Making Sense of Marx* (repr. 1986), Cambridge University Press.

Epstein, Scarlett (1962), 'Economic Development and Social Change in South India', Manchester University Press.

—— (1967), 'Productive Efficiency and Customary Systems of Rewards in Rural South India', in Raymond Firth (ed.), *Themes in Economic Anthropology,* Tavistock Publications, London.

Eswaran, Mukul, and Kotwal, Ashok (1985), 'A Theory of Two-Tier Labour Markets in Agrarian Economies', *The American Economic Review,* 75(1) (Mar.), 162–77.

Étienne, Gilbert (1968), *Studies in Indian Agriculture: The Art of the Possible,* University of California Press, Berkeley and Los Angeles.

Frankel, Francine R. (1971), *India's Green Revolution: Economic Gains and Political Costs,* Princeton University Press, Princeton.

Fredericks, L. J. (1977), 'Some Socio-Economic Characteristics of Rural Labour with Special Reference to a Rice Double-Cropping Area in Selangor, West Malaysia', in Hirashima (ed.) (1977).

Georgescu-Roegen, N. (1976), 'Process in Farming Versus Process in Manufacturing: A Problem of Balanced Development', in N. Georgescu-Roegen, *Energy and Economic Myths: Institutional and Economic Essays,* Pergamon, Oxford.

Ghodake, R. D., and Ryan, James G. (1981), 'Human Labour Availability and Employment in Semi-Arid Tropical India', *IJAE,* Conference Number, 36(4) (Oct.–Dec.), 31–8.

Ghose, Ajit Kumar (1980), 'Wages and Employment in Indian Agriculture', *World Development*, 8, 413–28.

GOI (1974), *Report of the Commissioner for Scheduled Castes and Scheduled Tribes for the Years 1971–72 and 1972–73*, Ministry of Home Affairs, New Delhi.

—— (1979), *Draft Sixth Five Year Plan 1978–83 (Revised)*, Planning Commission, New Delhi.

GOTN (1972), *Report on the Bench Mark Survey of the Agro-Economic Conditions of the Marginal Farmers and Agricultural Labourers Development Project Areas and Agro-Socio-Economic Conditions of Sub-Marginal and Marginal Farmers and Landless Agricultural Labourers*, Director of Statistics, Madras.

—— (1977), *World Agricultural Census, 1970–71, Tamil Nadu: District, Taluk and Panchayat Union Tables* [for each district except Madras, Kanniyakumari, and Nilgiris], Director of Agricultural Census, Madras.

—— (1983), *The Tamil Nadu Compulsory Labour Act (I of 1858): The Civil Court Manual (Tamil Nadu Acts)*, GOTN, Madras.

—— *Season and Crop Report*, successive volumes from 1947–48 to 1978–79, Director of Statistics, Madras.

—— (1986), *Season and Crop Report, 1981–82*, Director of Statistics, Madras.

Gough, Kathleen (1963), 'The Social Structure of a Tanjore Village', in Srinivas (ed.) (1963).

—— (1975), *Class Developments in South India*, Working Paper No. 1, Centre for Developing-Area Studies, McGill University, Montreal, Aug.

—— (1981), *Rural Society in Southeast India*, Cambridge University Press.

—— (1987), 'Socio-Economic Change in Southeast India, 1950s to 1980s', *Journal of Contemporary Asia*, 17(3), 276–92.

Guhan, S. (1983), *Palakurichi: A Resurvey*, Working Paper No. 43, MIDS, Madras, Nov.

—— (1984), *Irrigation in Tamil Nadu: A Survey*, Working Paper No. 49, MIDS, Madras, June.

—— and Bharathan, K. (1984), *Dusi: A Resurvey*, Working Paper No. 52, MIDS, Madras, Dec.

—— and Mencher, Joan P. (1983), 'Iruvelpattu Revisited', *EPW*, 4 and 11 June.

Gulati, Leela (1975), 'Female Work Participation: A Study of Inter-State Differences', *EPW*, 11 Jan.

—— (1976), 'Unemployment among Female Agricultural Labourers', *EPW*, 27 Mar.

Gunasinghe, Newton (1975), *Social Change and Disintegration of a Traditional System of Exchange Labour in Kandyan Sri Lanka*, Ceylon Studies Seminar Paper No. 6, Peradeniya Campus.

Habib, Irfan (1963), *The Agrarian System of Mughal India*, Asia Publishing House, Bombay.

—— (1982), *The Peasant in Indian History*, Presidential Address to the 43rd Session of the Indian History Congress, Kurukshetra.

Hall, J. F. (1930), *Report on the Result of the Investigation Directed by Government in G.O. No. 2734–I, dated 29.9.1929*, Collector's Office, Madurai, 12 May.

Hammond, J. L. and Barbara (1913), *The Village Labourer, 1760–1832: A Study in the Government of England before the Reform Bill*, Longmans, Green, London.

Hara, Yonosuke (1981), 'Institutional Changes in a Central Thai Village', *The Developing Economies*, 19(4) (Dec.), 305–13.

Harriss, Barbara (1987), 'Poverty in India: Micro-Level Evidence', draft paper for the Workshop on Poverty in India, Queen Elizabeth House, Oxford, Oct.

—— and Watson, Elizabeth (1987), 'The Sex Ratio in South Asia', in Janet Henshall Momsen and Janet G. Townsend (eds.), *Geography of Gender in the Third World*, Hutchinson, London.

Harriss, John (1982), *Capitalism and Peasant Farming: Agrarian Structure and Ideology in Northern Tamil Nadu*, Oxford University Press, Bombay.

—— (1986*a*), 'Agriculture/Non-Agriculture Linkages and the Diversification of Economic Activity in Rural Asia', Norwich.

—— (1986*b*), 'Knowing about Rural Economic Change: Problems Arising from a Comparison of the Results of "Macro" and "Micro" Research in Tamil Nadu', Norwich.

—— (1986*c*), 'Population, Employment and Wages: A Comparative Study of North Arcot Villages, 1973–1983', Norwich.

Hart, Gillian (1986), *Power, Labour and Livelihood: Processes of Change in Rural Java*, University of California Press, Berkeley and Los Angeles.

Haswell, M. R. (1967), *Economics of Development in Village India*, Routledge & Kegan Paul, London.

Hatti, Neelambar, and Raagard, Svend (1985), *Agricultural Labourers and Small Cultivators: A Study in Sirsi Taluk, Karnataka*, CDR Project Papers A.85.3, Centre for Development Research, Copenhagen, July.

Hayami, Yujiro (1981), *Understanding Village Community and the Direction of Rural Change in Asia*, Institute of Economic Growth Occasional Papers, NS No. 1, Hindustan Publishing Corporation, Delhi.

Heidrich, Petra (1988), *Agricultural Labour in Indian Society*, Studien über Asien, Afrika und Lateinamerika No. 39, Akademie Verlag, Berlin.

Hirashima, S. (1977), 'Hired Labour in Rural Asia: Problems and Issues', in Hirashima (ed.) (1977).

—— (ed.) (1977), *Hired Labour in Rural Asia*, Institute of Developing Economies, Tokyo.

Hjelje, Benedict (1967), *Slavery and Agricultural Bondage in South India in the Nineteenth Century*, Joint Reprint Series No. 1, The Scandinavian Institute of Asian Studies, Copenhagen.

Hobsbawm, E. J., and Rudé, George (1973), *Captain Swing*, Penguin University Books, Harmondsworth.

Horii, Kenzo (1977), 'Farming Technique and Rural Labour of Krian Rice Producing Area, Malaysia: Cases of Kampong Sungei Megat Aris and Kampong Parit Lebai Akir', in Hirashima (ed.) (1977).

Hossain, Mahabub (1988), *Nature and Impact of the Green Revolution in Bangladesh*, Research Report No. 67, International Food Policy Research Institute (IFPRI) in collaboration with the Bangladesh Institute of Development Studies, Washington, July.

IJAE (undated), *Conference Number, Volume III: 1948*, Researchco Reprints, Delhi.

ILO (1979), *Guide to Health and Hygiene in Agricultural Work*, ILO, Geneva.

—— (1985), *International Labour Conventions and Recommendations, 1919–1981*, ILO, Geneva.

ILO-ARTEP (1979), *Employment Expansion in Indian Agriculture: Proceedings of a National Seminar held in Bangalore*, ILO-ARTEP, Bangkok.

—— (1981), *Women in the Indian Labour Force: Papers and Proceedings of a Workshop,* ILO-ARTEP, Bangkok, Jan.

Indian School of Social Sciences, Calcutta (1976), *Bonded Labour in India,* with an introduction by M. K. Pandhe, India Book Exchange, Calcutta.

International Labour Conference (1954), *Fifth Item on the Agenda: Migrant Workers (Underdeveloped Countries),* Report 5(1), 38th Session, 1955, ILO, Geneva.

—— (1955*a*), *Fifth Item on the Agenda: Migrant Workers (Underdeveloped Countries),* Report 5(2), 38th Session, 1955, ILO, Geneva.

—— (1955*b*), *Sixth Item on the Agenda: Forced Labour,* Report 6(1), 39th Session, 1956, ILO, Geneva.

—— (1956*a*), *Sixth Item on the Agenda: Forced Labour,* Report 6(2), 39th Session, 1956, ILO, Geneva.

—— (1956*b*), *Fourth Item on the Agenda: Forced Labour,* Report 6(1), 40th Session, 1957, ILO, Geneva.

—— (1957), *Fourth Item on the Agenda: Forced Labour,* Report 6(2), 40th Session, 1957, ILO, Geneva.

—— (1958), *Seventh Item on the Agenda: Revision of the Plantations Convention (No. 110) and Recommendations (No. 110), 1958,* Report 7(1), 68th Session, 1982, ILO, Geneva.

Ishikawa, Shigeru (1976), *Labour Absorption in Asian Agriculture: An 'Issues' Paper,* ILO-ARTEP, Bangkok.

Jain, Devaki (1980), *Women's Quest for Power,* Vikas Publishing House, Ghaziabad.

Jose, A. V. (1974), 'Trends in Real Wage Rates of Agricultural Labourers', *EPW,* Review of Agriculture, 30 Mar.

Kandaswamy, A. (1964), *A Study of the Pattern of the Demand for and the Supply of Hired Labour in Agriculture in the Lower Bhavani Project Area, Madras State,* M.Sc. (Agri.) dissertation, Agricultural College and Research Institute, Coimbatore, Tamil Nadu.

Karanth, G. K. (1987), 'New Technology and Traditional Rural Institutions: Case of *Jajmani* Relations in Karnataka', *EPW,* 19 Dec.

Karashima, Noboru (1984), *South Indian History and Society: Studies from Inscriptions: A.D. 850–1800,* Oxford University Press.

Kikuchi, Masao (1981), 'Changes in Rice Harvesting Systems in the Philippines and Indonesia: Village Structure and Choice of Contractual Arrangements', *The Developing Economies,* 19(4) (Dec.), 291–304.

—— Cordova, V. G., Marciano, E. B., and Hayami, Yujiro (1979), 'Changes in Rice Harvesting Systems in Central Luzon and Laguna', IRRI Research Paper Series No. 31, International Rice Research Institute, Los Baños, July.

—— Hafid, Anwar, and Hayami, Yujiro (1984), 'Changes in Rice Harvesting Contracts and Wages in Java', in Binswanger and Rosenzweig (eds.) (1984).

Kitahara, Atsushi (1977), 'The Forms of Rural Labour and their Role in the Rural Economy of Thailand', in Hirashima (ed.) (1977).

Komoguchi, Yoshimi (1984), *A South Indian Village: Its Agriculture and Community,* Socio-Cultural Change in Villages in Tiruchirapalli District, Tamil Nadu, India, pt. 2, Modern Period No. 3, Institute for the Study of Languages and Cultures of Asia and Africa, Tokyo.

Kotovsky, Grigory (1964), *Agrarian Reforms in India,* People's Publishing House, New Delhi.

Krishnaji, N. (1971), 'Wages of Agricultural Labour', *EPW,* 25 Sept.

——(1980), 'Agrarian Structure and Family Formation: A Tentative Hypothesis', *EPW*, 29 Mar.

——(1983), 'Poverty and Fertility: A Review of Theory and Evidence', *EPW*, Annual Number.

——(1984), 'Family Size, Levels of Living and Differential Mortality in Rural India: Some Paradoxes', *EPW*, 11 Feb.

Krishnakumar, Asha (1989), 'An Economic Analysis of Groundnut Cultivation in Tamil Nadu', Madras.

Krishnamurty, J. (1984), 'The Occupational Structure', in D. Kumar (ed.), *The Cambridge Economic History of India*, ii, Orient Longman (in association with Cambridge University Press), Delhi.

Kumar, Ajoy (1986), *From Slavery to Freedom: The Tale of Chhatisgarh Bonded Labourers,* Struggle for Justice Series, 3, Indian Social Institute, New Delhi.

Kumar, Dharma (1965), *Land and Caste and South India*, Cambridge University Press.

Kurian, N. J. (1986), 'Economics of Rice Production in India: An Interregional and Intertemporal Analysis', *Journal of Quantitative Economics,* 2(1) (Jan.), 51–76.

Kurien, C. T. (1980), 'Poverty in Tamil Nadu: An Assessment', *MIDS Bulletin*, Dec.

——(1981), *Dynamics of Rural Transformation: A Study of Tamil Nadu 1950–1975,* Orient Longman, Madras.

Kuznets, Simon (1971), *Economic Growth of Nations,* Harvard University Press, Cambridge, Mass.

——(1980), *Size of Households and Income Disparities,* Economic Growth Centre Discussion Paper No. 342, Yale University, New Haven, Mar.

Lal, Deepak (1976), 'Agricultural Growth, Real Wages, and the Rural Poor in India', *EPW,* Review of Agriculture, June.

——(1984), *Trends in Real Wages in Rural India, 1880–1980,* World Bank, Washington, DC.

Lenin, V. I. (1972*a*), *The Development of Capitalism in Russia*, in *Collected Works*, iii, Progress Publishers, Moscow.

——(1972*b*), *On the Question of Our Factory Statistics (Professor Karyshev's New Statistical Exploits),* in *Collected Works*, iv, Progress Publishers, Moscow.

——(1972*c*), *The Purpose of Zemstvo Statistics,* in *Collected Works*, xx, Progress Publishers, Moscow.

——(1974), *Preliminary Draft Theses on the Agrarian Question,* in *Collected Works,* xxxi, Progress Publishers, Moscow.

Lipton, Michael (1986), 'Seasonality and Ultrapoverty', *IDS Bulletin,* 17(3) (July), 4–8.

Longhurst, Richard, Chambers, Robert, and Swift, Jeremy (1986), 'Seasonality and Poverty: Implications for Policy and Research', *IDS Bulletin*, 17(3) (July), 67–70.

Ludden, David (1985), *Peasant History in South India,* Princeton University Press, Princeton.

Mackenzie, A. T. (1962), *The Periyar Project,* Government Press, Madras (reprint).

Madras Institute of Development Studies (1980), *Structure and Intervention: An Evaluation of DPAP, IRDP and Related Programmes in Ramanathapuram and Dharmapuri Districts of Tamil Nadu,* MIDS, Madras.

Madurai Collectorate Records (1808), volume 1255, 'Report of Mr Hodgson on the Dindigul District', 28 Mar.

Mao Zedong (1967), *Selected Works*, 2nd edn. (1st edn. 1965), i, Foreign Languages Press, Beijing.

Marla, Sarma (1981), *Bonded Labour in India*, National Survey on the Incidence of Bonded Labour, Final Report, January 1981, sponsored by the Gandhi Peace Foundation and the Academy of Gandhian Studies, Biblia Impex, New Delhi.

Marx, K. (1973), *Grundrisse*, Penguin, Harmondsworth.

—— (1977), *Capital*, i, Progress Publishers, Moscow.

Mellor, John W. (1976), *The New Economics of Growth: A Strategy for India and the Developing World*, Cornell University Press, Ithaca.

—— and Johnston, Bruce F. (1984), 'The World Food Equation: Interrelations among Development, Employment and Food Consumption', *Journal of Economic Literature*, 22 (June), 531–74.

—— and Lele, Uma (1973), 'Growth Linkages of the New Foodgrain Technologies', *IJAE*, 23(1) (Jan.–Mar.), 35–55.

Mencher, Joan (1978), *Agriculture and Social Structure in Tamil Nadu: Past Origins, Present Transformation and Future Prospects*, Allied Publishers, New Delhi.

Menon, Parvathi (1986), 'The Agrarian Economy of the Carnatic in the Seventeenth and Eighteenth Centuries', Ph.D. thesis, Aligarh Muslim University, Centre for Advanced Study in History, Aligarh.

Menon, Saraswathi (1983), 'Social Characteristics of Land Control in Thanjavur District during the Nineteenth Century: A Sociological Study', Ph.D. thesis, Jawaharlal Nehru University, New Delhi.

Mies, Maria (1986), *Indian Women in Subsistence and Agricultural Labour*, ILO, Geneva.

Moerman, Michael (1968), *Agricultural Change and Peasant Choice in a Thai Village*, University of California Press, Berkeley.

Mukhopadhyaya, Swapna (1981), 'Women Workers in India: A Case of Market Segmentation', in ILO-ARTEP (1981).

Mundle, Sudipto (1979), *Backwardness and Bondage: Agrarian Relations in a South Bihar District*, Indian Institute of Public Administration, New Delhi.

Nagaraj, K. (1981), 'Structure and Inter-Relations of the Land, Labour, Credit and Product Markets of South Kanara: A Study of a Backward Agricultural Economy', Ph.D. thesis, Indian Statistical Institute, Calcutta.

—— (1985), *Towns in Tamil Nadu, Karnataka and Andhra Pradesh: A Study of Population and Spatial Configurations, 1961–1981*, Working Paper No. 54, MIDS, Madras, Mar.

—— (1986), 'Infant Mortality in Tamil Nadu', *MIDS Bulletin*, Jan.

—— (1989), 'Female Workers in Rural Tamil Nadu: A Preliminary Study', Madras.

Nair, K. Narayanan, Sivanandan, P., and Retnam, V. C. V. (1984), 'Education, Employment and Landholding Pattern in a Tamil Nadu Village', *EPW*, 16–23 June.

Nakamura, Hisashi (1976), *Accumulation and Interchange of Labour: An Inquiry into the Non-Market Economy in a South Indian Village*, Occasional Papers Series No. 14, Institute for Developing Economies, Tokyo.

—— (1982), *Disintegration and Re-integration of a Rural Society in the Process of Economic Development: The Second Survey of a Tank-Based Village in Tamil Nadu*, Studies in Socio-Cultural Change in Rural Villages in Tiruchirapalli District, Tamil Nadu, India, 5, Institute for the Study of Languages and Cultures of Asia and Africa, Tokyo.

Narasimhan, V. S. (1959), 'Agricultural Labour in Southern Taluks of Coimbatore District', M.Litt. thesis, University of Madras.

Nayyar, Rohini (1977), *Wages, Employment and Standard of Living of Agricultural Labourers in Uttar Pradesh*, Working Paper No. 5, World Employment Programme, Rural Employment Policy Research Programme, ILO, Geneva; published in *Poverty and Landlessness in Rural Asia*, ILO, Geneva.

—— (1987), 'Female Participation Rates in Rural India', *EPW*, 19 Dec.

NSS (n.d.), 'Tables with Notes on Differential Fertility and Mortality Rates in India', Report No. 175, 18th Round, Feb. 1963 to Jan. 1964, Department of Statistics, New Delhi.

Osmani, S. R. (1987), *The Food Problems of Bangladesh*, Working Paper No. 29, World Institute for Development Economics Research, Helsinki, Nov.

Palacpac, Adelita C. (1982), *World Rice Statistics*, International Rice Research Institute, Los Baños.

Parthasarathy, G. (1977), 'Employment, Wages and Poverty of Hired Labour within Indian Agriculture: Evidence on Selected Issues from Micro-Surveys in West Godavari District, Andhra Pradesh, India', in Hirashima (ed.) (1977).

—— and Pothana, V. (1983), 'Changing Agrarian Structure in Andhra Pradesh', *Journal of Rural Development*, Nov., pp. 581–95.

—— and Rao, G. Dasaradharama (1981), 'Women in the Labour Force in India', in ILO-ARTEP (1981).

Patel, Arun S. (1981), 'Irrigation: Its Employment Impact in the Command Areas of Medium Irrigation Projects in Gujarat', *IJAE*, 36(4) (Oct.–Dec.).

Patel, Surendra J. (1952), *Agricultural Labourers in Modern India and Pakistan*, Current Book House, Bombay.

Patnaik, Utsa (1972), 'Economics of Farm Size and Farm Scale, Some Assumptions Re-examined', *EPW*, Special Number, Aug.

—— (1976), 'Class Differentiation within the Peasantry: An Approach to Analysis of Indian Agriculture', *EPW*, Review of Agriculture, Sept.

—— (1985*a*), Introduction, in Patnaik and Dingwaney (eds.) (1985).

—— (1985*b*), *Reflections on the Agrarian Question and the Development of Capitalism in India*, Daniel Thorner Memorial Lecture, New Delhi.

—— (1987*a*), *Peasant Class Differentiation: A Study in Method with Reference to Haryana*, Oxford University Press, Delhi.

—— (1987*b*), 'Some Aspects of Development in the Agrarian Sector during Forty Years of Independent India', mimeo, New Delhi, Nov.

—— and Dingwaney, Manjari (eds.) (1985), *Chains of Servitude: Bondage and Slavery in India*, Sangam Books, Madras.

Poduval, R. N. (1987), *Foodgrain Economy of Tamil Nadu: Problems and Prospects*, Emerald Publishers, Madras.

Ponomarev, B. N., *et al.* (eds.) (1980), *The International Working Class Movement: Problems of History and Theory*, Progress Publishers, Moscow.

Prahladachar, M. (1983), 'Income Distribution Effects of the Green Revolution in India: A Review of Empirical Evidence', *World Development*, 11(11) (Nov.).

PWD (Public Works Department, Uthamapalayam) (1981), Records with Assistant Executive Engineer, PWD, Uthamapalayam Subdivision, Periyar Division, Uthamapalayam.

Rahman, Hussain Zillur, and Das, Samar Prasad (1982), *The Rural Labour Market in Noakhali*, CDR Project Paper No. A.82.14, Centre for Development Research, Copenhagen, Sept.

Ramachandran, V. K. (1980*a*), 'Agricultural Labourers in the Working Population of Tamil Nadu: Some Results from the Censuses of India 1961 and 1971 and the World Agricultural Census 1970–71', *MIDS Bulletin*, Mar.

—— (1980*b*), *A Note on the Sources of Official Data on Land Holdings in Tamil Nadu*, Data Series No. 1, MIDS, Madras, Dec.

—— (1983), *Inequality in the Distribution of Land Holdings and Assets among Households in Gokilapuram Village: A Report from a Field Survey*, Working Paper No. 31, MIDS, Madras.

—— (1984), *Irrigated Agriculture and Irrigation Labour: A Study of the Cumbum Valley, Madurai District, with Particular Reference to Gokilapuram Village*, Working Paper No. 47, MIDS, Madras.

—— (1986), 'Socio-Economic Characteristics of Agricultural Labourers in a Vanguard Agrarian Region: A Case Study of Gokilapuram Village, Madurai District', Ph.D. thesis, University of Madras.

—— and Balaji, M. (1980), 'A Report on the Implementation of the Food for Work Programme and Employment Guarantee Scheme in Ramanathapuram and Dharmapuri Districts', mimeo, Madras.

Rao, C. H. H. (1976), 'Factor Endowments, Technology and Farm Employment: Comparison of East Uttar Pradesh and Punjab', *EPW*, Review of Agriculture, Sept.

Rao, G. Hanumantha (1975), 'Caste and Poverty', Ph.D. thesis, Andhra University, Waltair, Andhra Pradesh, Apr.

Rao, Sudha V. (1984), 'Rural Labour: A Case Study of a Karnataka Village', *EPW*, 5 May.

Rastyannikov, V. G. (1975), 'The Agrarian Evolution of Indian Society in the 50s and 60s of the 20th Century', in V. I. Pavlov, V. G. Rastyannikov, G. K. Shirokov, and R. A. Ulyanovsky, *India: Social and Economic Development (18th–20th Centuries)*, Progress Publishers, Moscow.

—— (1976), 'Social Types of Hired Labourers in the Indian Countryside (1950s–1960s)', in Erik Komarov (ed.), *New India Studies by Soviet Scholars*, Problems of the Contemporary World, 33, *Social Sciences Today* Editorial Board, USSR Academy of Sciences, Moscow.

—— (1981), *Agrarian Evolution in a Multiform Structure Society: Experience of Independent India*, Routledge & Kegan Paul, London.

Reddy, C. Rammanohar (1985), *Rural Labour Market in Varhad: A Case Study of Agricultural Labourers in Rain-Fed Agriculture in India*, Working Paper No. 75, World Employment Programme, Rural Employment Policy Research Programme, ILO, Geneva, Oct.

Reddy, M. Atchi (1978), 'Official Data on Agricultural Wages in the Madras Presidency from 1873', *Indian Economic and Social History Review*, 15(4), 451–66.

—— (1979), 'Wages Data from the Private Agricultural Accounts, Nellore District, 1893–1974', *Indian Economic and Social History Review*, 16(3), 301–21.

—— (1983), 'Female Agricultural Labourers of Nellore, 1881–1981', *Indian Economic and Social History Review*, 20(2), 67–79.

Reserve Bank of India (1976), *Assets of Rural Households*, All India Debt and Investment Survey, 1971–72, Reserve Bank of India, Bombay.

Resettlement Report, Madura District: 'Proposals for a Resettlement of Madura District', BP 117 (Sett.) April 29, 1915 (Pres.), available at Collector's Office, Madurai.

RLE 1974–75 (1978), *Final Report on Indebtedness among Rural Labour Households*, Labour Bureau, Ministry of Labour, GOI, Chandigarh.

—— (1981), *Final Report on Employment and Unemployment of Rural Labour Households*, pts. 1 and 2 (supplement), Labour Bureau, Ministry of Labour, GOI, Chandigarh.

—— (1983), *Final Report on Income and Consumption Expenditure of Rural Labour Households*, Labour Bureau, Ministry of Labour and Rehabilitation (Department of Labour), GOI, Chandigarh.

Rodgers, Gerry, and Rodgers, Janine (1984), 'Incomes and Work among the Poor of Rural Bihar, 1971–81', *EPW*, Review of Agriculture, Mar.

Rosenzweig, Mark R. (1984), 'Determinants of Wage Rates and Labour Supply Behaviour in the Rural Sector of a Developing Country', in Binswanger and Rosenzweig (eds.) (1984).

Rudolph, Lloyd I., and Rudolph, Susanne Hoeber (1987), *In Pursuit of Lakshmi: The Political Economy of the Indian State*, The University of Chicago Press, Chicago.

Rudra, Ashok (1982), *Extraeconomic Constraints on Agricultural Labour: Results of an Intensive Survey in Some Villages near Santiniketan, West Bengal*, ILO-ARTEP, Bangkok, Aug.

—— and Bardhan, Pranab K. (1983), *Agrarian Relations in West Bengal: Results of Two Surveys*, Somaiya, Bombay.

Rukmani, R. (1989), 'An Analysis of Settlement Patterns in the Cumbum Valley', MIDS, Madras.

Ryan, James G., and Ghodake, R. D. (1984), 'Labour Market Behaviour in Rural Villages in South India: Effects of Season, Sex and Socioeconomic Status', in Binswanger and Rosenzweig (eds.) (1984).

St Croix, G. E. M. de (1988), 'Slavery and Other Forms of Unfree Labour', in Leonie J. Archer (ed.) (1988).

Saldanha, Indra Munshi (1989), 'Attached Labour in Thane: A Historical Overview', *EPW*, 20 May.

Sarvekshana (1981), 5(1–2), July–Oct.

—— (1982), 6(1–2), July–Oct.

—— (1986), 9(4), Apr.

Sastri, K. A. Nilakanta (1976), *A History of South India from Prehistoric Times to the Fall of Vijayanagar*, 4th edn., first pub. 1955, Oxford University Press.

Sen, Amartya (1987), *Freedom of Choice: Concept and Content*, Alfred Marshall Lecture, European Economic Association, Working Paper No. 25, World Institute for Development Economics Research, Helsinki, Aug.

Sinha, J. N. (1978), 'Rural Employment Planning: Dimensions and Constraints', *EPW*, Annual Number.

Sivertsen, Dagfinn (1963), *When Caste Barriers Fall: A Study of Social and Economic Change in a South Indian Village*, Universitets Forlaget, George Allen and Unwin, Oslo.

Smith, Joyotee, Cordova, Violeta, and Herdt, R. W. (1981), 'Trends in Labour Absorption and Earnings: The Case of Rice Production in the Philippines', abstract of paper in *IJAE*, Conference Number, Oct.–Dec.

Srinivas, M. N. (1963), Introduction, in Srinivas (ed.) (1963).

—— (ed.) (1963), *India's Villages*, Asia Publishing House, Bombay.

—— (1964), 'Village Studies and their Significance', in M. N. Srinivas, *Caste in India and Other Essays,* Asia Publishing House, Bombay.

Srinivasan, M. (1957), 'Pattern of Employment of Hired Labour in Agriculture in Certain Villages of Coimbatore Taluk of Madras State', *IJAE,* Apr.–June.

Srinivasaraghavaiyangar, S. (1893), *Memorandum on the Progress of the Madras Presidency during the Last Forty Years of British Administration,* 2nd edn., Madras.

Subramanian, S. (1983), 'Tank Irrigation in Tamil Nadu: Some Macro and Micro Perspectives', report submitted to the ILO, mimeo., MIDS, Madras, Mar.

—— (1985), 'Poverty and Inequality in Tamil Nadu', MIDS, Madras.

—— and Ramachandran, V. K. (1982), 'Agricultural Labourers in the Working Population of Tamil Nadu', *MIDS Bulletin,* June.

Sundaram, K. (1977), 'The Structure of the Work Force in Rural India: 1950–51 to 1971', *Indian Economic Review,* Apr.

Sundarayya, P. (1977), 'Class Differentiation of the Peasantry: Results of Surveys in Andhra Pradesh', *Social Scientist,* 56 and 57 (Mar. and Apr.).

Swaminathan, Madhura (1985), 'On to Pastures New: A Study of Migration', *Frontline,* Madras, 7–20 Sept.

—— (1986), *Rural Credit in Two Villages in Madurai District,* Final Research Report No. 1, Credit for Rural Development in Southern Tamil Nadu, Tamil Nadu Agricultural University and University of Reading, Sept.

—— (1989), 'Inequality and Economic Mobility: An Analysis of Panel Data from a South Indian Village', D.Phil. thesis, University of Oxford.

Swenson, Clyde Geoffrey (1973), *The Effect of Increases in Rice Production on Employment and Income Distribution in Thanjavur District, South India,* Ph.D. thesis, Michigan State University, University Microfilms, Ann Arbor.

Takahashi, Akira (1977), 'Rural Labour and Agrarian Change in the Philippines', in Hirashima (ed.) (1977).

Thompson, E. P. (1970), *The Making of the English Working Class,* Pelican, Harmondsworth.

—— (1974), 'Time, Work-Discipline, and Industrial Capitalism', in M. W. Flinn and T. C. Smout (eds.), *Essays in Social History,* Oxford University Press, London.

Thorner, Daniel and Alice (1962), 'Employer–Labourer Relationships in Agriculture', (first pub. 1957), in D. and A. Thorner, *Land and Labour in India,* Asia Publishing House, Bombay.

Trapeznikov, S. P. (1981), *Leninism and the Agrarian and Peasant Question,* 2 vols., Progress Publishers, Moscow.

Tribe, Keith (1979), 'Introduction to Weber', *Economy and Society,* 8(2) (May), 172–6.

Ushiyama, Keiji (1977), 'Rural Labour in Pre-War Japan: Case Study of Kambara-Gun, Niigata Prefecture', in Hirashima (ed.) (1977).

Uthamapalayam Agricultural Division (1985), *A Short Note,* Uthamapalayam.

Uthamapalayam Taluk, Revenue Records (1980–1), available at the Tahsildar's Office, Uthamapalayam, Tamil Nadu.

Vaidyanathan, A. (1984), 'Impact of Development on the Condition of Wage Labourers in Indian Agriculture', paper presented to a seminar at the University of Oxford.

van Schendel, Willem, and Faraizi, Aminul Haque (1984), *Rural Labourers in Bengal, 1880 to 1980,* Comparative Asian Studies Programme (CASP), Erasmus University, Rotterdam.

Venkataraman, Sridhar (1987), 'Agricultural Wages in Tamil Nadu: 1950/51–1980/81', M.Phil. dissertation, Jawaharlal Nehru University, June.

Vidyasagar, R. (1985), 'Debt Bondage in South Arcot District: A Case Study of Agricultural Labourers and Handloom Weavers', in Patnaik and Dingwaney (eds.) (1985).

Visaria, Pravin (1980), *Poverty and Unemployment in India: An Analysis of Recent Evidence,* Staff Working Paper No. 417, World Bank, Washington, Oct.

Vyas, V. S. (ed.) (1964), *Agricultural Labour in Four Indian Villages*, Sardar Vallabhbhai Vidyapeeth, Vallabhvidyanagar, Gujarat.

Weber, Max (1979), 'Developmental Tendencies in the Situation of East Elbian Rural Labourers', trans. Keith Tribe (first pub. 1894), *Economy and Society,* 8(2) (May), 177–205.

Wickramasekara, Piyasiri (1977), 'Aspects of the Hired Labour Situation in Rural Sri Lanka: Some Preliminary Findings', in Hirashima (ed.) (1977).

Yanagasawa, Haruka (1983), 'Socio-Economic Changes in a Village in the Paddy Cultivating Area in South India', Socio-Cultural Change in Villages in Tiruchirapalli District, Tamil Nadu, India, pt. 2, Modern Period No. 2, Institute for the Study of Languages and Cultures of Asia and Africa, Tokyo, Mar.

Yeshwanth, T. S. (1966), *Resurvey of a Tanjore Village: Madigai,* AERC, Madras.

—— (n.d.), *Uppattur Village,* Village Survey No. 47, AERC, Madras.

INDEX OF NAMES

SUBJECT INDEX

The manufacturer's authorised representative in the EU for product safety is Oxford University Press España S.A. of El Parque Empresarial San Fernando de Henares, Avenida de Castilla, 2 - 28830 Madrid (www.oup.es/en or product.safety@oup.com). OUP España S.A. also acts as importer into Spain of products made by the manufacturer.
Printed and bound by CPI Group (UK) Ltd, Croydon, CR0 4YY
14/09/2026
02216290-0005